BRADNER LIBRARY
SCHOOLCRAFT COLLEGE
18600 HAGGERTY ROAD
LIVONIA, MICHIGAN 48152

KF 3133 .B56 B64 2007
Bohrer, Robert A.
A guide to biotechnology law
 and business

A Guide to Biotechnology Law and Business

A Guide to Biotechnology Law and Business

Robert A. Bohrer
CALIFORNIA WESTERN SCHOOL OF LAW

CAROLINA ACADEMIC PRESS
Durham, North Carolina

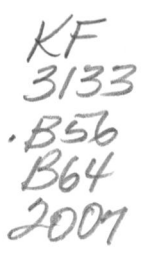
KF
3133
.B56
B64
2007

Copyright © 2007
Robert A. Bohrer
All Rights Reserved.

Library of Congress Cataloging-in-Publication Data

Bohrer, Robert A.
  A guide to biotechnology law and business / by Robert A. Bohrer.
    p. cm.
  Includes bibliographical references and index.
  ISBN-13: 978-1-59460-087-6 (alk. paper)
  ISBN-10: 1-59460-087-2 (alk. paper)
  1. Biotechnology industries--Law and legislation--United States.
2. Biotechnology--Law and legislation--United States. I. Title.

KF3133.B56B64 2007
343.73'0786606--dc22

                                      2007005410

Carolina Academic Press
700 Kent Street
Durham, North Carolina 27701
Telephone (919) 489-7486
Fax (919) 493-5668
www.cap-press.com

Printed in the United States of America.

For My Parents
Ira Bohrer of Blessed Memory
and Charlotte Bohrer
Who Shared Their Love of Learning With Me

And

To the Memory of Vince Frank,
Who Would Have Coauthored This Book

Contents

Table of Cases	xvii
Acknowledgments	xix

Chapter 1	**An Introduction to Biotechnology Law and Business**	**3**
§1.0	Biotechnology: Infinite Promise and Infinitely Challenging	3
§1.1	The Purpose of This Book	4
	§1.1(A) An Overview of This Book	4
§1.2	What is Biotechnology?	5
	§1.2(A) The Scope of the Biotechnology Industry	6
	§1.2(B) Applications of Biotechnology: Human Health	7
	§1.2(B)(1) Human Therapeutics	8
	§1.2(B)(2) Diagnostics for Human Disease	11
	§1.2(B)(3) Vaccines	12
	§1.2(C) Applications of Biotechnology: Agriculture	12
	§1.2(D) Industrial Applications of Biotechnology	13
§1.3	Summary: The Life Cycle of Biotechnology and the Lawyer's Role	14

Chapter 2	**A Primer on the Basic Science and Applications of Biotechnology**	**19**
§2.0	Modern Molecular Biology and Immunology	19
§2.1	DNA and the Genetic Code	19
	§2.1(A) From DNA to RNA to Protein: The Central Dogma of Molecular Biology (Now Known to be False!)	20
	§2.1(B) Genes: The Recipe for a Protein	21
	§2.1(C) Genes, Proteins, and Splice Variants	21
	§2.1(D) Putting It All Together: Genomes, Genotypes, and Phenotypes	22
§2.2	Prokaryotes, Eukaryotes, and Viruses	23
	§2.2(A) Viruses	24
	§2.2(B) Bacteria (Prokaryotes)	24

§2.2(C) Eukaryotes 25
§2.3 Introductory Genetic Engineering for Beginning Non-Biologists 25
  §2.3(A) Gel Electrophoresis 26
  §2.3(B) DNA Sequencing 27
  §2.3(C) Cloning Genes 28
  §2.3(D) Polymerase Chain Reaction 29
§2.4 An Introduction to Immunology (and Monoclonal Antibodies) 29
§2.5 Applications of Genetic Engineering in Health Care 34
  §2.5(A) Genetically Engineered Vaccines 34
  §2.5(B) Diagnostic Gene Probes 35
  §2.5(C) Therapeutic Applications of Genetic Engineering 36
  §2.5(D) Genetic Engineering and Human Gene Therapy 38
§2.6 Applications of Monoclonal Antibody Technology in Health Care 40
  §2.6(A) Therapeutic Uses of Monoclonal Antibodies 41
§2.7 Genetic Engineering Applications in Agriculture and Industry 42
§2.8 Summary 43

Chapter 3 Technology Transfer: The University-Industry Connection 45
§3.0 Introduction: Universities and Research Institutes—Where Biotechnology Is Born 45
§3.1 An Introduction to the Bayh-Dole Act: Transfer of Technology Developed with Federal Research Funding 46
§3.2 The Requirements of Bayh-Dole for Universities 48
  §3.2(A) Assignment of Rights 48
  §3.2(B) Sharing Proceeds 50
§3.3 The University Office of Technology Transfer (OTT): Functions and Models 51
  §3.3(A) Disclosure 51
  §3.3(B) Evaluation 52
  §3.3(C) Marketing 53
  §3.3(D) Licensing 55
  §3.3(E) Monitoring and Enforcement 55
§3.4 Alternative Models of Technology Transfer: Scripps, the NIH Guidelines, and Post-NIH Guidelines Developments 56
  §3.4(A) The Scripps Research Institute (TSRI) Significant Relationship Model 56
  §3.4(B) The NIH Guidelines: Developing Sponsored Research Agreements: Considerations for Recipients of NIH

                Research Grants (59 FR 32997-02, Monday, June 27,
                   1994) ("NIH Sponsored Research Guidelines")   58
        §3.5 Licensing a University Invention (Also a Few Tips on
                Negotiation)   60
            §3.5(A) Licensed Technology   62
            §3.5(B) Sublicensing   63
            §3.5(C) Field of Use, Geographic Territory, and Exclusive
                Versus Non-Exclusive   64
            §3.5(D) Royalties, Licensing Fees, and Milestone Payments   66
            §3.5(E) The Bottom Line on University Licensing   67
        §3.6 Sponsored Research Agreements   67
            §3.6(A) The Description of the Research   67
            §3.6(B) Time for Review before Publication   68
            §3.6(C) Sponsor's Rights to Inventions Resulting from the
                Sponsored Research   68
        §3.7 Conclusion: University Technology and Biotechnology   69

Chapter 4   Intellectual Property (IP) Protection for Biotechnology   71
        §4.0 Introduction: The Need for Patents and the Patentability of
            Biotechnology   71
        §4.1 Patents and Trade Secrets: Choosing a Method for Protection   73
        §4.2 Differences Between U.S. and Foreign Patent Law   75
        §4.3 Patentable Subject Matter   77
        §4.4 New and Useful: The Requirement of Utility   80
        §4.5 Novelty and Publication   82
        §4.6 Nonobviousness   84
            §4.6(A) Secondary Evidence of Nonobviousness   88
        §4.7 Written Description, Enablement, and Best Mode: The
                Requirements of §112   89
        §4.8 Other Special Problems for Biotechnology Patents   93
            §4.8(A) Protection from Foreign Competition   93
            §4.8(B) Biotechnology Products Derived from Human Tissue   95
            §4.8(C) Patent Protection for Stem Cell Research and
                Development   96
        §4.9 Biotechnology Patents: Moral and Ethical Issues   97
            §4.9(A) The Rights of Indigenous Peoples   97
            §4.9(B) Patent Protection for Drugs Derived from Traditional
                Remedies   99

§4.9(C) Patenting Animals: Beyond Patentable Subject Matter to
 Agricultural Policy and Animal Rights 99
§4.10 Broad Patent Claims to Drug Targets 100
§4.11 Patent Infringement and the Doctrine of Equivalents 102
 §4.11(A) Reverse Equivalents 104
 §4.11(B) Unexpected Results 105
§4.12 Conclusion: Future Directions for Biotechnology Patent Law 105
 §4.12(A) International Patent Issues 106
 §4.12(B) Life-Saving Pharmaceuticals in Developing Countries 106
 §4.12(C) Scientific Progress, the Human Genome Initiative,
 and Changes in Patent Law 106

Chapter 5 Biotechnology Business Strategy 109
§5.0 Introduction (Including a Brief History of Biotech Business
 Models) 109
§5.1 Strategic Planning for the New "Hybrid" Biotechnology 112
 §5.1(A) The Role of Markets in Strategic Planning 113
 §5.1(B) Minimizing Risk in Clinical Development: Animal
 Models, Clinical Trial Costs (Sizes and Duration), Clear
 Pathways, and Unmet Medical Needs 115
§5.2 Integrating Regulatory Strategy and Market Issues 119
§5.3 Financing Biopharmaceutical Development: Considerations
 and Strategy 120
§5.4 Summary: The Multiple Strands of a Biotech Business Plan 125

Chapter 6 Financing the Process of Development 129
§6.0 Introduction 129
§6.1 A Brief Introduction to Corporations and Securities: (The
 MBAs and Lawyers May Wish to Skip This Section) 130
§6.2 Selling Securities: Caveat Vendor 132
§6.3 In the Beginning: Is There an Angel in the House? 132
§6.4 Venture Capital Financing Issues 134
 §6.4(A) Due Diligence 135
 §6.4(B) Valuation of the Investment (and the Company) 137
 §6.4(C) Capital Structure: Authorized Shares 139
 §6.4(D) Stock Options Plans 140
 §6.4(E) Stages of Venture Capital Investment 141
 §6.4(F) Anti-Dilution 142

§6.5 Beyond the First Round: Licensing and Corporate Strategic
    Alliances   142
  §6.5(A) The Framework for a Strategic Alliance   144
  §6.5(B) Timing of Alliances and the Impact on Valuation   145
  §6.5(C) Foreign Partnerships   148
§6.6 Introduction to the Public Offering   149
  §6.6(A) The Decision to Go Public   149
  §6.6(B) Preparing the Public Offering: The Letter of Intent   151
  §6.6(C) Going Public: The Team   151
  §6.6(D) Going Public: Due Diligence   152
  §6.6(E) Going Public: Filing the Registration Statement   153
  §6.6(F) Going Public: The Effective Date and the Closing Date   154
  §6.6(G) Going Public: The Quiet Period   155
  §6.6(H) Going Public: Lock-Up Periods and the PSLRA "Safe
      Harbor"   156
§6.7 Conclusion: Biotechnology Finance as a Four-Dimensional
    Rubik's Cube   158

Chapter 7 An Overview of the Regulation of Biotechnology   161
§7.0 Introduction   161
§7.1 History of the Regulation of Genetic Engineering   163
§7.2 Regulation Under the NIH Recombinant DNA Advisory
    Committee Guidelines (NIH-RAC)   164
§7.3 The FDA and Human Health Care Applications of
    Biotechnology   168
  §7.3(A) The FDA Approval of Human Therapeutics and
      Diagnostics   168
  §7.3(B) The FDA and New Animal Drugs   171
§7.4 The EPA and Agricultural Biotechnology   172
  §7.4(A) EPA's FIFRA Procedures for Field Testing Genetically
      Engineered Organisms   173
  §7.4(B) The FIFRA Standard for the Approval of Pesticides   174
§7.5 EPA'S Regulation of Non-Agricultural Biotechnology—TSCA   175
  §7.5(A) TSCA's Scope, Procedures, and Standards   176
  §7.5(B) Requirement of a Premanufacture Notice (PMN)   177
§7.6 The USDA and Agricultural Biotechnology   178
  §7.6(A) USDA's Regulation of Veterinary Biological Products—
      VSTA   179

§7.6(B) USDA's Regulation of Genetically Engineered
  Microorganisms and Plants—PPA ... 181
§7.6(C) The Scope, Standards, and Procedures of the PPA ... 182
§7.7 The Regulation of Food Products from Genetically Engineered
  Plants ... 183
  §7.7(A) Adulterated: The First Concept of Food Safety ... 183
  §7.7(B) Food Additive: Added Substances That Do Not
    Adulterate ... 184
  §7.7(C) Generally Recognized as Safe (Between Additive and
    Adulteration) ... 185
  §7.7(D) Foods Derived from New Plant Varieties Created by
    Genetic Engineering—the FDA Statement of Policy ... 187
  §7.7(E) Food Plants Engineered to Produce a Biopesticide or
    Treated with a Biotechnology Derived Biopesticide ... 192
  §7.7(F) Conclusion: Risk Perception and Food Biotechnology ... 193

**Chapter 8 Special Regulatory Issues—FDA Regulation of Drugs,
  Biologics, and Devices** ... 197
§8.0 Introduction ... 197
§8.1 Regulatory Strategy: An Introductory Example ... 199
§8.2 The FDA Structure and Statutory Authority for Drugs and
  Biologics (With a Brief Word about Devices) ... 200
  §8.2(A) A Brief Excursion into the Distinction between Drugs
    and Biologics ... 201
  §8.2(B) A Brief Word about Devices ... 205
§8.3 Stages of Drug Development and Drug Development Strategy ... 206
  §8.3(A) From Preclinical Decision Making to NDA ... 209
    §8.3(A)(1) The Five Key Parameters of Strategic Success:
      FDA Approval ... 210
    §8.3(A)(2) Optimal Indication ... 210
    §8.3(A)(3) The Shortest Possible Time ... 210
    §8.3(A)(4) Conserving Scarce Resources: Time and Money ... 210
    §8.3(A)(5) Planning for Reimbursement ... 211
  §8.3(B) Strategic Issues in the Development Process and the
    Terms of Success ... 211
    §8.3(B)(1) The Preclinical Development Process and the
      Optimal Indication ... 212
    §8.3(B)(2) Conserving Time and Money ... 214
    §8.3(B)(3) Planning for Reimbursement ... 216

§8.3(C) Drug Development Strategy: Learning from the
Mistakes of Previous Companies   216
§8.3(D) Conclusion: Planning the Clinical Development of a
New Therapeutic Begins at the Early Preclinical Stage   220
§8.4 A Practical Guide to Clinical Trials and the NDA   220
§8.4(A) The New Drug Application or Product License
Application   222
§8.4(B) Beyond Approval: Post-Marketing Issues   225
§8.4(C) Reporting Adverse Drug Experiences (ADEs)   226
§8.5 Accelerated Approval and Early Access to Unapproved Drugs:
Drugs for Severely Debilitating and Life Threatening Illnesses   227
§8.5(A) Fast Track, Accelerated Approval, and Priority Review   227
§8.5(B) Early Access to Experimental Drugs   229
§8.6 Product Exclusivity and the FDA: Orphan Drugs, Patent
Term Extension, and Pediatric Study Extensions   231
§8.6(A) The Orphan Drug Act   231
§8.6(B) Patent Term Restoration   233
§8.6(C) Pediatric Study Patent Extensions   235
§8.7 "Generic Biopharmaceuticals"   237
§8.8 Advertising and Promoting Prescription Drugs (Herein Also a
Brief Word about Dietary Supplements-DSHEA)   241
§8.8(A) The Legal Background of the FDA's Marketing and
Promotion Regulation   241
§8.8(B) Marketing to Physicians and Formulary Committees   243
§8.8(C) Direct-to-Consumer (DTC) Advertising in Print and by
Electronic Media   245
§8.9 DSHEA—Anything Goes (Almost)   247
§8.10 Postscript: International Pharmaceutical Regulation and
Harmonization   253

**Chapter 9 Ethical Perspectives on New Ethical Dilemmas for
Biotechnology**   257
§9.0 Introduction   257
§9.1 The Scientific Advances and Ethical Challenges to
Biotechnology Research and Development   258
§9.2 Ethical Frameworks and *A Theory of Justice*   260
§9.3 Applying the Frameworks to the Problems   263
§9.3(A) Bioterrorism and Restraints on Research and
Publication   263

§9.3(B) Balancing Incentives for Pharmaceutical Innovation against Affordability and Access for Persons with Medical Needs 265
§9.3(C) Echazabal and Workplace Genetic Testing: The Law and Ethics of Protecting Workers against Risks 266
§9.3(C)(1) The Legality of Protecting Workers against Risks to Their Own Health 267
§9.3(C)(2) *A Theory of Justice* and Workplace Genetic Testing 269
§9.4 Embryonic Stem Cell Research 274

Chapter 10 Special Regulatory Issues: Human Gene Therapy 277
§10.0 Introduction to Gene Therapy: At the Edge of a New Era in Medicine 277
§10.1 What is Gene Therapy? 278
§10.1(A) Candidate Diseases for Gene Therapy 280
§10.1(B) Selecting Appropriate Target Cells 281
§10.1(C) Methods for Delivering Therapeutic Genetic Material 281
§10.2 Risks of Gene Therapy 284
§10.2(A) The Risk of Producing Replication-Competent Viruses 285
§10.2(B) The Pathogenicity of Viral Vectors 286
§10.2(C) Summary: Risks of Gene Therapy 288
§10.3 The Regulatory Framework for Gene Therapy 289
§10.3(A) Notification of the NIH-RAC 289
§10.3(B) Appendix M and the NIH's Principal Concerns 290
§10.3(C) FDA Review of Gene Therapy Protocols 291
§10.4 The Ethical Debate over Gene Therapy 292
§10.4(A) General Limitations on Somatic Cell Gene Therapy 293
§10.4(B) Ethical Issues Surrounding Germ-Line Gene Therapy 294
§10.4(C) The Ethical Problem of Access to Gene Therapy 297
§10.5 Conclusion 298

Chapter 11 Liability for Biotechnology Products 299
§11.0 Introduction 299
§11.1 A Basic Overview of Negligence 300
§11.1(A) Violation of a Statute or Regulation as Negligence *Per Se* 301
§11.1(B) The Relevance of Custom to Negligence 302
§11.1(C) General Evidence of Reasonable Care 303
§11.2 An Introduction to Strict Product Liability 304

§11.2(A) Defective Condition Unreasonably Dangerous — 305
§11.2(B) Manufacturing Defect — 305
§11.2(C) Design Defect — 306
§11.2(D) Failure to Warn — 308
§11.2(E) Preemption of Liability by FDA Approval:
  A Complicated Tale — 309
§11.3 Legal Cause: A Problem in Both Negligence and Strict
  Liability — 311
  §11.3(A) Expert Testimony in Product Liability Cases — 312
§11.4 The National Vaccine Injury Compensation Program:
  42 U.S.C. §300aa-1 *et seq.* (2006): One Answer to the
  Causation Conundrum — 315
§11.5 Drugs, Vaccines, and Other Human Therapeutics as
  Unavoidably Unsafe: Restatement (Second) of Torts
  §402A Comment k — 318
§11.6 Restatement of Torts (Third) and Product Liability Reform — 320
§11.7 Liability for Biotechnology in Other Contexts: Beyond Product
  Liability for Drugs, Vaccines, and Medical Devices — 322
  §11.7(A) Liability in the Workplace — 322
  §11.7(B) Liability for Biotechnology Pesticides — 327
  §11.7(C) Liability for Food Products Produced by Biotechnology — 328
  §11.7(D) Liability for Other Biotechnology Products in Industry
    and the Environment — 331
§11.8 Conclusion — 332

Index — 333

Table of Cases

"Agent Orange" Product Liability Litigation, 312
Amgen v. Chugai, 93
Andrulonis v. U.S., 322
Ariad Pharms., Inc. v. Eli Lilly & Co., 101
Bates v. Dow Agrosciences, 327
Brown v. Superior Court, 308, 332
Burr v. Duryee, 102
Chevron USA, Inc. v. Echazabal, 267
Daubert v. Merrell Dow Pharmaceuticals, Inc., 313
Diamond v. Chakrabarty, 46, 73
Enzo Biochem v. GenProbe, 90
Escola v. Coca-Cola Bottling Co. of Fresno, 305
Festo Corp. v. Shoketsu Kinzoku Kogyo Kabushiki Co., 103
Frye v. United States, 311
Genentech, Inc. v. Wellcome Found. Ltd., F.3d 1555 (Fed. Cir. 1994), 87
Graham v. John Deere Co., 88
Graver Tank, 102, 104
Harvard College v. Canada, 78
Hilton Davis Chem. Co. v. Warner Jenkinson Co, Inc., 103
Hormone Research Foundation, Inc. v. Genentech, 102
Hybritech Inc. v. Monoclonal Antibodies Inc., 88
In Re Allen, 77
In re Bergy, 79
In re Dillon, 919 F.2d 688 at 692, 105
In re Graeme I. Bell, 85
In re Hogan, 104–105
In re Mayne, 105
In re Thomas F. Deuel, 85
Johns Hopkins v. Cellpro, 92
Jurgens v. McKasy, 102
Loctite Corp v. Ultraseal Ltd., 102
Medtronic Inc., v. Lohr, 310
Mexicali Rose v. Superior Ct., 329
Moore v. Regents of University of California, 95
NVE Inc. v. Thompson, 436 F.3d 182 (3rd Cir. 2006), 252
Papas v. Upjohn, 327
Parke Davis & Co. v. H.K. Mulford & Co., 72
Penwalt Corp. v. Durand-Wayland Inc., 104
Pharmanex v. Shalala, 221 F.3d 1151 (10th Cir. 2000), 250
Platzer v. Sloan-Kettering, 50
Richardson v. Richardson-Merrell, 312
Sanitary Refrigerator, 102
Scripps Clinic & Research Found. v. Genentech Inc., 104
The T.J. Hooper, 303
Thompson v. Western States Medical Center, 535 U.S. 357 (2002), 242
Warner-Jenkinson Co., Inc. v. Hilton Davis Chem. Co., 103
Washington Legal Foundation v. Henney, 202 F.3d 331, (D.C. Cir. 2000), 242
Yong Cha Hong v. Marriot Corp., 329

Acknowledgments

A great many people have helped me with this book in different ways. There are too many to list them all, but I shall list several in different categories.

First are those who specifically lent their time to reading drafts of this book and offering helpful suggestions. These include John Mendeloff of the University of Pittsburgh; Dale Busch, former General Counsel of the Salk Institute for Biological Studies; Jasmin Patel of Novartis; Jacob Handy of Morrison and Foerster; and, my wife Karen.

Second, there are many scientists who have contributed to my efforts to understand the science of biotechnology. These include the late Clifford Grobstein, of the University of California, San Diego (UCSD), who was my first guide to the field of biotechnology policy; Don Helinski, of UCSD, who gave me the opportunity to work and learn at UCSD's Center for Molecular Genetics; Ron Brown, a cofounder of Syntro and former CEO of Octamer, Inc., who first introduced me to the extraordinary world of biotechnology as a business; and, the late John O'Brien, of UCSD, who extended to me the great privilege of working with him for several years in the development and commercialization of his science. Mario Bourdon of the La Jolla Institute for Molecular Medicine provided me with the opportunity to learn about the world of nonprofit biomedical research during my years as a member of the Board of that Institute. Ami Loewenstein of Technion's Dimotech and the faculties of Medicine and Biology at the Technion-Israel Institute of Technology in Haifa also were very generous in allowing me to work and learn about their science, research, and efforts in technology transfer

Third, there are those who have helped me develop as a lawyer and legal scholar. John Cribbet, the former Dean of the University of Illinois College of Law, strove to impart the importance of "the big picture" to generations of law students. Dean Cribbet, credit for my effort to see the big picture in this book is yours. The late Dean Albert Sachs, of Harvard Law School, helped me understand the value of legal scholarship. My colleague John Noyes has been a sounding board for my ideas for over twenty years. Dean Steven Smith of California Western School of Law has encouraged me in all of my scholarship and particularly in my efforts to complete this book.

I have been very fortunate to teach Biotechnology Law to many wonderful students over the past decade or so and I owe much to them for their contributions to my class, which have helped expand my understanding of the issues and my efforts to communicate both technical and legal material. To name just a few I would mention: Lisa Haile of DLA Piper Rudnick Gray Cary; Christine Gritzmacher of GenProbe; Chris Dayton of BiogenIdec; Jayshree Gerken of Fish and Richardson, and, Rena Patel of Bristol Meyers Squibb.

My sons Matthew and Nathaniel have been very patient and supportive during the many hours I was working on this book that I otherwise would have spent with them. Biotechnology is about the science of life. You have taught me about its meaning.

I owe more than I can say to my Editor-For-Life and wife, Karen Bohrer, without whose help and support I could never have finished this project.

Thanks.

Bob Bohrer
San Diego, California
August 2006

AN GUIDE TO BIOTECHNOLOGY LAW AND BUSINESS

CHAPTER 1

An Introduction to Biotechnology Law and Business

§1.0 Biotechnology: Infinite Promise and Infinitely Challenging

Imagine a world without cancer, heart disease, or arthritis. Or imagine that the same world could produce food crops able to flourish in salt water and disease-resistant poultry that grow twice as big on the same amount of food. Biotechnology research and development could very well lead to such a world. The potential of biotechnology is enormous. Sales of biotechnology products exceeded $25 billion in the year 2000, a three-fold increase in an eight-year period. Biotechnology is seen by many as the next great technological revolution, driven by extraordinary progress in the engineering and manufacturing of life itself.

Although still in its infancy (or perhaps toddlerhood), biotechnology does indeed seem likely to produce a true technological revolution. It is spawning new industries in the fields of health care, agriculture, and industrial chemistry. It is enabling people in those fields to work in new and more efficient ways. Physicians now run sensitive new diagnostic tests in minutes in their offices and farmers are beginning to produce more food at less cost. Of course, the extent to which biotechnology will produce changes in our social structure and daily lives remains to be seen, but the possibility that it will do so seems very real. A society free of significant disease, efficiently producing abundant food in an environmentally sound way would be very different from that which we know today.

Biotechnology, the technology based on the great advances in modern molecular biology, is infinitely promising. It promises great improvements, even dramatic changes in our lives, as dramatic as the change from pre-industrial feudal society to the twentieth century market economy of the United States. As with any such change, it also offers the promise of fame and fortune to the innovators and entrepreneurs who succeed in harnessing the power of biotechnology to provide important new products and services. The rush of capital and entrepreneurs to biotechnology, with its largest center in Northern California, has been likened to the Gold Rush of 1849, when the adventurous souls of an earlier era were lured from far and wide by the siren call of a new frontier offering vast riches. It will almost certainly be the case that many more will fail than will succeed. The challenge of turning the most spectacular scientific advances into successful products is as formidable as it is potentially rewarding. Biotechnology is indeed a technological, social, and business revolution that is infinitely promising and equally challenging.

§1.1 The Purpose of This Book

This book is intended to provide "the big picture" to everyone interested or involved in the biotechnology industry. It is a book about law and business because it is fair to say that no prior technological change was so shaped at every stage by law and regulation, from the basic intellectual property rights that separate biotechnology from molecular biology to the complex financial and regulatory environment in which biotechnology companies compete. Many people have worked in the biotechnology industry for years, but within just one of its many specialties, such as patent law, business development, or regulatory affairs. This book is intended to provide even experienced biotechnology lawyers or executives a useful introduction to the areas with which they may be less familiar. Patent lawyers may not feel compelled to read the chapter on intellectual property protection while technology transfer administrators may not need to read the chapter on technology transfer, but each may wish to know more about all the rest of the life-cycle of biotechnology.

§1.1(A) An Overview of This Book

This book is intended to provide an essential guide for those who work in biotechnology. What is required in order to have a chance to survive and flourish in the biotechnology industry? An understanding of the basic science and a desire to learn more is essential. Chapter 2 is brief introduction to the science upon which the biotechnology industry is based. Much of that science originates in federally funded research in universities and research institutes such as Stanford University, the Salk Institute, and The Scripps Research Institute. Chapter 3 is about technology transfer and university licensing, the process by which valuable university and not-for-profit research institute inventions are licensed to for-profit companies. Chapter 4 is a guide to intellectual property issues in biotechnology, which is how scientific discoveries can be provided with the property rights essential to raising the money that a biotechnology company needs.

The two greatest talents a biotechnology executive needs are the ability to change course and the ability to raise enormous amounts of money. Chapter 5 is about business planning for biotechnology and Chapter 6 is about the basic mechanisms for financing biotechnology, including a discussion of joint ventures and strategic alliances. Intellectual property, money, and a good business plan are just the start. The long journey from the laboratory to the marketplace requires navigating a maze of regulatory requirements. Chapter 7 provides an overview of the regulation of biotechnology, with special emphasis on the regulation of food products produced from genetically engineered crops. Chapter 8 focuses on the FDA, the principal regulatory agency for the great majority of biotechnology companies. From the start, ethical issues have played a significant role in the financial and regulatory environment for biotechnology and have significantly impacted the rate at which biotechnology is developed. Chapter 9 provides an introduction to ethical reasoning and discusses of some of the most significant recent ethical issues raised by biotechnology. Chapter 10 is about the special issues raised by the most challenging and potentially most powerful application of biotechnology, human gene therapy. Unfortunately, not every biotechnology product that makes it to the market will turn out as anticipated. There are certain to be lawsuits over injuries allegedly caused by unreasonably dangerous biotechnology products. Chapter 11 concludes this guide to biotechnology business and law with an overview of the legal structure for personal injury liability as it is likely to be applied to biotechnology products.

It is fair to say that a business plan for any biotechnology product requires an international strategy and that international law is of great significance to the biotechnology industry in a number of areas, from technology transfer limitations to foreign patent protection and pharmaceutical regulation. Foreign patent issues are discussed in Chapter 4, foreign joint ventures are discussed in Chapter 6, and international pharmaceutical regulation is discussed in Chapter 8.

§1.2 What is Biotechnology?

Biotechnology is the technology based on modern molecular biology. Modern biology may well have begun in 1953 when Frances Crick and James Watson discovered the structure of DNA, the molecule that comprises the genetic code of every living thing.[1] That central discovery of modern biology became a technology in 1973 when scientists first learned to manipulate DNA through an array of techniques known as recombinant DNA or genetic engineering. The ability to manipulate genes was followed in 1975 by Kohler and Milstein's discovery of a basic technique for manipulating immune system cells *in vitro*. The pace of discovery in biology has accelerated, and with it our power to manipulate life grows ever greater. This introduction describes a number of biotechnology applications and their related legal issues. Chapter 2 provides a more detailed scientific and technological base for those applications and is intended primarily for those readers without a strong technical background. This §1.2 is intended to provide an overview of the industry and technology for newcomers to the biotechnology industry with varied levels of technical sophistication.

§1.2(A) The Scope of the Biotechnology Industry

It is somewhat misleading to discuss the "biotechnology industry." Although there are a great many companies founded specifically to utilize biotechnology processes or develop biotechnology products (these are sometimes referred to as "dedicated biotechnology companies" or dbcs), biotechnology is a set of tools which can be and are used by large, long-established companies which do not have a focus solely on biotechnology. Thus Monsanto has made major investments in the development of biotechnology products, as have Merck, Novartis, and virtually all of the other established pharmaceutical companies. In fact, while a great deal of the development of biotechnology in the United States is being done by dedicated biotechnology companies, in Europe and Japan there are fewer such dbcs, and a greater percentage of biotechnology research and development is being done by existing pharmaceutical and chemical companies.

The United States is undoubtedly the world leader in biotechnology. A major reason for that leadership is the fact that the United States funds by far the largest amount of university and not-for-profit institute research in molecular biology. In biotechnology, in contrast to other "high tech" industries such as semi-conductors and telecommunications, there is a very fine, sometimes almost non-existent line between basic university research and potentially valuable commercial research. While most university research in particle physics or astronomy has only a very remote possibility of turning into a

---

1. A slight oversimplification: RNA, DNA's sister molecule, provides the genetic instructions for a fair percentage of viruses.

marketable invention, much of the work done in molecular biology has commercial potential. Researchers interested in the genes that cause disease or control development are doing research that is appropriate for a university context, but may well have significant commercial application. Thus both of the two building blocks of biotechnology, genetic engineering and the production of monoclonal antibodies, were first developed in university laboratories (the former in Northern California, the latter in England). More recently, such major biotechnology developments as the first genetic test for breast cancer susceptibility[2] and the first efforts at human gene therapy originated in university labs or the laboratories of the National Institutes for Health (NIH).

According to the Biotechnology Industry Organization, there were almost 1,500 biotechnology companies in the United States in 2001 with total industry revenues reaching $28.5 billion. Biotechnology is very much an international endeavor. Even very early-stage biotechnology companies are often involved in collaborative research and development arrangements with one or more foreign companies, or "partners." Sales are also very much an international effort. For pharmaceuticals the U.S. market is approximately half of the world market, while the EU and Japan represent the lion's share of the rest of world sales. This trend can only continue with the integration of the European Community into the world's largest single market for medical products, the opening of Eastern Europe to Western investment and sales, and cost containment pressures rising in the U.S. domestic market.

§1.2(B) Applications of Biotechnology: Human Health

The greatest investment in biotechnology has been in the development of human health care products: therapeutics, vaccines, and diagnostics. An Office of Technology Assessment Report, Biotechnology in a Global Economy (1991), predicted that "Biotechnology is likely to be the principal scientific driving force for the discovery of new drugs and therapeutic[s] ... as the industry enters the 21st century (p.7)." It would be difficult to describe all of the ways

---

2. U.S. Patent No. 5,747,282 (Skolnick, et al., assignee U. of Utah, et. al.). A brief word about the footnotes in this book is in order. I have tried to limit the number and length of footnotes in this book in order to make the text itself more readable. Therefore I use footnotes sparingly and for two purposes: first, where to guide the reader to the location of a source, and second, where the footnote will lead the reader to much more detailed and extensive information on the subject. For the non-lawyer reader, I have followed the general citation style of legal materials. Therefore periodicals are cited with the volume of the periodical preceding the periodical name and the page number following the periodical name: e.g. 354 New England J. of Med. 993 (2006).

in which biotechnology is being used to develop human health care products but it is possible to highlight the principal directions being taken at this time.

§1.2(B)(1) Human Therapeutics

In the development of therapeutics, the first efforts of biotechnology companies were aimed at finding and using the genes responsible for producing important human proteins that were difficult to isolate from natural tissue. Thus human insulin and human growth hormone were the first two biotechnology products approved for therapeutic use by the FDA. In each case, scientists had isolated the genes responsible for the desired protein, inserted the genes into another organism or cell-type referred to as the "expression system." Common expression systems include E. coli, a bacteria, yeast, or, more recently, suitable mammalian cell cultures such as Chinese hamster ovary (CHO) cells. After the newly engineered recombinant cells begin producing the desired protein, the protein is separated from the cell and its media and purified. The technique used is often referred to as fermentation, for the process is analogous to the process of using bacteria or yeast to produce cheese or wine.

Many approaches are being used to develop subsequent generations of biotechnology therapeutics. Chief among these are rational drug design, combinatorial library screening, antisense oligonucleotides, monoclonal antibodies, gene therapy, and proteomics (the systematic study of a cell's proteins) and genomics (the study of a cell's genes) to discover new targets for high-throughput screening. Rational drug design brings the power of the new biological techniques to the synthesis and screening of potential therapeutic compounds. It generally involves identifying the precise target for therapeutic intervention, such as an enzyme necessary for a virus to reproduce or a cellular receptor.[3] Once the target substance is identified, the gene responsible

---

3. Receptors are molecules that appear on the surface of cells and, when bound to their ligand (the biological trigger for the receptor), signal the cell to begin a particular process or response. The affinity for one protein or polypeptide to bind in a specific way to another protein or non-protein ligand is an important phenomenon that has implications for a great deal of molecular biology. The relationship between the protein features of "foreign" organisms such as viruses and bacteria and the immune system of mammals is discussed in Chapter 2, §2.3. It is fundamental to a basic understanding of biotechnology, however, that proteins (long chains of amino acids) fold into complex three-dimensional shapes and that a protein's function is determined by its three-dimensional shape. Any change in the three-dimensional folded structure may impair or prevent function. These structural changes can result from a change of even a single amino acid among hundreds, or by heating, changes in ph, or, as discussed below, by alterations in the attached sugar molecules.

for producing it may be cloned and significant quantities of the target are then available. Rational drug design then proceeds to determine the precise shape of the target and predict the structure and composition of compounds that can bind to the target and alter its function in a therapeutic way.

The screening of combinatorial libraries against a target substance is increasingly being used to avoid the often-difficult structure determination phase of rational drug design. In this approach, the purified target substance (such as a cell receptor or viral enzyme) is used to screen a very large number of randomly generated small molecules. The very large number of molecules tested against the screen makes it reasonably likely that one or more of the combinations will bind to the target. Those combinations that do bind, or show an affinity for the target, then become "hits" which, if they are successful in additional cell-based or animal model assays, become "leads" that can be tested in humans as therapeutic drug candidates.

Small, interfering RNAs, known as RNAis, are one of the very latest discoveries of molecular biology to be put to therapeutic use. Genes are composed of long sequences of nucleic acids known as DNA (more fully described in Chapter 2) and genes are transcribed into messenger RNA (mRNA) before being translated into proteins. Every cell in an organism contains the same cells, but not all genes are expressed (turned "on") in any particular cell at any particular time. Genetic diseases can be caused by the expression of genes at an abnormally high or undesirable level (such as the many cancers resulting from the over-expression of cell-growth promoting or regulating genes known as oncogenes). RNAi is a short synthetic strand of nucleic acid designed to bind to a target strand of RNA to prevent it from being translated or expressed. RNAi is already finding wide use in the laboratory study of genetic pathways and is capable of silencing a wide variety of genes and their corresponding pathways. If the problem of delivering RNAi to its target in humans can be solved, RNAi has almost unlimited potential as a new form of human therapeutic.

Antibodies are the substances produced by the immune system to bind to foreign substances and target them for destruction by other components of the immune system (see Chapter 2). Monoclonal antibodies are highly specific, highly homogenous antibodies which all bind to exactly the same target. Monoclonal antibodies are being developed for therapeutic use in at least two different ways. Some monoclonal antibodies are designed to bind to a pathogenic target, either disabling the pathogen or supplementing the body's own immune response, while other antibodies are designed to carry chemotherapeutic or radio-therapeutic molecules precisely to a target, such as a tumor. A number of antibodies have now been successfully developed and are in the marketplace. Centocor's Remicade, an antibody to TNF, has been approved in Crohn's disease and Rheumatoid Arthritis, while Genentech and IDEC have both developed tumor-targeting antibodies used in treating cancer. IDEC's Zevalin is the first radio-isotope-carrying antibody approved by the FDA and is used in the treatment of non-Hodgkin's lymphoma.

Human gene therapy has been one of the most difficult and far-reaching applications of biotechnology to the development of human therapeutics. Gene therapy is an attempt to treat disease by altering the characteristics of the diseased cells at the genetic level. Thus, while the first generation of biotechnology products, such as human insulin, attempted to treat disease by providing a replacement for the gene product, gene therapy would treat disease by providing cells with the genetic ability to produce appropriate quantities of insulin inside the body itself. Because there are many diseases that have a large genetic component (including cancer, cardiovascular disease, and Alzheimer's Disease) the long-range potential of gene therapy is enormous. At this time there have been numerous experimental trials of gene therapy. The pace of technological development in the field has been much faster than would have been believed possible only a few years ago. Thus, while substantial hurdles remain, it is increasingly clear that human gene therapy will be a major advance in medicine. In Chapter 10, Special Regulatory Issues: Human Gene Therapy, the technology of gene therapy, the legal issues, and the ethical issues it raises are considered in greater detail.

Stem cell research is the one of the most recent major areas of biotechnology research and commercialization. Stem cells are cells that are formed early in an organism's development.[4] The process by which a multicellular organism, whether a snail or a human develops, begins with a single cell which is wholly "undifferentiated" and therefore, through multiple stages of "differentiation," can develop into more mature cell types, such as neurons, blood cells, and skin cells. Stem cells can be derived from early-stage embryos (specifically, from the inner-cell mass of three- to seven-day-old blastocysts containing approximately 100 cells), in which case they are almost completely undifferentiated and are capable, at least in the ordinary process of development, of becoming any later cell type. Such cells are "totipotent." In the course of development, different lineages of later stem cell types are created. For example, hematopoietic stem cells, found in the bone marrow of vertebrates, are capable of becoming any later type of blood cell, not necessarily any other type of mature cell. Such later lines of stem cells are "pluripotent."

---

4. *See generally* http://stemcells.nih.gov/info/basics/basics1.asp and Wikipedia, *Embryonic Stem Cell*, http://en.wikipedia.org/wiki/Embryonic_stem_cell_research.

The advances of the past decade have been in techniques to isolate and purify stem cells, techniques that allow such cells to be maintained in a stable, undifferentiating state, and methods of encouraging stem cells to differentiate into particular desired cell types. The promise of such developments is in the power to create specific cell types to treat particular diseases. For example, in neurodegenerative diseases, such as Parkinson's disease, a source of replacement dopaminergic neurons created from stem cells might provide a significant therapeutic advance, as would a source of pancreatic B-islet cells for persons suffering from type 1 diabetes. However there are, as yet, no actual clinical trials that have been undertaken using human embryonic stem cell products. In §4.8(C) there is a brief discussion of intellectual property protection for stem cell research and development.

§1.2(B)(2) Diagnostics for Human Disease

There are two primary ways in which biotechnology is being used to create diagnostics for human disease: genetic probe technology and monoclonal antibody technology. Gene probes are based on the same principle as antisense oligonucleotides, described above. The key to the development of a gene probe is determining a sufficient portion of the sequence of a gene associated with the particular disease. These disease gene targets can be a viral gene, bacterial gene, or a human gene, such as one of the variant or mutant genes responsible for cystic fibrosis. Once the target gene sequence is known, a probe can be generated which is designed to bind only to the target sequence. By use of appropriate labeling and signal amplification techniques, a blood sample or tissue sample can be searched for the presence of the disease-associated gene sequence with great specificity and in a relatively short period of time.

Monoclonal antibody-based diagnostics are based on the same principle as therapeutic monoclonal antibodies, described more fully in §2.4. However, it is often even easier to produce a diagnostic antibody than a gene probe or therapeutic antibody. To produce a gene probe, a target gene sequence must be determined. To produce a diagnostic antibody, all that is necessary is a purified quantity of the desired target, for instance a strain of disease-causing bacteria. Once the target is isolated, it can be used to generate monoclonal antibodies specific for that target by the procedures described in Chapter 2. These monoclonal antibodies are then also labeled and used to determine whether or not a specific disease-causing agent (or other substance of interest, such as the hormones produced in early pregnancy) is present in a blood, tissue, or urine sample. In addition, antibodies attached to radioactive isotopes have been approved as *in vivo* diagnostics to locate and image the tumors to which the particular antibodies react.

§1.2(B)(3) Vaccines

Vaccines have traditionally been used to prevent, rather than treat, disease. Vaccines accomplish this purpose by training the immune system to recognize a disease-causing agent, enabling the "primed" immune system to overwhelm the virus or bacteria before it is sufficiently established to cause disease. To train the immune system to recognize the infectious organism, traditional vaccines have relied on techniques that either present a "killed" organism that still presents its characteristic shape (e.g. the Salk Polio Vaccine) or attenuated organisms that share some of the pathogen's structural features but not its virulence (e.g. the Sabin Polio Vaccine and the traditional small pox vaccine).

Both recombinant DNA technology and monoclonal antibody technology are being used to produce vaccines. Genetic engineering can create attenuated organisms far more precisely than the selective breeding techniques formerly used. In addition, if key "recognition" features (epitopes or antigens) of a virus or bacteria can be identified along with the genes that produce those features, genetic engineering can produce a vaccine that consists only of the recognition features, without the rest of the organism (these are called "subunit" vaccines and dominate the effort for an AIDS vaccine). DNA encoding these features can be used as a vaccine by itself, a technique known as "naked" DNA vaccination.[5]

§1.2(C) Applications of Biotechnology: Agriculture

Biotechnology has significant potential for contributing to almost every aspect of agriculture. Genetic engineering can produce new animal drugs, diagnostics, and vaccines in precisely the same ways outlined for human health care products in the preceding subsection. In addition, genetic engineering techniques can be used to modify plants and animals themselves in ways that are not possible through traditional crossbreeding and hybridization. Desired traits can be very precisely manipulated, both within the array of genetic material belonging to the organism (its "genome") and between species, even from plants to animals. In addition to a wide array of animal health care products, agricultural biotechnology companies are developing transgenic plants that are pest resistant (or, in some cases, pesticide tolerant), slower to rot,

---

5. http://www.vical.com/company/dnatech.htm.

drought tolerant, higher in nutrients, or even some that produce valuable drugs.[6] Transgenic animals are also being developed to be leaner, grow faster, and resist disease. Finally, agricultural biotechnology is already marketing biopesticides, which are produced using genetic engineering techniques to isolate substances found in nature that are "natural" pesticides and genetically modify crops to produce those proteins themselves.

Agricultural biotechnology is developing rapidly but faces its own particular problems. In addition to the general business problem of addressing smaller markets that are difficult to penetrate, agricultural biotechnology faces substantial legal hurdles in the form of environmental and food safety concerns, particularly in Europe. An overview of agricultural biotechnology regulation is included in Chapter 7.

§1.2(D) Industrial Applications of Biotechnology

Biotechnology is less than 30 years old and biotechnology processes are still relatively expensive. Thus, biotechnology is least advanced in developing products for general industrial production, which is often low-margin and extremely cost-sensitive. The primary applications of biotechnology in industry are in the areas of hazardous waste remediation, biosynthetic materials, the development of bioengineered catalysts for industrial processes, and the creation of "biosensors." In hazardous waste remediation, microorganisms are being developed which can metabolize and detoxify a variety of common toxic wastes such as PCBs (poly-chlorinated biphenyls) and dioxin. In developing biosynthetic materials, efforts are directed at using biotechnology to produce natural polymers such as silk.

Still another industrial use of biotechnology is in the development of new biocatalysts for industrial chemistry. Catalysts are substances that speed chemical reactions by bringing together the reacting substances and presenting them in the appropriate orientation. Enzymes are the catalysts of biological reactions, and recombinant DNA technology can be used to produce and improve enzymes that can serve in a wide range of industrial processes, from digesting organic stains in detergents to substituting for the calves' intestinal product rennin in the production of cheese.

Biosensors can be thought of as the industrial extension of monoclonal antibody diagnostics, described in §1.2(B)(2). In the case of biosensors for industrial application, the antibodies are not labeled with a color-changing chemi-

---

6. *See* http://www.bio.org/health care/pmp/keypoints.asp.

cal; instead, antibodies that are bound to their target (such as an industrial effluent or toxic waste product) are detected by sensitive electrical equipment calibrated to display the concentration of the antibody's target molecule.

As in agricultural biotechnology, the most important legal issues confronting the development of industrial biotechnology may be in the field of government regulation. The regulatory framework for industrial biotechnology is surveyed in Chapter 7.

§1.3 Summary: The Life Cycle of Biotechnology and the Lawyer's Role

The premise of this book is that law and lawyers play an indispensable role in facilitating the development of biotechnology at virtually every stage. Biotechnology product development has a relatively uniform "life cycle" and the remainder of this introduction traces the life cycle of a typical biotechnology product's development, from laboratory to formulary, briefly noting the law and lawyers involved at each stage.

Research and Technology Transfer (Laboratory): Most biotechnology products have their inception in NIH-funded research at a university or not-for-profit research institute (such as The Salk Institute or Whitehead Institute). Although such research would seem to be far removed from any legal considerations, many basic parameters of the university research environment are the result of laws or government regulation. A university's right to inventions made by university scientists is governed by both statutes and contracts. Research involving recombinant DNA is subject to the NIH guidelines for such research. Research involving animals is subject to a variety of state and federal regulations and to the review of an institutional board dealing with animal welfare, mandated by federal law. An even more important set of issues, from the standpoint of the university researcher, surrounds the rights to biological materials often exchanged with other researchers. The procedures for transferring technology from the university to the private sector are regulated both by statute and the NIH.

Intellectual Property Protection: Although there is a significant legal framework surrounding university research, there is little contact with lawyers during the basic research process. The involvement of lawyers is likely to come at the successful completion of a research project with commercial potential. At that stage, the university is likely to rely on patent lawyers to protect the university's claim to any inventions and, possibly, to assist in licensing the technology to the private sector for commercial development.

Financing: It is not at all unusual for a company to be formed specifically to exploit a particularly valuable or enabling technology licensed from a university. In some cases, the university itself desires to transfer the technology to a start-up company, with the university's compensation being in large part the stock of the new company. The Salk Institute facilitated the creation of SIBIA, the La Jolla Cancer Research Foundation helped found Telios, and Boston University gave birth to Seragen. Whether the technology is transferred to a university-spawned start-up company or to an independent, small dbc, the capital needs for the commercial development of a biotechnology product are likely to be in excess of $50 million up to $200 million and lawyers in the field of corporate transactions, often with a subspecialty in venture capital law, play a key role in biotechnology commercialization. Lawyers are also certain to be involved in the future arrangements for corporate partnerships that will further fund the venture and in the initial public offering of the company's stock that will in many cases be necessary to bring in sufficient capital to complete development of the company's first product.

Regulatory Approval: If all goes well in the early years of biotechnology product development, from both a technological and financial perspective, the next major set of legal issues to enter the picture centers on government regulation of the testing and marketing of biotechnology products. Whether a biotechnology product is aimed at the human health care market, the agricultural market, or the industrial market, there will be substantial government regulatory hurdles to surmount before the product can be sold. The complex matrix of government regulation of biotechnology was outlined in The Coordinated Framework for the Regulation of Biotechnology, published by the Office of Science and Technology Policy on June 26th, 1986 (51 Fed. Reg. 23302). The key government agencies regulating the development of biotechnology are: the Food and Drug Administration (FDA), for human health care products, new animal drugs, and many food products affected by biotechnology; the Environmental Protection Agency (EPA), for most industrial applications of biotechnology and all issues surrounding the development of biopesticides or pest-resistant species; and, the U.S. Department of Agriculture (USDA), for the development of new animal vaccines and the environmental safety of new, genetically engineered plant species (other than for pesticide or pest-resistant purposes).

Marketing and Reimbursement (Formulary): While regulatory approval for testing and marketing a product may be the biggest hurdles for a biotechnology product's development, the legal issues certainly do not stop with the product's first sales. Marketing of biotech products, especially human health care products, involves close regulation by the FDA and delicate issues of third-party payor reimbursement. Formularies are lists of drugs, maintained by health insurers and providers, which are preferred for use in treating particular diseases or are reimbursed at preferable rates. Not all drugs for any particular condition, such as depression or elevated cholesterol, receive the additional marketing advantage of being the preferred drug included in most formularies. Making the case for formulary inclusion is part of the drug development and marketing process.

Like other drugs, a biotechnology drug may often have utility in disease conditions, or indications, beyond that for which it was tested and approved by the EPA. For example, Amgen's G-CSF was approved for chemotherapy-induced neutropenia and NeoRx's GM-CSF for use in connection with autologous bone marrow transplants, yet each may have utility for the other's indication. The delicate problem of complying with the FDA's rules for such "off-label" uses and other marketing issues is considered in Chapter 8.[7]

Post-Formulary (Liability): Unfortunately, for many biotechnology products, regulatory approval and successful entry into the marketplace may not end the issues requiring the guidance of lawyers. It is inevitable that biotech products, especially biotech pharmaceuticals, vaccines, and diagnostics, will give rise to claims of personal injury and product liability. The first wave of lawsuits for injuries due to alleged injuries from biotech products were those claiming injury from StarLink[8]-contaminated corn.

Although the development of a biotechnology product over time does move through the general phases described here as the product life-cycle, from basic research through post-market liability, at any particular time in the development of biotechnology the process may well more closely resemble a Rubik's Cube than an orderly, linear progression of events. To the entrepreneur planning a biotechnology company, or to the venture capitalist making an early investment in the company, the various future stages of development may be viewed as interlocking parts that must fit together in the plan of development. The problem of developing and proving the technology must be aligned with the costs of that development (and of course if it is not proprietary, or other-

---

7. While FDA regulation of marketing and promotion is covered in Chapter 8, the issues of managed care, Medicare, Medicaid, and reimbursement by private insurers are beyond the scope of this book.

8. Andrew Harris, *Corn Rows: Genetically Engineered Corn That Creates Its Own Pesticide Prompts Almost 30 Lawsuits*, MIAMI DAILY BUS. REV., September 18, 2002 at A8:
> In the two years since it was revealed that StarLink, a strain of corn genetically engineered to create its own pesticide, had drifted into the human food chain, nearly 30 lawsuits have been filed against the corn's creators, growers and distributors by farmers, consumers and fast-food franchises.

wise able to be protected as intellectual property, it is unlikely to be developed at all). A financing strategy must account not only for the time and expense of initial R&D, but also for the time and money costs expended in the regulatory process. That regulatory process largely circumscribes the projected market. Within the target market, pricing and reimbursement problems ultimately determine the financial return on the project, connecting back to the financing issues.

Viewed over time, lawyers play a part in virtually every major phase of biotechnology commercialization. Viewed from the vantage point of a biotech executive faced with the extremely complex problem of fitting the various pieces of the puzzle together, knowledge of the legal issues involved is indispensable. Law may or not be the glue that holds the parts together, but it is beyond any doubt that law and lawyers unquestionably help shape and define the key pieces. We hope this book will help those involved in solving the problems of biotechnology commercialization understand the legal environment in which biotechnology is developing.

CHAPTER 2

A Primer on the Basic Science and Applications of Biotechnology

§2.0 Modern Molecular Biology and Immunology

Chapter 1 provides an overview of the biotechnology industry and the life cycle of biotechnology products from the research laboratory to the marketplace. At different stages in this life-cycle, the degree of scientific expertise required of a business executive or lawyer varies significantly, with a great deal of scientific expertise required to make early stage decisions about intellectual property protection but much less to make decisions about alternative financing arrangements. However, a certain minimal understanding of the science upon which biotechnology is based is essential for anyone involved in managing or advising a biotechnology business, from the CFO to the corporate counsel. This chapter provides an overview of the basic science on which most of biotechnology is based: modern molecular biology and immunology.

§2.1 DNA and the Genetic Code

In many ways, modern biology may be said to have begun in 1953 when Watson and Crick discovered the three-dimensional structure of DNA, the basic genetic material which determines the heredity of all plants and animals (although some viruses use RNA as their basic genetic material). Watson and Crick's discovery that DNA takes the shape of a double helix (spiraling ladder) is familiar to virtually every high school graduate. The backbone of the helix, or the sides of the ladder, is composed of sugar (deoxyribose) and phosphate molecules, while the rungs of the ladder are varying sequences of interlocking nucleotides, known as Adenine (A), Guanine (G), Thymidine (T), and Cytosine (C). The great power of the structure is derived from the invariant way in which the four nucleotides on the two strands will pair up, an A always pairs with a T and a G always pairs with a C. Two complementary, interlocked nucleotides are known as a "base pair." In this way, each strand of DNA contains the information necessary to generate the other strand, with each nucleotide along a single strand requiring that its complementary nucleotide be placed on the opposite strand. Thus, if the helix is unwound and a new complementary stretch is synthesized for it, any particular stretch will always replicate its original complementary sequence. This natural pairing of the four nucleotides is the basis for the faithful replication and preservation of the hereditary message of the DNA genetic code.

The information stored in the DNA is built on the four-letter (A, G, T, C) alphabet of the nucleotides. When the helix is unraveled so that one strand (the "sense" strand) can be "read," the four letters are read in blocks of three-letter "words" or codons. Sixty-four possible three-letter codons can be generated from the four-letter alphabet. Each of the 64 possible codons represents one of the 20 amino acids or a signal to start or stop "reading." Since there are many more (64) codons than there are naturally occurring amino acids (20), most amino acids can be represented by more than one codon (scientists refer to such a code as "degenerate"). For example, the amino acid leucine is encoded by, among others, the combinations CTT, CTC, and CTA. However, there is absolute specificity in the other direction—that is, CTA represents leucine and only leucine. Those few amino acids for which there is one and only one codon are said to have "unique" codons (for example, tryptophan is encoded only by TGG). These properties of uniqueness and degeneracy play a major role in the patentability of new gene sequences, as discussed in Chapter 4, §4.6 Nonobviousness.

§2.1(A) From DNA to RNA to Protein: The Central Dogma of Molecular Biology (Now Known to be False!)

A stretch of DNA containing some number of codons is transcribed into messenger RNA[1] and translated by t-RNA into a sequence of the correspon-

---

1. RNA, or ribonucleic acid, is very similar to DNA in both structure and function. For purposes of this overview it is sufficient to know that the principle differences are that the sugar used in the backbone of RNA is ribose, rather than DNA's deoxyribose, and that one of the four nucleotide bases of DNA, Thymidine, is replaced by a similar nucleotide, Uracil, in RNA.

ding amino acids. Such chains of amino acids are known as polypeptides, and proteins are simply complete polypeptides of a substantial length. (A short polypeptide or peptide of five amino acids may have biological activity, but it would not ordinarily be thought of as a protein.) Although there are three broad categories of biological molecules—carbohydrates, lipids, and proteins—the principal determinants of the basic structure and function of any living organism are the proteins coded for in its DNA. The proteins are used in the metabolic and synthetic pathways that produce the other biological molecules.

§2.1(B) Genes: The Recipe for a Protein

One useful way of thinking about the meaning of the term "gene" is by thinking of a gene as a stretch of DNA which contains the complete set of instructions (in codons) for the construction of a particular protein (chain of amino acids with a particular function). Furthermore, in a complex, or multicellular organism, whether it is an octopus or a human, virtually every cell of any single creature contains exactly the same genetic information, or DNA, as every other cell of that creature. The difference between a skin cell and a liver cell is determined by which genes are expressed and in what proportions.

§2.1(C) Genes, Proteins, and Splice Variants

The relatively straightforward process of the central dogma, DNA to RNA to Protein, one gene to one protein, has been rendered slightly more complicated in recent years. It has long been known that there are two basic categories of cells (see also §2.2, below). Prokaryotic cells, such as bacteria, have no organized nucleus and contain only "coding" DNA in their genetic material. Eukaryotic cells contain an organized nucleus with chromosomes and have long stretches of non-coding DNA, known as "introns," separating the coding regions, known as "exons." The eukaryotes include all higher organisms from yeast to man. While it was previously thought that the instructions for a protein, i.e. a gene, consisted of particular exons that were always linked together in the same way to produce a protein, it is now known that exons can be rearranged, or even recombined with other exons, to form different proteins. These different variations of the exon elements of a gene are known as splice variants. The result is that one eukaryotic "gene" can produce several different proteins, each a splice variant of that gene. It is through the process of different splicing and rearrangement of exons that the total number of human genes, now thought to be in the neighborhood of 30,000, can produce something on the order of five times that many human proteins.

Many proteins are involved in the process by which the expression of individual genes is determined, or regulated. Such regulatory proteins play a role in starting and stopping the process by which a particular segment of DNA is "transcribed" into "messenger" RNA (m-RNA) and then "translated" into proteins. The proteins that control the expression of a gene are called "transcription factors" and the pathway that activates a transcription factor is often referred to as a "signal transduction and activation of transcription" or "STAT" pathway. The "signal" that activates a transcription factor may be initiated by the binding of a messenger chemical outside the cell to a receptor that extends through the cellular membrane on the cell's surface (a transmembrane receptor), or it may be generated by an internal signaling molecule and its intracellular receptor. A molecule that binds to a receptor is referred to as its "ligand."

§2.1(D) Putting It All Together: Genomes, Genotypes, and Phenotypes

With this basic set of principles, the genetic processes of the cell (whether a human liver cell or a fruit fly neuron) can be understood. DNA is the repository of the genetic instructions that define and control the individual cell. All the cells of any individual organism contain the same genetic material.[2] The set of genetic instructions that define a species (and are contained within each cell of every "normal" member of that species) is the "genome" of that species. The Human Genome Project has now completed the sequencing of the approximately three billion base pairs (the majority of which are non-coding introns, see §2.3(C)) that comprise the 23 pairs of human chromosomes. Humans, like all sexually reproducing organisms, have two copies of every chromosome, one maternally inherited and one paternally inherited. These three billion base pairs specify between 30,000 and 40,000 genes strung along each of the twenty-three (paired) chromosomes.

A chromosome is simply a very large bundle of DNA on which particular sets of genes are found. For instance, a gene involved in a common form of leukemia is found on chromosome 11. For any particular gene within a species, there may be a number of different DNA variations in the exons, or coding portions, of the gene. Such different versions of genes are known as alleles (the different genes which can occupy a given location in a species

---

2. It is a slight simplification to say that all the cells of an organism contain the same genetic material. In sexually reproducing organisms, the germ cells produce sex cells (sperm and eggs) that contain only half the amount of genetic material of a non-sex cell.

genome). At any particular point in time for a particular gene (such as that which determines eye color in humans) there exists a finite number of alleles found in that species, while the process of mutation and recombination (changes in the genetic material of the reproductive cells of an organism and thus heritable) can create still other alleles. The genetic variations found within an individual organism are referred to as the "genotype" of that organism, while those observable differences produced by genetic variations, such as hair color or as manifest genetic disease, are referred to as the organism's "phenotype."

For many genes, there are numerous "normal" alleles, such as those that determine eye color or blood type. In the case of some genes, only some of the alleles may be considered "normal" because other variations would cause disease. For example, there are several different mutations that have occurred over time to give rise to different pathogenic alleles of the cystic fibrosis gene (which encodes the cystic fibrosis transmembrane protein), but just one of those variants accounts for approximately 75 percent of cystic fibrosis cases. However, one of the most important contributions of the last 10 years or more of genetic research has been to underscore the importance of the interaction among many genes in determining the extent to which any one allele is likely to affect phenotype. Two individuals with exactly the same cystic fibrosis mutation may have very different degrees of disease because of the interplay with other genes. Similarly, mutations in the first genes known to be associated with breast cancer, BRCA1 and BRCA2, produce disease in only some of the women with any particular mutation, underscoring the importance of other genes and possibly environmental factors. The extent to which a gene affects phenotype, or is observable, is known as penetrance. Much public concern over genetics issues is caused by a misinformed impression that genetic variations are determinative of outcomes, rather than more or less strongly (or even very weakly) statistically linked to outcomes.

§2.2 Prokaryotes, Eukaryotes, and Viruses

Biologists divide organisms into prokaryotes, eukaryotes, and viruses. Prokaryotes—bacteria and archaea (one-celled organisms that have bacteria-like structure, but some of the biochemical features of more complex eukaryotes)—have no organized membrane-defined "organelles" and no nucleus in which the cell's DNA is contained. Eukaryotes have a membrane-bounded, organized nucleus, and all "higher" organisms, from yeast to man, are eukaryotes.[3] Viruses are the simplest organisms and generally are constructed simply with an outer protein envelope, or "coat" protein, enclosing the viral genetic material. In the case of RNA viruses and retroviruses the genetic material is in the form of RNA.

§2.2(A) Viruses

Because a virus has only genetic material surrounded by protein, it is an "obligate parasite"—it needs a host to reproduce and must rely on its host's apparatus for carrying out much of its life-cycle. DNA viruses use the host's m-RNA and t-RNA. RNA viruses use the host's t-RNA. Retroviruses, such as HIV (the AIDS virus), encode RNA that produces a protein enzyme known as reverse transcriptase (RT). The retrovirus's RT then "reverse transcribes" the viral RNA into DNA. The new DNA version of the retroviral genome is then, with the assistance of other proteins coded by the virus, integrated (inserted) into the host's nuclear DNA. After integration, the viral DNA is transcribed and translated by the host cell.

Viruses are generally specific both for particular hosts and for particular cell-types within a host. Recently, however, the ability of viruses to adapt and be transmitted to new hosts is believed to be responsible for outbreaks of new diseases, such as SARS. The general specificity of viruses is derived both from the "fit" between molecules on the outer surface of the target cell and the envelope protein of the virus and from the suitability of the particular target cell's own machinery for performing the necessary functions for the (necessarily parasitic) virus. One example of the relation between virus and host is the HIV or AIDS virus. The HIV virus infects human T-cells because a protein known as GP-120 on the viral envelope, or coat, binds to a protein known as CD4 on the surface of the targeted class of T-cells.[4]

§2.2(B) Bacteria (Prokaryotes)

Bacteria are the major group of prokaryotes. Bacteria are many orders of magnitude larger and more complex than viruses. (There are viruses, known as phage, which target and infect bacteria.) Bacterial DNA is found throughout the cytoplasm of the bacteria, often in one long strand of bacterial DNA

---

3. Again a slight oversimplification, because in eukaryotes some cell types may lack an organized nucleus. In vertebrates, for example, red blood cells do not have a nucleus.
4. CD4 is not the only human cell surface protein required for HIV entry into cells; CCR5 is also necessary.

and in the form of "plasmids," which are circular rings of double-stranded DNA. For a given strain of bacteria, each plasmid contains a fixed set of bacterial genes in a particular sequence around the circular chain. Bacterial plasmids, particularly that of E. coli, have been studied extensively and were the first commonly used system for producing genetically engineered proteins.

E. coli has been one of the most important prokaryotes in biotechnology. There are different strains of E. coli, most of which are non-pathogenic and many of which can be found flourishing in the intestines of human beings, usually without any ill effect. In part because it is common, harmless, and easy to grow and maintain, E. coli was used in most of the first experiments involving genetic engineering. Indeed, it was Paul Berg's 1971 proposal to transfer genetic material from SV-40 (a simian virus) into E. coli that led first to the first great controversy over genetic engineering, then to the Asilomar meeting, and ultimately to the development of the NIH-RAC to regulate experimentation with recombinant DNA (see Chapter 7).

§2.2(C) Eukaryotes

Eukaryotic cells are significantly more complex than prokaryotic cells. In addition to a defined nucleus containing the cell's genetic material organized by chromosome, eukaryotic cells have significantly more complicated structures for manufacturing, transporting, and secreting biological molecules, specialized structures for producing energy, and many times more genes. Yeasts are among the least complicated eukaryotic organisms and yeast biology has therefore provided one of the most useful model systems for understanding eukaryotic cell biology and studying eukaryotic genetics. Other eukaryotic model systems of significance in biological research include C elegans (a roundworm), drosophilae (fruit flies), and mice. The production of proteins by genetic engineering is now commonly done in eukaryotic cell systems and frequently relies on Chinese Hamster Ovary (CHO) cells.

§2.3 Introductory Genetic Engineering for Beginning Non-Biologists

E. coli has been an important tool in the development of biotechnology. Perhaps the most important use of E. coli was in the development of the first system for the production of recombinant or genetically engineered human proteins. Although it may be sufficient for many lawyers to conceive of genetic engineering simply as the cutting and pasting of genes within and between species, it should be obvious that such cutting and pasting requires biological scissors and paste. The biological scissors used in genetic engineering are known as restriction enzymes, each of which has the ability to cut DNA at a precise location as determined by the adjacent sequence of nucleotides. The first restriction enzymes discovered were found in E. coli. For example, the restriction enzyme EcoR1 recognizes the sequence GAATTC and cuts DNA between the G and the A when it "finds" that sequence. The resulting DNA fragments are, for reasons which go beyond the scope of this text, said to have "sticky" ends to which a desired DNA sequence can be attached using the biological "paste" known as DNA ligase (another protein enzyme). In this way, the DNA that encodes the protein known as human insulin can be inserted into an E. coli plasmid, the plasmid reinserted into the bacteria, and human insulin produced by the resulting "genetically engineered," or recombinant, bacteria.[5] "Cutting and pasting" DNA into eukaryotic cells, including mammalian cells, involves more complicated techniques.

E. coli were used as the first system in which to produce many of the early recombinant, or genetically engineered proteins, such as human insulin and human growth hormone. The E. coli plasmid known as pBR322 was one of the first significant stretches of a genome to be completely mapped and sequenced, that is, the precise normal order of every one of its base pairs determined, as well as the genes that are encoded for by the various segments of the plasmid. A substantial number of different restriction enzymes have known, precise effects in cleaving the pBR322 plasmid, adding to the utility of E. coli as a host system for genetic engineering.

§2.3(A) Gel Electrophoresis

In addition to the ability to cut and paste DNA, genetic engineering requires the ability to manipulate DNA in several other ways. The first additional skill needed is the ability to sort out the pieces or fragments of DNA that result when a restriction enzyme is used. This is easily accomplished by gel electrophoresis, which is a broadly useful technique for sorting a variety of different molecules by size, or molecular weight. Gel electrophoresis is

---

5. An important limitation on such uses of E. coli is the inability of the bacteria to glycosylate, that is, add sugar molecules, to human proteins in the same way as occurs in the human cells. For some proteins the glycosylation does have a significant impact on the protein's folding, or three-dimensional shape, and therefore on the protein's function. In such cases a yeast system or mammalian cell culture may be necessary to produce the desired protein in functional form.

based on the principle that, in an electrical field, charged molecules will migrate towards the poles, through a gel, at different rates determined by their size and charge. This is relevant because DNA carries a slight electrical charge, as do proteins. The separated fragments of DNA can then be detected by the addition of short pieces of radioactive (or otherwise labeled) DNA "probes," which find complementary sequences in the target DNA and "hybridize" or bind, allowing the various hybridized fragments to be detected and the distances between them used to generate estimates of their size. Gel electrophoresis has been dramatically enhanced in recent years by the addition of capillaries (small tubes) to the process to allow much greater sensitivity for larger starting samples.

§2.3(B) DNA Sequencing

DNA sequencing was greatly improved during the course of the Human Genome Project. The core of the process is the cutting of the DNA to be sequenced into smaller fragments, separated into four batches where each set of fragments is replicated, but where a precisely calculated fraction of the nucleotides used to make the new complementary fragments are labeled "terminator" nucleotides of either A, T, G, or C. Since only a fraction of the "Adenosines" in the "A" batch are labeled terminators, strands of varying length will be produced before one of the labeled "terminator" fragments is used in replication and the replication stopped with the labeled terminator at one end. Thus in one batch there are all fragments of varying lengths ending in C, in another batch fragments ending in A, and so on.

Sequencing machines then sort all of the fragments from all four batches by length and, by reading the labels, determine the order of the As, Ts, Gs, and Cs along the entire length of the original DNA (the shortest fragment might end in an A, the next shortest in a C, thus indicating that in the original DNA the A was followed by a C, and so on). With the help of automated sequencing machines, a strand of DNA several thousand base pairs in length can be sequenced (have the order of its nucleotides determined) in a relatively short period of time and, with technological improvements continuing, the sequencing can be done with increasing speed. This automation of the Sanger process of gene sequencing using labeled terminator nucleotides enabled the Human Genome Project to be completed ahead of schedule.[6]

---

6. For those interested in a lengthier explanation of sequencing that is still aimed at a general audience, see *Facts About Genome Sequencing*, http://www.ornl.gov/sci/techresources/Human_Genome/faq/seqfacts.shtml.

§2.3(C) Cloning Genes

In addition to the cutting and pasting techniques and sequencing methods described above, research and development involving recombinant DNA, or genetic engineering, requires the ability to "fish" out genes' (DNA or RNA sequences) coding for proteins of particular interest and the ability to make copies of those genes. This process of searching for and duplicating genes of interest is what is usually known as gene cloning. While the details of the process are certainly beyond the scope of this book, a very basic understanding of the principles should be helpful to lawyers involved in the business and regulatory aspects of biotechnology. (Biotechnology patent lawyers may safely skip this section. Others should note its importance for the discussion of nonobviousness in Chapter 4, §4.6.) As stated in §2.1 above, the four-letter alphabet of DNA is used to encode the amino acid sequences of proteins, and the DNA message is transcribed into m-RNA and then translated by t-RNA into the proteins themselves. In the process then there are m-RNA copies of every DNA sequence or gene that is being expressed (activated, or used in the manufacture of protein at that time in that cell). The m-RNA must travel from the DNA to the site in the cell where proteins are assembled. In the process, long "junk" stretches of the DNA sequence, known as introns, are edited out, leaving a shorter m-RNA set containing only the nucleotide sequence (the exons) that encodes the amino acids of the protein.

The process of fishing for a desired mammalian gene has generally focused on these edited m-RNAs. At any given time, any cell would contain at least one m-RNA strand for each protein being produced by that cell. After extracting the various strands of m-RNA from a particular cell or cells and sorting them by gel electrophoresis, each resulting m-RNA "band" may be converted into DNA through a two-step process. The enzyme reverse transcriptase translates the m-RNA into a single-strand of DNA and the enzyme DNA polymerase converts the single stranded DNA into a double stranded DNA (known as a "cDNA"). Each cDNA generated by this process can then be inserted into E. coli or another expression system (as described above) and the resulting protein products assayed to determine which, if any, of the cDNAs is producing the desired protein. Implicitly, one of the major difficulties in this process is finding an m-RNA for a particularly rare protein. If a protein is produced in very small amounts, by a limited number of cells, only when needed, then the chances of finding an m-RNA for that protein in any given group of cells might be very low. It is that difficulty which rendered the search so competitive and expensive for genes for such rare proteins like the colony stimulating factors (which stimulate the proliferation of various types of blood cells). Since one of the key difficulties in the search for particular genes might be the minute amount of m-RNA present for a rare, transiently-produced protein, it is now greatly facilitated by the technique of DNA amplification known as Polymerase Chain Reaction, or PCR (discussed in the next subsection).

§2.3(D) Polymerase Chain Reaction

The last basic tool of genetic engineering discussed in this chapter is the ability to make many copies of any particular DNA sequence of interest. Although a number of techniques have been developed for this purpose, the first method to be developed, the polymerase chain reaction (PCR), is still the most widely used. PCR earned its inventor, Kerry Mullis, the Nobel Prize and was invented by him while at Cetus, a company that is now part of Chiron.[7] Hoffman-LaRoche and Perkins-Elmer ultimately acquired the rights to the invention. PCR involves the use of a form of DNA polymerase, an enzyme that, in the presence of DNA and a supply of nucleotides, will make a strand of DNA complementary to any "primed" single-stranded DNA. The new double-stranded DNA is then treated so that it separates into two single strands, more "primer" is added, and the DNA polymerase turns the two single strands into two double strands. In this way, repeated cycles of doubling, separation, priming, and redoubling produce a very large number of copies of the original DNA within a very short period of time.

§2.4 AN INTRODUCTION TO IMMUNOLOGY (AND MONOCLONAL ANTIBODIES)

Genetic engineering is one of the two foundational technologies upon which the biotechnology industry is built. The other involves the production of monoclonal antibodies. An understanding of monoclonal antibodies requires a brief overview of the some of the basic principles of immunology, just as understanding genetic engineering required an introduction to the molecular biology of DNA, RNA, and proteins.

The basic function of the immune system is to recognize foreign (non-self) "invaders" and to respond by disabling or destroying the invader before it is

---

7. *See* Saiki, Bugawan, Horn, Mullis and Erlich, *Analysis of enzymatically amplified beta-globin and HLA-DQ alpha DNA with allele-specific oligonucleotide probes*, 324 Nature 163 (Nov. 13–19, 1986).

able to produce any toxic or pathogenic effect. Thus, there are two distinct functions for the immune system: recognition and response. The recognition function of the immune system extends to any foreign substance larger than a small organic molecule, from polypeptides and polysaccharides at the smallest end, up through bacteria and foreign cells at the larger end. The response function of the immune system involves a variety of mechanisms, from the targeting of cytotoxic proteins known as complement, to the operation of a variety of types of cells that can lyse (rupture or break open) foreign cells or digest foreign proteins.

To stimulate an immune response, one or more features (chemical structures) of the foreign substance must be recognized as non-self. Substances recognized by the immune system are called antigens, and the particular feature or features recognized are known as epitopes or antigenic determinants. Thus a larger antigen, such as a bacterium, may present a great many antigenic features (such as each of the glycoproteins and polysaccharides attached to the bacterial cell wall) and a smaller antigen, such as a three or four amino acid polypeptide may present only one antigenic determinant or epitope.

The recognition function of a healthy immune system is extraordinarily specific, both in the ability to distinguish self from non-self and to distinguish between the various antigenic determinants or epitopes of different foreign substances. The ability to distinguish among antigens is the reason why immunity is so specific: exposure to measles provides an immune response and protection against the measles virus but not against the hepatitis virus, which has different epitopes or antigenic determinants.

The immune system of all vertebrates involves white blood cells known as lymphocytes. There are two broad categories of lymphocytes: T-cells and B-cells. The two categories of lymphocytes give rise to two different (although somewhat interdependent) types of immune response: cellular or cell-mediated (T-cell) and humoral or antibody-dependent (B-cell). B-cells produce antibodies, the proteins that bind to antigens and stimulate further immune response. T-cells are involved in either stimulating or suppressing the action of B-cells or other T-cells (so-called helper and suppressor T-cells) or in actually breaking down or digesting foreign invaders (so-called cytotoxic T-cells). While the structure and function of T-cells is of great importance to research involving inflammatory and autoimmune diseases, it is the production of antibodies by B- cells that has produced the second major component of the biotechnology industry and which is discussed in greater detail in this subsection.

There are millions of B-cells in a mature adult mammal, and each B-cell produces one and only one antibody, or intricately folded protein. Antibodies are Y-shaped structures, consisting of two long, "heavy-chains" which form the basic Y structure (see figure 2.1) and two shorter, "light chains" which are on the outside of and attached to the two arms of the "Y."[8] The two heavy-chains and the two light chains are identical, and each consists of a constant region, a variable region, and a hyper-variable region. The constant regions are at the base of the chains, and the variable and hyper-variable sections run close to the tips of the two arms. The variable and hyper-variable regions of the light chain are parallel to and interface with the variable and hyper-variable regions of the heavy-chains. It is the uniquely configured space between the variable and hyper-variable regions of the light chain and the variable and hypervariable regions of the heavy chain that form the "lock" of each antibody and within which the antibody's specific antigenic determinant or epitope is bound. Because of the underlying mechanism for antibody gene rearrangement and mutation, millions of different configurations of light and heavy chain variable regions can be created, each with an affinity for one and only one antigenic determinant or epitope.

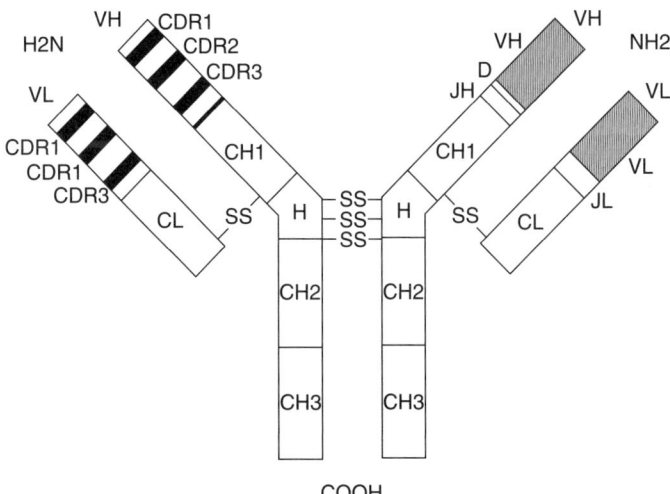

This figure is reproduced from U.S. Patent 5,229,272. (Note: this patent diagram shows an antibody that was patentable because it was genetically engineered to bind a different antigen with each arm, whereas ordinary antibodies bind the same antigenic determinant with both arms.) The hyper-variable

---

8. *See* Fig. 2.1.

regions involved in binding are shown as CDR1, CDR2, and CDR3 on the "left" arm of the antibody, and ordinarily are mirrored on the "right" arm of the antibody. The "light chain" is the shorter "arm" of the antibody, while the "heavy chain" is the longer, inside arm of the Y, and continues into the constant regions labeled "CH2" and "CH3" (constant heavy 2 and constant heavy 3). The constant, or non-variable region of the light chain is labeled "CL" on both arms of the antibody.

When an antibody encounters its matching antigen, there are several important consequences. First, if the epitope to which it binds is one that is essential to the pathogenicity or biological activity of the antigen, pathogenicity or biological activity may be blocked. An example of this would be the binding of an antibody to a viral epitope that is necessary for the virus to bind to and enter its target cells, or to a region of the inflammatory substance TNF that prevents it from stimulating inflammation. Such antibodies are often referred to as "neutralizing" antibodies. Second, when the antibody binds to the antigen, it signals for the production of complement (toxic proteins which can destroy the antigen) or cytotoxic T-cells, or both. Third, through the process of clonal selection, the B-cell that produces an antibody that binds to an antigen is stimulated to produce many more antibodies and to reproduce itself many times. Thus the population of B-cells that produce a particular antibody will grow quickly upon exposure to antigen. Fourth, through a process of cell "memory," the genetic instructions for that antibody are stored in "memory" B-cells, which allows much more rapid production of that B-cell line and its antibody should that antigen be encountered again.

The work of Kohler and Milstein in 1975 enabled scientists to harness much of the power of B-cells and antibodies. For a very long time prior to Kohler and Milstein, it was known that mammals inoculated with an antigen would produce antibodies in response to the antigen and that those antibodies could be isolated from the serum of the inoculated animals. However, because any virus or bacteria will have numerous antigenic determinants, and because any one antigenic determinant may bind in slightly different fashion to several different antibodies, the collection of antibodies from the serum of inoculated animals produces a varied assortment of so-called "polyclonal" antibodies.

Kohler and Milstein's breakthrough was in isolating individual B-cells from an inoculated animal and fusing the individual B-cells with myeloma cells (immortal cell-lines of bone-marrow tumor origin). Each product of the fusion of a B-cell and a myeloma cell is an immortalized (capable of indefinite growth and reproduction in cell-culture) cell-line, called a "hybridoma," which produces only one specific type or "monoclonal" antibody—the antibody that was being produced by the original B-cell fusion partner. In this way, a "library" of hybridomas can be screened to single out those producing antibodies with the greatest affinity ("tightness" of binding) and specificity for the desired target. The selected hybridoma can then be propagated to produce large quantities of the desired monoclonal antibody, each antibody with precisely the same affinity for precisely the same antigenic determinant. This ability to produce monoclonal antibodies has been of enormous importance in the area of medical diagnostics and holds promise for therapeutic and vaccine development as well.

Although the production of monoclonal antibodies through the fusing of lymphocytes with myeloma cells into hybridomas remains the dominant method for antibody production, a few other technological developments deserve mention. First, most of the earlier monoclonal antibodies were generated by exposing mice to an antigen, with the resulting product being a murine antibody. Because of subtle differences between murine antibodies and human antibodies, murine antibodies sometimes have less specificity for particular antigens than is desired or, if injected into a human, cause an immune response to the foreign (mouse) protein comprising the antibody. This immune response, a reaction known as HAMA (human anti-mouse antibody), can be so severe as to preclude repeated administration of the antibody. A variety of techniques have been developed to overcome these problems. First, for some applications, it is possible to use only the binding portion (variable region) of the murine antibody, clipping off the antibody "tail." Such antibody fragments, or "Fabs," may have the necessary affinity without the same degree of allergenicity or HAMA. Second, it is possible to produce "chimeric" antibodies, which preserve the essential antigen-binding portions of the murine variable region and use a human or humanized constant region, which also reduces allergenicity. Finally, some companies are producing fully human hybridomas and monoclonals, using B-cells from the lymph tissue of volunteers who have been diagnosed as positive for a particular pathogen or disease and fusing those B-cells with human immortalized cell-lines.

In addition to a variety of approaches to modifying antibodies to reduce their allergenicity, it is now also possible to produce antibodies by genetic engineering, rather than hybridoma fusion. To do this, the several genes that encode the various regions of an antibody are cloned and inserted into another host or expression system. Using bacterial viruses, or phage, researchers at The Scripps Research Institute Clinic and Stratagene, Inc. have successfully developed a system for the bacterial production of antibodies, and other expression systems are being developed as well. In fact, one technique being developed would genetically engineer the production of antibodies in plants.

Such "plantibodies" may have much greater yield and lower cost than the current technology for antibody production.

§2.5 Applications of Genetic Engineering in Health Care

With this basic introduction to the science and techniques of genetic engineering and monoclonal antibodies, most biotechnology applications can be better understood. The first application, already mentioned in the course of explaining the techniques of genetic engineering, is to the large-scale production of otherwise rare or expensive proteins. Most of the first applications of genetic engineering have been of this type, particularly in the field of human health. The first biotechnology-derived therapeutics approved by the FDA were all genetically engineered versions of naturally occurring human proteins that were believed to have therapeutic use. These include human insulin, human growth hormone, tissue plasminogen activator (tPA), erythropoietin, interleukin-2 (Il-2), alpha interferon, interferon beta, granulocyte macrophage colony stimulating factor (gm-csf), and granulocyte colony stimulating factor (g-csf). Other such recombinant human proteins are continuing to be of interest, such as GDNF and the proteins involved in the promotion and inhibition of angiogenesis (new blood vessel formation).

§2.5(A) Genetically Engineered Vaccines

Genetic engineering also has led to a major improvement in the development of vaccines. All vaccines, whether genetically engineered or produced by traditional means, work by "priming" the immune system of the vaccinated person to recognize the infectious agent's structure so that when the natural, or wild-type, infectious agent is encountered, it is quickly overwhelmed by an accelerated or heightened immune response (of B-cells or T-cells or both, as described in §2.4, above). Traditional vaccines utilized either attenuated (less pathogenic) or "killed" forms of the virus or bacteria to challenge the immune system with the antigenic determinants of the target pathogen. Successful examples of the traditional vaccine technology include the smallpox vaccine (using the less pathogenic cowpox virus), the Salk vaccine (using chemically "killed" polio virus strains), and the Sabin oral polio vaccine (using attenuated strains of the polio virus).

One of the key problems in traditional vaccine methodology was the limited power to attenuate or kill particular pathogens. Although the Salk vaccine succeeded in inactivating the poliovirus without changing its shape, and the Sabin vaccine succeeded in attenuating the virus without diminishing its "immunogenicity," other viruses and pathogenic bacteria were not always susceptible to such manipulation by the techniques available before genetic engineering. Thus genetic engineering can be used to produce safe vaccines in at least two ways. First, by making specific alterations in the genes of the target organism, the techniques of genetic engineering provide a much greater power to attenuate pathogenicity without reducing immunogenicity. Second, by identifying and cloning the genes responsible for producing particularly immunogenic features of the target organism, such as the envelope proteins of the virus, those features can be produced without any of the rest of the organism creating "sub-unit" vaccines that trigger immune system recognition but that are wholly incapable of causing disease. Finally, the injection of DNA (so-called "naked DNA") encoding particular antigens has been shown to cause the DNA to be taken up, the encoded proteins produced, and an immune response generated.[9]

§2.5(B) Diagnostic Gene Probes

Genetic engineering also provides powerful tools for creating diagnostics and therapeutics. The primary forms of genetically engineered diagnostics are DNA or RNA gene probes. Gene probes are created by identifying a stretch of DNA or RNA from a target pathogen (such as a virus or bacteria) or an identified disease-causing mutation (such as BRACA1, which indicates an increased risk for breast cancer). The target gene sequence of 18–30 bases (enough for specificity) is then used as a template for the creation of complementary DNA or RNA segments. The complementary segments can then be attached to different kinds of labeling molecules and then used as probes to assay a blood or other tissue sample. If the complementary DNA or RNA is added to a properly prepared blood or tissue sample and "hybridizes" or binds to a gene segment in the sample, the hybridization confirms the presence of the target gene sequence. The most recent development in this area has been the technology that allows thousands of such gene probes to be placed on a "chip" array so that, if a sample of genetic material extracted from blood or other tissue results in hybridization to any one of them, the result is an elec-

---

9. http://www.vical.com/company/dnatech.htm.

tronic signal that identifies the particular "hit." Such miniaturized, large-scale arrays allow thousands of genetic tests to be done in one pass.

§2.5(C) Therapeutic Applications of Genetic Engineering

Genetic engineering is also used to create therapeutics in a number of different ways beyond the production of rare, naturally occurring proteins. The most direct method of using genetic engineering to produce therapeutic products may be the development of so-called "antisense" therapeutics. Much like the gene probes described above, antisense therapeutics are designed to be complementary to and hybridize to a target sequence. The target sequence can be a segment of a viral gene, or a disease-producing gene of the cell itself, such as a cancer-causing oncogene. Because "natural" RNA is rapidly broken down by the cell, research into antisense RNA therapeutics has largely centered on modified or synthetic RNA that is more resistant to breakdown by the cell. The theory of antisense therapeutics is that when the antisense RNA reaches its target cell, it will bind to its target disease sequence, preventing further transcription into m-RNA or translation into protein, shutting down the disease process. The first antisense therapeutic to be the subject of a New Drug Application, ISIS Pharmaceuticals' Vitravene(tm) for Cytomegalovirus Retinitis in AIDS patients, was approved by the FDA in 1998. However, antisense has run into significant obstacles, primarily in achieving therapeutic concentrations at the target site (i.e. delivery) and may require new technological advances to succeed. A related and more recent scientific discover, of small interfering RNAs (RNAi) now seems to be more promising. RNAis were discovered to be a natural mechanism for turning off or inhibiting the production of proteins by an antisense-like mechanism. It will likely be a focus of significant research for years to come.

Genetic engineering techniques also play a major role in another process of creating therapeutics, known as rational drug design. Rational drug design is based on the premise that knowledge of the three-dimensional structure of particular proteins involved in a disease can greatly increase the speed with which compounds can be found to block the disease process. Disease-related proteins which can serve as targets for rational drug design can range from enzymes involved in viral reproduction (such as the reverse transcriptase enzyme of the HIV or AIDS virus) to particular protein receptors on the surface of the disease affected cells (such as the cellular receptors targeted by immune cells in an autoimmune disease such as psoriasis).

Rational drug design is made possible, in most cases, by the power of genetic engineering to produce large, pure quantities of such rare proteins as a viral enzyme or cell receptor. Once a quantity of the pure protein is available, rational drug design efforts employ a variety of methods, including x-ray crystallography and nuclear magnetic resonance spectroscopy, to determine the three dimensional structure of the protein. Even with substantial quantities of pure protein, determining the structure of a protein can be extremely difficult. Knowing the amino acid sequence of the protein is, at the present time, insufficient to allow reliable prediction of the way that the protein will fold into its normal three-dimensional shape. Proteins can be hard to grow in crystalline form, which is necessary for the x-ray diffraction technique primarily used to determine structure. However, as with sequencing, this is another area where the synergy of computers, robotics, and biochemistry has recently produced significant advances. The process of crystallizing proteins and obtaining their three-dimensional structures has been automated and robotized, permitting the thousands of attempts often required to obtain the right conditions for crystallizing a protein, an effort that often took years, to be performed in a matter of days. This has led to "high-throughput" x-ray crystallography, the core technology of biotechnology companies such as Syrrx (now part of Takeda Pharmaceuticals).[10]

Once a structure has been determined, rational drug design centers on the search for small molecules with a shape complementary to key regions of the target protein. For example, once the three dimensional structure of a viral enzyme is known, efforts at drug development would center on small molecules which would be predicted to fit precisely within the enzyme's active site. These drugs would then prevent the enzyme from performing its normal function (binding to its usual target "substrate") and thus would stop viral replication. To further illustrate this point, if a viral enzyme's normal function is to cut a large precursor protein into the small subunits that comprise the viral envelope, blocking the interaction between enzyme and precursor protein with a small molecule that binds to the enzyme would stop the viral replication at that juncture. The most recent advances in rational drug design are programs that can predict, with increasing power, protein shape from protein sequence and automated x-ray crystallography equipment that can generate three-dimensional structures in a "high throughput" system.

---

10. Syrrx performed "several million" crystallization experiments in its first eighteen months of operation. See D. Hosfield et al. 2003, *A fully integrated protein crystallization platform for small-molecule drug discovery*, available at http://www.atcg3d.org/PDF/Hostfield_JSB2003.pdf.

§2.5(D) Genetic Engineering and Human Gene Therapy

The "final frontier" in the application of genetic engineering to human health care is the field of human gene therapy. Human gene therapy is currently in the experimental stage, with hundreds of trials of various gene therapy protocols either underway or already completed at a number of institutions around the United States and around the world. Gene therapy involves the attempt to treat disease at the genetic level. Although there may be as many as 4,000 human diseases that have a substantial genetic component (including heart disease, all cancers, diabetes, and arthritis), until recently the concept of treating disease at the genetic level was relegated to the realm of science fiction. Type I diabetes, for example, is treated by providing exogenously produced insulin to replace the endogenous insulin that is absent due to the death of the insulin secreting cells of the pancreas. A gene therapy approach to Type I diabetes might attempt to provide an insulin gene and appropriate regulatory sequences to some other group of cells, which could then secrete insulin as needed.

The extraordinarily rapid development of the ability to identify genes, determine their function, and manipulate genetic material is turning science fiction into science fact. Although the field of human gene therapy raises legal and technical issues of sufficient complexity and number to require more detailed treatment in a separate chapter (see Chapter 10, Special Regulatory Issues—Human Gene Therapy), the rapid development of gene therapy merits its inclusion in this overview of the science of biotechnology.

An important distinction in the field of gene therapy is made between somatic cell gene therapy and germ-line gene therapy. In somatic cell gene therapy, which includes all currently approved protocols and trials, the aim is to change the genetic operation of target cells other than cells involved in the creation of germ cells for sexual reproduction. Thus the changes resulting from somatic cell therapy are non-heritable. In germ-line therapy genetic changes would be made in germ cells, which would then be heritable by any progeny of the treated individual. For example, somatic cell gene therapy for Huntington's disease might be aimed at central nervous system cells of an individual at risk for the disease without affecting the probability that the treated person's offspring would be at risk for Huntington's. In germ-line therapy, which might well be performed on early pre-embryos in conjunction with *in vitro* fertilization, germ cells would be affected and those changes would be inherited by future generations.[11]

---

11. Research has suggested the possibility of an alternative, quite possibly more efficient means of performing germ-line gene therapy. Dr. Ralph Brinster of the University of

Gene therapy requires two general functions: first, identification of the genetic basis for a disease; and, second, delivery of genetic material to the target cells in such a way that the new genetic material will be expressed and regulated at appropriate levels. The identification of the genetic basis of disease involves the molecular biology techniques of gene cloning and sequencing discussed in §2.3 above.

The problem of delivering genetic material to target cells has been the major technical barrier for gene therapy, and current protocols involve a variety of approaches in two categories, *in vivo* and *ex vivo*. In *in vivo* gene therapy the genetic material, in whatever vector or system, is injected into the patient. In *ex vivo* gene therapy, tissue or cells are removed from the patient or tissue donor, the new genetic material is introduced into cells in cell culture, and the transduced cells are then injected or implanted in the patient. Autologous transplantation of transduced cells involves the removal of a group of target cells, transfection (the insertion of additional, functioning genetic material), and the return of the transduced cells to the patient.[12] Transduction of the removed cells usually involves incubation with an appropriate viral vector. In the future, autologous transplantation may be replaced by the implantation of genetically engineered tissues that have been rendered non-immunogenic.

Commonly used techniques for *in vivo* gene therapy include genetically engineered viruses (called viral vectors), fat-like coatings called liposomes, and even the injection of DNA sequences (referred to as "naked" DNA). Viruses in nature perform the essential functions of gene therapy, that is, they effec-

---

Pennsylvania reported that he was able to select and remove sperm stem cells from the testes of mice (sperm stem cells are early-stage sperm-producing cells), insert a foreign gene into the sperm stem cells, and reimplant them. The reimplanted, genetically engineered sperm stem cells then produced sperm that carried the new gene. *See* Gina Kolata, *Gene Technique Could Shape Future Generations*, N.Y. TIMES, Nov. 22, 1994, at A1.

12. The preferred cells for autologous transplantation will often be stem cells of one type or another. As with sperm stem cells, *supra* note 19, a variety of different stem cells are found in the body. Each type of stem cell is an early-stage cell that is long-lived and divides numerous times over its lifespan, giving rise to one or more different populations of mature cell types. For example, marrow (hematopoietic) stem cells are the progenitors of a number of types of immune system cells, and genetic engineering of marrow stem cells could result in the repopulation of the immune system with genetically engineered mature cells. One frequently discussed application of gene therapy for marrow stem cells is as a treatment for AIDS. If uninfected stem cells could be removed and engineered to be resistant to the AIDS virus (for example with an antisense sequence for a key viral gene), the resulting mature cells, including the CD4 cells that are the major target of the virus, would also be resistant to the virus. Embryonic stem cells are discussed in the text accompanying note 4.

tively target cells and insert a genetic message (albeit not a therapeutic one) into those cells. Viral vectors take advantage of these characteristics of viruses and are constructed to carry the therapeutic gene sequence, to be non-pathogenic, and to be incapable of reproducing. At this time, however, autologous transplantation and viral vector transduction are the most common techniques employed in gene therapy. These first efforts at genetic alteration of human cells are a significant milestone in modern medicine. However, the difficulty of achieving sufficient expression of the desired gene product in the desired cells for a significant period of time remains an enormous scientific and technical barrier to successful gene therapy.

As of this writing, both *in vivo* and *ex vivo* gene therapy experiments have produced adverse reactions, including death. In a widely publicized tragedy at the University of Pennsylvania, the direct injection of an adenoviral vector into the liver of one subject, Jessie Gelsinger, led to liver inflammation and death. In a less widely publicized series of events in France, at least two patients in a trial of gene therapy for x-linked severe combined immunodeficiency disease (X-SCID) contracted leukemia, apparently as a result of the viral vector's effect on the transduced cells. After the French reports, the FDA placed a moratorium on gene therapy trials, yet has since lifted it. Recent reports have also indicated potential for an *in vivo* gene therapy protocol aimed at stimulating the growth of vasculature to the heart of persons with coronary artery disease. Gene therapy will undoubtedly be a powerful medical tool in the future, but when that future will arrive is still unclear.

§2.6 Applications of Monoclonal Antibody Technology in Health Care

As is the case with genetic engineering, there is a broad range of health care applications for monoclonal antibody technology. To date, by far the most prevalent has been the development of monoclonal antibody-based *in vitro* diagnostics. Because any single line of monoclonal antibodies has a specific affinity for and binds to one and only one antigen, monoclonal antibodies are a powerful tool for determining whether that antigen is or is not present in a particular sample. Their specificity makes monoclonal antibodies extraordinarily useful as diagnostics.

The single most widely used product of biotechnology may well be the monoclonal antibody-based home pregnancy test, which is a good example of antibody diagnostic technology in general. In the home pregnancy test, the antigen that provides the diagnostic indicator of pregnancy is the hormone human chorionic gonadotropin (HCG), found in the urine of pregnant women soon after conception. When mice are inoculated with HCG, they produce antibodies to the hormone, and the inoculated mice can be used to create monoclonal antibodies as discussed in §2.4 above. One or more of these antibody lines can be linked to an appropriate marker or label molecule, attached to an appropriate instrument, and the result is a very sensitive assay for HCG as an indicator of pregnancy.

A second diagnostic use of monoclonal antibodies is to create *in vivo* imaging agents using antibodies attached to radioactive isotopes. Researchers have begun to identify proteins that are strongly associated with cancerous cells or produced in greater quantities in cancerous cells. By using those proteins as antigens in the process of making monoclonal antibodies, the result is a monoclonal antibody that can be injected into the body of a cancer patient and will bind to those sites where the cancer associated protein is found. By attaching radioactive isotopes to the antibodies and using an external imaging medium, the resulting film can clearly locate concentrations of the cancerous cells. The FDA has already approved monoclonal antibody imaging agents for use in colon cancer, ovarian cancer, prostrate cancer, and small cell lung cancer.

§2.6(A) Therapeutic Uses of Monoclonal Antibodies

Although the earliest health care applications of antibody technology to reach the market were diagnostics, monoclonal antibodies are increasingly being approved for therapeutic use.

The ability of monoclonal antibodies to bind to a specific antigen can also be of therapeutic value, depending on the role of the target antigen in the disease process. One very successful therapeutic antibody is Centocor/Eli Lilly's ReoPro. ReoPro binds to a target receptor on the surface of platelets, which are key blood cells involved in clotting and thrombosis. Administration of ReoPro during angioplasty and other coronary artery surgery keeps the platelets from sticking to one another ("aggregating") and reduces the risk of clotting and restenosis (re-narrowing) of the coronary arteries. Similarly, antibodies to TNF and IL-1, which prevent the binding of those inflammation-promoting substances to their receptors, have been approved for several inflammatory conditions.

The use of monoclonal antibodies as immunotherapeutics, particularly for cancer, has been discussed since the beginning of monoclonal technology. It took many years for the concept to become a reality, but in the last several years it has been successfully demonstrated, with the FDA approval of IDEC

Pharmaceutical's Rituxan for non-Hodgkin's lymphoma, the first monoclonal antibody cancer therapeutic to be approved by the FDA.[13] The general concept is that if appropriate antibodies that recognize unique, tumor-specific antigens could be produced, those antibodies could be injected into a patient, hone in on the cancerous cells, and stimulate a general immune response that would be of therapeutic effect on the cancer. One major obstacle has been the search for such tumor-specific antigens that are nevertheless shared by all patients with a particular cancer. Such tumor-antigens are being discovered and the field of cancer antibodies (as well as cancer "vaccine" therapeutics) holds great promise.

§2.7 Genetic Engineering Applications in Agriculture and Industry

Although by far the greatest share of commercial investment in biotechnology to date has been in the human health care applications described in §2.5 and §2.6, biotechnology is being used to develop a wide range of agricultural and industrial applications. In agriculture, the techniques of genetic engineering have been used to create genetically engineered plants and animals with desirable characteristics, produce plants with improved characteristics (such as pest resistance or pesticide tolerance), and produce recombinant proteins that are in themselves biological pesticides. Together with monoclonal antibody technology, the same techniques that have been applied to human health are producing new animal drugs, diagnostics, and vaccines.

In industry, genetic engineering techniques are used to create microorganisms that are capable of breaking down hazardous or toxic wastes into harmless constituent molecules. The ability of monoclonal antibodies to bind in highly specific ways to particular substances has a variety of industrial applications. First, monoclonal antibodies can be used as a "diagnostic" for the presence of a variety of organic molecules, for which other assays may be more time-consuming or costly. Secondly, the antibodies can be used as biocatalysts, which hold their antigens in a position that increases their reactivity with another compound. Finally, both the techniques of genetic engineering and

---

13. IDEC's antibody targeted CD20, a cell surface marker on mature B cells. The temporary depletion of B cells, which are proliferating at a pathological rate in non-Hodgkin's B Cell Lymphoma, has proved to be a valuable therapeutic strategy in those patients. Rituxan was the first antibody to be approved that uses the immune system's own power to kill the targeted tumor cells and benefit the cancer patients.

monoclonal antibodies add substantially to the power of industry to produce a variety of "specialty" chemicals, including amino acids, vitamins, and biopolymers (such as silk), and to purify or separate the products of chemical manufacture.

§2.8 Summary

Biotechnology, the application of modern molecular biology and immunology, is generally based on the technologies of genetic engineering and monoclonal antibody production. Genetic engineering allows scientists to unlock and manipulate the genetic code for the protein building blocks of living organisms. By "fishing" out or cloning the genes that control the production of particular proteins, scientists can produce large quantities of heretofore rare or unobtainable substances, such as human insulin or human growth hormone, many of which have direct value as therapeutics. The techniques of genetic engineering also provide the basis for new generations of vaccines and diagnostics, as well as open the door to gene therapy, or the direct treatment of disease at the genetic level.

Monoclonal antibody technology is based on the manipulation of mammalian immune system cells, allowing for the production of essentially unlimited quantities of a single antibody type, each such antibody type having a unique specificity and affinity for a particular antigen or target molecule. Monoclonal antibodies, currently in wide use as diagnostics, also have the potential to allow for a new generation of therapeutics for treating a wide range of diseases. Both monoclonal antibody technology and genetic engineering are being used for industrial and agricultural applications as well as human health applications.

CHAPTER 3

Technology Transfer: The University-Industry Connection

§3.0 Introduction: Universities and Research Institutes—Where Biotechnology Is Born

Where does biotechnology come from? The answer is that it comes from breakthroughs in the life sciences, many of which take place in the leading universities and not-for-profit research institutes in the United States. The federal government, through the NIH, put more than $16 billion dollars into research at universities and research institutes in the fiscal year 2002. That is a large percentage of overall research funding at United States universities and research institutes and results in a considerable amount of discovery and invention.

According to the Association of University of Technology Managers (AUTM), university-based research (which hereafter shall also mean research at independent research institutes such as The Salk Institute, The Scripps Research Institute, and the Whitehead Institute for Biomedical Research) produced approximately 13,000 invention disclosures in the fiscal year 2000. In that same year, approximately 6,300 new patent applications were filed (equal to just under 50 percent of the disclosures) and about 3,800 U.S. patents were issued to universities based on applications filed in previous years. While not all of that activity is in the life sciences or related to biotechnology, the great majority of those disclosures and inventions are in life sciences, because the majority of the funding is in those fields.[1] That is a lot of science and a lot of potential biotechnology.

---

1. According to the AAAS analysis of the federal research budget, the NIH received $15.8 billion in 2005, which was 57 percent of all federal spending on basic research (http://www.aaas.org/spp/rd/prel07tb.htm#tb2).

Many people believe that the Supreme Court's decision in *Diamond v. Chakrabarty* (447 US 303 (1980)) was the great legal event that enabled the growth of the biotechnology industry. It is equally possible to argue that the midwives who assisted in the birth of the biotechnology industry were a political "odd couple" named Birch Bayh and Bob Dole. The Bayh-Dole Act (P.L. 96-517, Patent and Trademark Act Amendments of 1980), co-sponsored by the liberal Democrat from Indiana and the conservative Republican from Kansas, may well deserve equal or greater credit for the phenomenal success of the biotechnology industry over the past 20 or more years. While *Chakrabarty* was certainly important for investor confidence, in truth, only a very small portion of the intellectual property in biotechnology consists of patents on whole living organisms, the issue in that case. On the other hand, an enormous amount of the intellectual capital in biotechnology originated in universities and made its way into commercialization through the structure provided by the Bayh-Dole Act.

This Chapter is intended to provide a brief introduction to Bayh-Dole (§3.1), an overview of the process of university technology transfer in its most common form (§3.2) and some interesting alternative forms (§3.3), and a guide to working successfully with university technology transfer offices to procure the rights to the technology needed (§3.4). Why license technology from universities? To paraphrase Willie Sutton's famous (though possibly apocryphal) quote: "Because that's where the science is."

§3.1 An Introduction to the Bayh-Dole Act: Transfer of Technology Developed with Federal Research Funding

The Bayh-Dole Act is, in its basic framework, brilliant. Consider the problem:

1. how to distribute the burden of discovering valuable inventions made in the course of federally funded research (and the burden of efficiently transferring the rights to those inventions to appropriate parties for their further development, manufacture, and distribution);
2. how to provide sufficient incentives to the parties on whom the burden is placed to increase the likelihood that they will undertake their roles in a reasonably diligent manner.

The choices, at the most basic level, are to put the burden of discovery and transfer on the federal government, the recipient research institution, or the inventor/researcher. In Europe, traditionally the rights and the burden belong to the inventor, which has proven frustratingly inefficient. Each inventor must reproduce the process of technology transfer from scratch. Before Bayh-Dole, the default rule was that the rights belonged to the federal government. By 1980 there was widespread conviction that allocation was also inefficient: the federal government was increasingly overwhelmed by the number of inventions and too removed from the inventor to operate efficiently. In order to adapt to an environment of increasing federal research budgets and increasing invention Bayh-Dole provided a simple solution: place the burden on the universities and give both the universities and the inventors sufficient reward to stimulate their performance.

There is debate over whether Bayh-Dole is primarily responsible for the revolution in university technology transfer since 1980 or is simply one among several factors. There is even debate about whether increasing technology transfer is a good thing. Despite these disagreements over cause and effect and economic theory, one fact is certain: the rapid growth in the federal research budget was accompanied by an even greater increase in the number of inventions that were produced by federally funded researchers, even during a period in which non-university research and development was producing fewer and fewer patents per research dollar. Universities are getting increasingly efficient at producing and licensing inventions. Bayh-Dole is the law under which they do it.

As with almost any major legislation, a diversity of political interests led to attachment of several additional requirements alongside the basic goal of making the process of technology transfer more efficient. Bayh-Dole provides that institutions that receive federal research funds may retain the rights to the inventions produced in the course of that research as long as they:

1. Diligently pursue commercialization;
2. Share proceeds with the inventor;
3. Favor, where possible, small business licensees; and,
4. Require manufacture of products to be performed substantially in the U.S.

These simple requirements are designed to accomplish the Act's purposes: locate the burden of uncovering inventions and transferring them at a level where those tasks can be efficiently undertaken while giving incentives to both the inventors and the universities to do their part. However, the additional requirements of favoring small businesses and domestic manufacture go beyond that to further other domestic political demands.

§3.2 The Requirements of Bayh-Dole for Universities

Virtually every major research university has taken three basic steps to implement the Bayh-Dole Act: instituted policies requiring assignments of invention and disclosure of inventions from all employees; instituted a policy setting forth the inventor's share of the proceeds of any invention; and, established a university office of technology transfer to administer the technology transfer process.

§3.2(A) Assignment of Rights

Even before Bayh-Dole, some major U.S. research universities required all employees to sign, as a condition of employment, an agreement that all inventions made by the employee would be assigned to the university and would be promptly disclosed to the university. However, the practice was by no means universal and at many universities inventors were left to deal with the issues of rights and technology transfer more or less on their own (Columbia was one such university[2]). After Bayh-Dole, such blanket invention assignment and disclosure agreements became universal practice. The standard language of such agreements reads more or less as follows:

> I acknowledge my obligation to assign inventions and patents that I conceive or develop while employed by University or during the course of my utilization of any University research facilities or any connection with my use of gift, grant, or contract research funds received through the University. I further acknowledge my obligation to promptly report and fully disclose the conception and/or reduction to practice of potentially patentable inventions to the Office of Technology Transfer or authorized licensing office. Such inventions shall be examined by University to determine rights and equities therein in accordance with the Policy. I shall promptly furnish University with complete information with respect to each.
> In the event any such invention shall be deemed by University to be patentable or protectable by an analogous property right, and University desires, pursuant to determination by University as to its rights

---

2. *See* David C. Mowery, Richard R. Nelson, Bhaven N. Sampat, and Arvids A. Ziedonis, *The Growth of Patenting and Licensing by U.S. Universities: An Assessment of the Effects of the Bayh-Dole Act of 1980*, 30 RESEARCH. POLICY 99 (2001). Mowery et. al. argue that the data shows that the Bayh-Dole Act was just one of the factors that contributed to the continuing acceleration of university technology transfer in the life sciences.

and equities therein, to seek patent or analogous protection thereon, I shall execute any documents and do all things necessary, at University's expense, to assign to University all rights, title, and interest therein and to assist University in securing patent or analogous protection thereon.[3]

The quoted language incorporates three separate obligations: the obligation to disclose, to assign, and to cooperate in the process of patenting and technology transfer. It is the assignment provision that can occasionally raise the thorniest issues, particularly in states such as California where state law limits blanket pre-assignments.[4] California Labor Code §2870 declares that such agreements:

> shall not apply to an invention that the employee developed entirely on his or her own time without using the employer's equipment, supplies, facilities, or trade secret information except for those inventions that either:
> (1) Relate at the time of conception or reduction to practice of the invention to the employer's business, or actual or demonstrably anticipated research or development of the employer; or
> (2) Result from any work performed by the employee for the employer.

The result of such policies may be messy disputes over the rights to inventions that were conceived fully and in a flash, like Minerva from Jupiter's head, while the employee is driving or otherwise away from the lab and on his own time (as the Nobel-prize-winning conception of PCR occurred to Kary Mullis).[5] In

---

3. University of California Patent Policy, http://www.ucop.edu/ott/patentpo.html (last visited May 11, 2006).

4. While only a small number of states have similar provisions, those states are among the leaders in biotechnology research and commercialization, including (in addition to California) Washington, North Carolina, Illinois, and Minnesota.

5. One account of Mullis's moment of inspiration is this one:
> The idea for PCR came to Mullis on an April night in 1983, as he made the 300-mile drive north from Berkeley to his weekend cabin in Mendocino County. At the time, Mullis was making oligonucleotides for Cetus Corp., one of the many biotech firms that sprang up in California in the early '80s. Oligonucleotides are short pieces of DNA which, when radioactively labeled, can be used to determine if a sample of genetic material contains a specific nucleotide sequence or gene. He had virtually automated Cetus' production of the material, and feared that he and his staff were far out-producing the demand.
> I was thinking of things that were related to what I was wanting to do—find some uses for oligonucleotides. And I was driving and it was quiet and something just went, Click!

http://gtalumni.org/news/magazine/sum94/click.html (visited May 11, 2006).

any state, however, there can be disputes as to the rights to inventions which the inventor claims were conceived shortly after leaving a particular place of employment.

In most cases, however, the assignment agreement is understood and respected by the researchers, but the obligation to disclose the invention is not necessarily fulfilled in a timely manner. In general, university scientists' first objective is rapid publication of important findings, as to the first to publish belongs the eternal fame and glory. While the disclosure and related technology transfer processes can be performed in an expedited way that does not interfere with the scientist's first priority, it sometimes simply just escapes the scientist's attention in the rush of discovery, preparing the paper, and, in some cases, preparing for the major scientific meeting at which the research will be discussed. This problem of timely disclosure is just the first difficulty faced by university offices of technology transfer and is discussed below in §3.3 The University Office of Technology Transfer.

§3.2(B) Sharing Proceeds

The Bayh-Dole Act recognized that if technology transfer is to function reasonably well, inventors need an incentive to participate in the process. The statute addresses this concern by requiring that institutions that elect to retain the rights to federally sponsored research must share the proceeds with the inventors (35 U.S.C. §202 (c) 7(B)). There appears to have been some effect of market forces that have led most major universities and research institutes to arrive at a reasonable, if not absolutely uniform, consensus as to what "share" means. Table 3.1 shows the revenue sharing policies in place at a number of leading research institutions. It appears that the general trend is for universities to provide inventors with either a fixed percentage in the neighborhood of 30–35 percent or a sliding scale that begins somewhat higher and then moves downward to 25 percent–33 percent. It would seem logical for universities to arrive at somewhat comparable policies, for non-salary compensation must be kept at competitive levels, just as salaries at such institutions must be competitive, in order to attract and keep top researchers.

Despite the gradual confluence of policies on revenue sharing, one interesting case did explore the question of whether the Bayh-Dole requirement set any lower limit on the amount that universities must share. In Platzer v. Sloan-Kettering (787 F. Supp. 360 (S.D.N.Y. 1992)), the inventors brought suit against their employer seeking, among other relief, a greater share of the revenues than the particularly miserly portion afforded them by the employer's patent policy. At the time, the policy was set forth in the case as being based on a sliding scale as follows:

Annual Gross Proceeds	Inventor's Share
$0–$50,000	25 percent
$50,000–$150,000	15 percent
$150,000–$300,000	10 percent
Over $300,000	5 percent

While the case also turned on procedural issues, the court was quite clear in rejecting any notion that the statutory requirement to "share" had any fixed minimum. Instead, the court found that Congress had clearly intended such policies to be "left to the supply and demand of the market" (Platzer at 368). The lesson of Platzer for scientists is clear: check the patent policies before you sign on.

§3.3 THE UNIVERSITY OFFICE OF TECHNOLOGY TRANSFER (OTT): FUNCTIONS AND MODELS

Every Office of Technology Transfer (OTT) has to perform the same essential functions, beginning with the initial disclosure of a possible invention and, in the best cases, continuing through monitoring and auditing the performance of a licensee. This section 3.3 briefly lists and discusses these essential functions.

§3.3(A) Disclosure

The process of university technology transfer cannot begin without possible inventions coming to the attention of the OTT. Since virtually every university has a policy, signed by all employees, requiring researchers to disclose possible inventions to the OTT, one might think that assuring timely disclosure is generally solved. However, as previously noted, many times researchers fail to disclose their inventions in a timely manner. Many universities sponsor periodic seminars for their scientists to reinforce the need for disclosure and to help educate scientists as to what kinds of research findings need to be disclosed. While that helps, it does not eliminate the problem of scientists who for various reasons fail to fill out and turn in timely disclosures of their research findings.

In smaller research institutes, timely disclosure is less likely to be a problem because the head of the OTT may well be familiar with the research programs of every scientist and can proactively keep abreast of research progress. At larger institutions, with hundreds of independent investigators, the challenge is far greater. Some partial solutions that fit within the traditional model of technology transfer can help. First, the OTT can actually use the grant funding system to help with the burden of keeping up. Simply reviewing the abstracts of funded grant applications can give the OTT a "heads-up" on at least some of the research that is likely to produce significant inventions. Unfortunately, in some really large universities, there are bureaucratic impediments to even that small step: the grant applications are managed by one office, the OTT is another, and the computer systems of the two may not be easily used to share information. Another potentially helpful addition to the traditional disclosure process is also not easy to implement. Simply requiring that copies of manuscripts sent out for publication review be simultaneously sent to the OTT would prevent at least some valuable rights being lost by default. In these days where copying e-mails with attachments is a trivial task, this would seem eminently doable. Yet again, it is not self-enforcing and only works if scientists (and their clerical support personnel) are committed to participating.

§3.3(B) Evaluation

Once disclosure does occur, the more difficult steps in the process begin, the first of which is evaluation. The evaluation stage is critical because it is often the stage at which a decision must be made as to whether or not to commit funds to intellectual property protection. The decision to file a patent needs to be made before the invention is disclosed to the public, whether by publication or presentation at a scientific meeting. In the high stakes academic battles for scientific "firsts," there is usually not more than a month or so to make a decision on whether or not to file at least a "provisional" application (see Chapter 4 §4.02). If a licensee cannot be assured within that time, then the university must make a decision as to whether or not to go "at risk" and expend funds for intellectual property protection without any assurance that a licensee will be obtained to reimburse the university for all of its patent costs.

Evaluating an invention properly requires significant expertise of a variety of types. The evaluation needs to address the questions regarding the patentability of the findings (where the scientist's own expertise can be of significant help), the commercial need for the technology, the commercial investment in similar or competing technology, and the identity of potential licensees. To compound the difficulty of this early, critical stage in technology transfer, the rapid pace of advances in many fields and the great range of research at major universities makes it almost certain that even the most knowledgeable technology transfer staff will be asked to evaluate new discoveries outside their core expertise. To a considerable extent, the difficulty of evaluation is inherent in the traditional technology transfer model in which each disclosure is dealt with individually and marketed and licensed independently. Whether this difficulty can be reduced by alternative models of technology transfer is discussed in §3. 3 below.

§3.3(C) Marketing

The marketing stage can be a part of the evaluation process. If there is enough time for "non-confidential" marketing of the invention prior to publication, there may be sufficient expression of interest (or lack thereof) from potential licensees to provide some indication of the market for or value of the invention. What is "non-confidential" marketing? It is marketing that initially relies on a description of the technology without the details that would make it a "publication" prior to patent filing; and, because it is lacking those details, it can be done without the need for confidentiality agreements to protect patent rights.

For example, a recent discovery hailed in the popular press was the activity of a particular protein in controlling metabolism and its potential value as an obesity drug.[6] A non-confidential marketing description of that invention might simply state, "a novel protein for which animal data had confirmed the protein's ability to reduce body-weight, independently of diet, by approximately 20 percent." This "teaser" description could be the basis on which interested potential licensees make the decision to sign a confidentiality and non-disclosure agreement (often referred to as a "CDA") to obtain the details of the invention and make a decision as to whether or not to negotiate for a license to the invention.

Marketing, even at the non-confidential stage, can either be relatively targeted or more widely broadcast. The easier, but likely less effective, route is to broadcast widely the non-confidential descriptions of inventions through such means as website postings of available inventions. The federal government itself has long done a form of this through its weekly publication of government-owned inventions available for licensing in the Federal Register. For example, the following was among the announcements in the February 8th 2006 Federal Register:

---

6. The Salk Institute Professor Ron Evan's article on PPAR (d) and weight-loss, *Peroxisome-Proliferator-Activated Receptor Activates Fat Metabolism to Prevent Obesity*, appeared in the journal CELL (April 18, 2003 at 159).

Multideterminant Peptides That Elicit Helper T-Lymphocyte Cytotoxic T-Lymphocyte and Neutralizing Antibody Responses Against HIV-1

Jay A. Berzofsky et al. (NCI)

U.S. Patent No. 6,294,322 issued 25 Sep 2001 (HHS Reference No. E-152-1991/1-US-01).

Licensing Contact: Robert M. Joynes, J.D.; 301/594-6565; *joynesr@mail.nih.gov.*

....

This invention described peptide constructs that may be of clinical importance in HIV/AIDS vaccine development. A vaccine for the prevention and/or treatment of HIV infection would ideally elicit a response in a broad range of the population. It would also have the capability of inducing high titered neutralizing antibodies, cytotoxic T lymphocytes, and helper T cells specific for HIV-1 gp 160 envelope protein. A vaccine based on synthetic or recombinant peptides has been developed which elicits these responses while avoiding the potential safety risks of live or killed viruses. Unlike previously developed vaccines this invention avoids those regions of gp 160 which may contribute to acceleration of infection or the development of immune deficiency. This invention provides peptides up to 44 amino acid residues long that stimulate helper T-cell response to HIV in a range of human subjects. Six multideterminant regions have been identified in which overlapping peptides are recognized by mice of either three or all four MHC types. Four of the six regions have sequences relatively conserved among HIV-I isolates. These multideterminant cluster peptides are recognized by T cells from humans of multiple HLA types, and have been found in a phase I clinical trial to elicit neutralizing antibodies, cytotoxic T cells, and helper T cells in at least some of the human subjects. These peptides are currently being tested in primates. Once delivery systems and a stronger mucosal response are induced, NCI plans to use these peptides in human clinical trials. Dated: January 30, 2006. 71 FR 6506.

This disclosure tells the world that the NIH has an invention that consists of several peptides that appear to be capable of eliciting a therapeutic immune response to most strains of HIV-1 virus and that therefore may be the basis of a breakthrough in HIV vaccine research. Equally important, it does not provide any sequence or other detailed information about the peptides, thus avoiding the problem of publication and loss of intellectual property rights.

More difficult than broadcast marketing is the limited dissemination to one or more prospective licensees that have been selected based on their existing research programs or needs. For example, using the example above, the effort would be to identify biotechnology and pharmaceutical companies that are particularly focused not simply on AIDS, but on immunological approaches to AIDS, possibly including anti-HIV antibodies but most specifically HIV vaccines. Furthermore, since the invention is a relatively uncommon form of vaccine platform, consisting of multiple peptides, it would also be useful to identify companies in the HIV vaccine field that are likely to have a vaccine development technology that would facilitate the development of multivalent vaccines to the NIH peptides. Obviously, the targeted marketing approach requires more work and more knowledge of the industry. Again, given the range and volume of science research at a major university, it is impossible for the OTT personnel to be familiar with everything or even have the time to do the research required to market every invention appropriately. The researcher's knowledge of the field can be invaluable in this approach, but it is still a difficult task for the OTT staff. A scientist-to-scientist contact can often be the best way to start such discussions.

§3.3(D) Licensing

If all goes well, the evaluation and marketing efforts lead to negotiations with a licensee. The main issues that are the subject of negotiations in licensing university technology are discussed in §3.4 below. For purposes of this section, it is enough to note that it is another of the very demanding tasks placed on the OTT during the technology transfer process. Not only must the OTT have technical sophistication and marketing ability, it must possess the ability to undertake sophisticated negotiations with very skilled business executives on the other side.

§3.3(E) Monitoring and Enforcement

While the goal of the OTT is to license the technology successfully, it is not the last stage in the process. Many biotechnology inventions are ten years or more from marketing approval when they are licensed. During that time, the OTT needs to continue the prosecution of the patent claims, monitor the progress of the licensee in meeting the milestones for diligence set out in the license, collect milestone payments and, in the best of all possible worlds, collect royalties (and, when necessary, examine the records of the licensee in accordance with the provisions of the license to assure that the amount of royalties is, in fact, properly calculated). In other, less happy cases, the OTT may be involved in negotiations of extensions of milestone deadlines with the licensee, dealing with the messy problem of terminating a license, or even participating in the process of reclaiming technology from a bankrupt licensee. The thrill of victory (executing a painstakingly negotiated license) may still be followed by the agony of defeat (when the licensee's performance stops).

§3.4 Alternative Models of Technology Transfer: Scripps, the NIH Guidelines, and Post-NIH Guidelines Developments

The previous section details the steps in the most common mode of technology transfer from universities to the private sector, in which each invention and each disclosure become the subject of their own process of evaluation, marketing, and negotiation. One might call this the "ad hoc" or "piece by piece" approach. It also attempts to convey some of the difficulties posed by the ad hoc approach, due to the difficulty of evaluating technology over a wide range of science, identifying the most likely licensees, marketing the inventions accordingly, and actually negotiating the license for each invention through negotiations with sophisticated pharmaceutical or biotech executives. There are, however, alternative modes of technology transfer, which circumvent or avoid some of these difficulties, although they may present other problems instead. The following subsections take a brief look at: a major alternative model for structuring the university technology transfer function, the TSRI "Master Agreement" model (§3.4(A)); and the NIH guidelines "Developing Sponsored Research Agreements: Considerations for Recipients of NIH Research Grants" (§3.4(B)).

§3.4(A) The Scripps Research Institute (TSRI) Significant Relationship Model

The Scripps Research Institute is the world's largest private, non-profit biomedical research institute, with 286 faculty, 800 post-docs, 164 graduate students, and over 1,500 support personnel housed in over one million square feet of space in La Jolla, California. In 2001, TSRI received over $193 million in sponsored research support, as shown in Table 3.2.

Table 3.2 (Source: http://www.scripps.edu/intro/facts.html)
Sponsored Research

	NIH	Other Federal Agencies	Other	Total
2004	$219,961,000	$12,028,000	$28,282,000	$265,221,000
2003	$196,304,000	$11,824,000	$34,321,000	$242,449,000
2002	$173,816,000	$11,716,000	$38,904,000	$224,436,000
2001	$144,273,000	$9,735,000	$39,400,000	$193,408,000
2000	$131,531,000	$8,173,000	$39,592,000	$179,296,000

One question raised by the data in Table 3.2 is what are the sources of the $28.2 to $39.4 million in "Other" funding? In 2001, $20 million of the "other" funding, or a little more than ten percent of total funding, came from Novartis, the global pharmaceutical company. The agreement for that funding was signed by Sandoz Pharmaceuticals, which was merged with Ciba-Geigy to form Novartis.

Originally, the Sandoz agreement was to have been a ten-year agreement for $30 million a year in exchange for a right of first-refusal on *all* TSRI technology on pre-arranged licensing terms (with different royalty rates and milestones for diagnostics and therapeutics). However, when that agreement was announced, it created a major national controversy (*see* Christopher Anderson, *Scripps-Sandoz deal comes under fire; National Institutes of Health to investigate contract between Scripps Clinic and Research Foundation and Sandoz Pharma*, 259 SCIENCE 889 (Feb. 12, 1993)). Congressman Ron Wyden complained that the taxpayer-funded research (federal funding averages about 80 percent of the total TSRI funding) was being given away to a foreign company that would then gouge the public for its products. The "foreign" identity of Sandoz may explain why the then-expiring agreement, between Scripps and Johnson & Johnson, which was to be replaced by the Sandoz agreement upon expiration, had passed without similar controversy.

The NIH Director at the time, Bernadine Healy, commissioned a special committee to look into such arrangements with two significant results: TSRI and Sandoz rewrote the terms of their agreement to make them more palatable to the public; and, the NIH recommended Guidelines, that purported to interpret the Bayh-Dole Act, to guide recipients of federal research funding in any future large-scale technology transfer arrangement with private industry. The rewritten terms of the Scripps/Sandoz agreement are discussed in the next subsection, the NIH Guidelines.

§3.4(B) The NIH Guidelines: Developing Sponsored Research Agreements: Considerations for Recipients of NIH Research Grants (59 FR 32997-02, Monday, June 27, 1994) ("NIH Sponsored Research Guidelines")

The furor raised by the announcement of the original agreement between Scripps and Sandoz led the special panel appointed by then-Director Bernadine Healy to issue a number of recommendations concerning the "problem" of long-term significant relationships between pharmaceutical companies and recipients of NIH funding. The resulting Guidelines justified the panel's various suggested policies and concerns by referring to the Bayh-Dole Act's *seven* purposes, as set forth in the NIH Sponsored Research Guidelines:

Specifically, the objectives of the Bayh-Dole are to:

- Promote collaboration between commercial concerns and nonprofit organizations, including universities;
- Promote the utilization of inventions arising from Federally supported research or development;
- Encourage maximum participation of small business firms in Federally sponsored research and development efforts;
- Ensure that inventions made by nonprofit organizations and small business firms are used to promote free competition and enterprise;
- Promote the commercialization and public availability of inventions made in the United States by United States industry and labor;
- Ensure that the Government obtains sufficient rights in Federally sponsored inventions to meet the needs of the Government and protect the public against nonuse or unreasonable use of inventions; and,
- Minimize the costs of administering policies in this area.

(NIH Sponsored Research Guidelines).

It is worth noting that these seven purposes may sometimes be in conflict. For example, promoting utilization might sometimes be best served by large pharmaceutical companies rather than a "small business firm." It also may be difficult to imagine that these purposes are of equal importance. For example, the panel itself recognized that promoting utilization was likely more important than promoting U.S. manufacture by providing that waivers of the U.S. manufacture requirement could be obtained where attempts to procure a U.S. manufacturer as licensee were unsuccessful or commercially impracticable. Nevertheless, the NIH found sufficient coherency in these purposes to issue the Guidelines, with both general and specific recommendations.

The panel's recommendations began with their general recommendations, which were expressly stated as being "UNIVERSAL POINTS FOR CONSIDERATION." These included imprecations to all institutions to strive to insure academic freedom and timely dissemination of research results as well as seek the development and utilization of inventions by selecting grantees with the ability to develop and market the end products diligently. These "universal" points note that seeking development and utilization requires timely patenting, monitoring, and enforcing licensee performance.

The NIH's panel was not satisfied with these general means of fulfilling Bayh-Dole's purposes, concluding that long-term significant relationships with pharmaceutical companies required special scrutiny and limits because of their potential to undermine the promotion of and transfer of technology to small businesses, one of those seven explicit goals of the Bayh-Dole Act. On that basis, the panel declared that "heightened scrutiny" was required whenever any of these "threshold criteria" apply:

> (a) the amount of financial support from the sponsor meets or exceeds $5 million in any one year, or, $50 million total over the total period of funding under the agreement;
> (b) the proportion of funding by the sponsor exceeds 20 percent of the Grantee's total research funding;
> (c) the sponsor's prospective licensing rights cover all technologies developed by a major group or component of the Grantee organization, such as a large laboratory, department or center, or the technologies in question represent a substantial proportion of the anticipated intellectual output of the Grantee's research staff; or
> (d) the duration of the agreement is for 5 or more years.

(NIH Sponsored Research Guidelines).

The panel, in a remarkable demonstration of inscrutable logic, reasoned that such substantial relationships were likely to *both* "adversely affect open commercial access, especially for small businesses, to a Grantee's Federally funded research activities" *and* "delay or impede the rapid development and commercialization of technology." This is a remarkable set of conclusions because it is certainly not clear how giving rights to inventions to a global pharmaceutical company, which has agreed to pay more than $50 million for those rights, would slow the commercialization of that technology in comparison with licensing to a small, often financially-strapped, biotechnology company. Empirical evidence for that proposition would be interesting, but it would be a difficult hypothesis to test. Nevertheless, these "threshold criteria" have generally served as red flags in the negotiation of university-industry relationships ever since. The result has been that Scripps rewrote its agreement with Sandoz (now Novartis) and there has been no other institution-wide agreement between a major research institution and a pharmaceutical company since the Guidelines have issued. While arguably that has increased the burden on technology transfer offices, it may also mean that there are even more opportunities for licensing individual technologies from major universities and research institutes. The negotiation of such licenses is the subject of the following section, §3.5.

§3.5 Licensing a University Invention (Also a Few Tips on Negotiation)[7]

The United States biotechnology industry has been seeded, in large part, by discoveries made in university labs. Subsections 3.5(A)–(E) examine the terms of a typical license for an invention and some of the issues surrounding the various terms and conditions that must be negotiated in a license of that type. Section 3.6 looks at a sponsored research agreement and discusses the terms and conditions typically included in an agreement of that type.

The most fundamental requirement for successfully negotiating any type of agreement, including a license with a university for a university invention, is to know what it is you really need and what it is the other party to the negotiation really needs. As long as what you really need (as opposed to what you might wish for or might like in the best of all possible worlds) and what the other party really needs are not in conflict, an agreement should be achievable. This basic rule has two corollaries. First, the objective of negotiation is to reach an agreement. Second, if the technology really succeeds, both parties will be successful almost regardless of the precise terms and, if it does not, the only terms that matter are the boilerplate dealing with the unpleasant subjects of termination, indemnification, and so forth. Information on those kinds of terms is readily available in all sorts of practice guides and does not need repetition here. With this in mind, let us start with the most basic concept, which is the concept of the license itself.

A license, by definition, is merely a grant to the licensee of the right to do something which otherwise would violate some right of the party granting the

---

7. This §3.5 covers the main licensing issues in a university to industry licensing agreement. Many of these issues are quite similar in the context of licenses from biotechnology companies. Additional issues that arise in the context of biotechnology company to pharmaceutical company licensing agreements are covered in §6.5.

license, known as the licensor. So, a license to an invention is a grant of permission to make use of the invention without being sued for infringement of patent rights that are claimed by the licensor. Just from that brief description a few basic points should be clear. First, a license is not a transfer of ownership; it is merely permission to use something that is still owned by the licensor. This is very different from an assignment, which can transfer ownership (and in fact is used in the university context to acquire the inventor's rights, as discussed above in §3.2(A)). Second, a technology license is premised on the notion that the licensor has a valid and enforceable right to the licensed technology. While in the case of a license from a company this might occasionally include a protected trade secret, in the case of a university it generally means the university's grant of a license assumes that the university has been issued or will be issued a patent that covers the licensed technology.[8] If, for example, the patent application is ultimately rejected or abandoned, then the licensor can license only its know-how, which provides the licensee with much less protection against competition, is likely to be independently developed, and therefore has much less value. However, it is often significantly more than a year after the signing of the license before it becomes clear what the likely scope of the patent rights will be. Often the patent rights that issue are significantly narrower than the patent rights that were originally sought. What is crucial to the licensee is that the patent rights that ultimately issue are of sufficient scope to cover the primary business use of the invention by the licensee. The license will undoubtedly require the licensee to pay the costs of patent prosecution. The best that the licensee can hope to negotiate is a right to be consulted about the choice of patent lawyers and patent-prosecution decisions.

Understanding the licensee's planned use for the technology is critical for both sides to the negotiation. In most cases, even when it seems relatively simple, such as when the proposed licensed technology is for a single promising pharmaceutical compound discovered in a university lab, it needs careful thought. When it is a "platform" technology that can serve as the drug discovery engine for a start-up biotechnology company it requires even more consideration. In the following parts of this subsection, the key terms of a license are discussed and the issues that most commonly arise are analyzed from the perspective of different possible licensee business plans.

---

8. It is possible to license "know-how" that relates to the licensed technology, as well as the rights which may be within the scope of any patent. This necessitates identifying the subject of the know-how and the royalty rate for its use if no patent issues.

§3.5(A) Licensed Technology

The first key term, which certainly requires understanding the business plan of the licensee, is "licensed technology." The licensed technology term requires very careful definition. It most likely begins by referring to the technology described in specific patent applications, but then requires further elaboration. For example, the licensee would like to add language that makes it clear that the licensed technology also includes any subsequent patent applications that are continuations in part or divisions of the specified initial applications, which in most cases is acceptable to the licensor. After all, the purpose of the agreement is to provide the licensee with the opportunity to develop a successful product, which may require the use of "improvements" to the original technology as claimed in continuations of the originally named applications and to provide the licensor with an income stream from the development of that successful product.[9]

If the licensed technology is a potential therapeutic, based on a particular new protein or chemical structure, then the business plan of the licensee is relatively straightforward: develop the therapeutic into a marketable product. While other issues may arise in the context of sublicensing provisions, the main issue for this kind of license that arises in the definition of licensed technology relates to any improvements made by the licensee rather than the licensor. The university certainly doesn't want to hand over technology to the licensee simply so that the licensee can design a product that is outside the scope of the ultimate patent claims and avoid any royalties. The licensor will usually want to avoid this by inserting a "reach through" provision in the description of licensed technology, e.g.:

> "Improvements" means all patentable and non-patentable inventions, discoveries, technology and information of any type whatsoever, including without limitation polymer compounds, methods, processes, technical information, knowledge, experience and know-how which utilize, incorporate, derive from, or are based on the Licensed Technology or could not be conceived, developed or reduced to practice but for the use of the Licensed Technology.

---

9. One notable exception to the general willingness of licensors to include improvements described in continuations in part of the original patent is when the inventor was funded as a Howard Hughes Medical Institute (HHMI) investigator. In such cases the licensee should try to negotiate an option to license subsequent continuation patents on the same terms as the original license, with milestone payments for any such additional license to be credited to all such licenses. The guiding principle of fairness ought to be that one product should yield one royalty and one set of payments.

This, or clauses like it, are not uncommon and ought to be acceptable to both sides where the licensee simply plans to develop and market a product based on the licensed technology. However, such a clause could be a disaster in the case where the licensed technology is intended to be used as drug discovery technology and the licensee's business plan includes making that technology available to big pharmaceutical companies in exchange for funding and some royalty of their own. In the case of a "discovery" technology license, much more care needs to be given to make sure that the licensor does realize the appropriate value from the power of the technology to develop leads, but that the licensee is able to leverage the technology appropriately as its business plan develops. That problem spills over into the next topic: sublicensing rights and the sharing of the sublicensee's payments.

§3.5(B) Sublicensing

Every licensee needs the right to sublicense. Even large pharmaceutical companies may need to divest some or all of their rights to a technology, but for biotech companies it is almost certain that a sublicense will be part of the business plan. Sublicensing thus needs to be something that can be done without much difficulty—a licensor may need to approve, but it may not withhold its consent without good cause and may take no longer than 30 days to respond. The more difficult issue is what to do about sublicensee payments. Here again, a common pitfall is to use one boiler plate for the two very different types of licenses: the compound license and the drug discovery license. A licensor university's principal concern is that it receive a reasonable return from its technology, which means that the return has to come either from the royalty and milestone payment stream from the product the licensee or sublicensee develops and markets, or from the fees and royalties that the licensee receives from the sublicensee, or some combination of both.

For a compound-"product-oriented" license, the licensor wants to make sure that it has the same royalty and that it gets a fair percentage of any fees that the sublicensee pays to the licensee, regardless of who ends up developing and selling the compound or product. A fair percentage depends on how much more developed the product or compound is when it is sublicensed. The more that the licensee has done to develop and "prove" the value of the licensed technology, the lower the percentage of the sublicensing fees to which the licensor should be entitled.[10]

---

10. For an excellent discussion of this point, *see* Knox Bell, *Win/Win Licensing: University to Biotechnology Company,* 22 BIOTECHNOLOGY L, REP. 9 (2003).

For the other type of license, the drug discovery license, the situation is vastly different. If the biotech company licensee develops a drug discovery technology into a sophisticated system for assaying pharmaceutical company chemical libraries, or even collaborating with pharmaceutical companies on the design and synthesis of libraries for assaying in the system, then it is much more difficult to determine what the licensor's share should be of the licensing fees and royalties paid by the pharmaceutical company. It is probably too low to simply apply a royalty rate of 3 percent to 5 percent and treat the sublicensee's payments as though they were income from sales, but it is probably too high to apply percentages in the range of 15–50 percent that can apply to sublicensing fees for compound-based product licenses. While no hard and fast formula is possible, the parties need to thoroughly understand the value of various scenarios, including any equity the university may have in the licensee, and attempt to agree on a percentage of licensing fees and downstream royalties that works both for the licensor and the licensee's business plan.

§3.5(C) Field of Use, Geographic Territory, and Exclusive Versus Non-Exclusive

Most discussions of licensing deal first with the issue of whether a license is exclusive or non-exclusive and in which geographic territory, and then deal with the issues of field of use. Exclusive licenses limit the licensee to entering into a single license agreement in the geographic territory covered by the license, such as North America, Asia, or the European Union. Non-exclusive licenses allow any number of licensees simultaneously in the territory covered by the license. Of course co-exclusive licenses allowing only two licensees, or in some rare cases some finite number more than two, are also possible, though not nearly as common as exclusive and non-exclusive licenses. The major exception to this is the common university practice of reserving for itself the right to use the invention for its own research in its own labs. This ought not to be a problem, so long as the right of the licensee to improvements is adequately dealt with in defining licensed technology, as discussed in §3.5(A).

However, the relationship of the definitions of the field of use, geographic territory, and exclusivity is inextricable. Any license, even an exclusive license, only protects against competition from other licensees in the field of use defined by the license and in the geographic territory defined by the license. Geographic limitations are also important to understand. A geographic territory in which a licensee has the exclusive license does not protect against the importation of competitive products from a licensee in another territory to the extent that importation of such goods is permitted. At this time, the question of whether or not drugs or biologics produced and marketed outside the United States can be imported by individual patients ordering from pharmacies outside the United States is an extremely controversial one.[11] There has been serious consideration of the passage of a bill that would remove any barriers to such imports. Furthermore, if the technology is an enabling one, it is fairly clear that discoveries made with that technology outside the geographic territory can be developed or sold within the licensed territory. In that event, an "exclusive" North American license would be practically co-exclusive with any non-North American licensee whose product is thus available in the U.S. market. So, the question of exclusivity and geographic territory are not only linked, the relationship may change if the restrictions are loosened on personal importation of pharmaceuticals. It is likely to make it increasingly difficult for licensors to carve up exclusive licenses within separate geographic territories.

This raises the issue of field of use terms in licenses. The field of use can be broad, narrow, or extremely narrow. For example, a license to a broad field of use of a protein would be for "human therapeutics," or more narrowly for "oncology," or extremely narrowly for "end-stage renal disease." The point to keep in mind is that just as an exclusive license to a geographic territory is only as exclusive as the territorial barrier is impermeable, an exclusive license to a field of use is only as exclusive as the actual product for that field of use is truly distinguishable from the products created for other fields of use. For example, in the last example of field of use given, Amgen, the developer of a drug, erythropoietin, which stimulates red-blood cell production (a treatment for anemia), licensed Ortho to sell the drug for all uses *other than* the treatment of patients whose disease was due to end stage renal disease, i.e. dialysis patients. Thus patients whose anemia was due to chemotherapy, for example, would be within Ortho's exclusive field of use. However, the two drugs are generally indistinguishable: the active ingredient and dosing is generally the same regardless of the cause of the anemia; and, the result is that the "exclusive" license is actually a somewhat competitive "co-exclusive" license. This rather difficult licensing relationship led, not surprisingly, to extensive litigation.[12] By comparison, diagnostic antibodies are not interchangeable with therapeutic ones; animal drugs are rarely substituted for human ones; but even different formulations (i.v. versus injectable) of a particular drug may, in some cases, be used interchangeably. So any division of

---

11. *See* John Carey, *Drug R&D: Must Americans Always Pay?*, Bus. Week, October 13, 2003 at 38.

12. *See e.g.* Ortho Pharm. Corp. v. Amgen, Inc., 882 F.2d 806 (3rd Cir. 1989).

the field of use needs to be made in such a way that other fields of use will not result in a competitive product if the licensee's reasonable expectations are to be fulfilled.

§3.5(D) Royalties, Licensing Fees, and Milestone Payments

If the licensor university and licensee can agree on whether or not the license is exclusive, the field of use, the geographic territory, and the core financial terms need to be agreed upon. It is useful to think of all of the payments as being interdependent. Whether a payment is made as a royalty on sales or as a milestone payment, or even as a return on the sale of equity in the licensee issued as part of the initial licensing fee, it is all part of the licensor's return on the technology. Thus, where a university accepts a substantial block of stock in a start-up company as part of the licensing agreement, the royalty payments and milestone payments should be correspondingly less. Otherwise the licensee would be borrowing from Peter to pay Paul by lowering the revenues of the company and thus the value of the university's shares in order to pay the royalties and milestones. It is also important for both sides to remember one cardinal fact: if the technology is truly successful, the returns will be substantial even if the royalty rate and milestone payments are a bit low. Conversely, if the technology is not successful, then the royalty rate is largely irrelevant and the milestone payments are a windfall to the licensee.

While it would be convenient if there were universally agreed upon standard royalty rates and milestone payments for human diagnostics, for therapeutics, and for drug discovery platforms, that is simply not the case. A research tool that is licensed non-exclusively may call for a .5 percent royalty, but a truly powerful and value-creating tool (such as a process for significantly increasing the yield of therapeutic antibody production) may command a higher rate. A potential therapeutic may garner a 3 percent royalty or a 10 percent royalty, depending on the therapeutic field, the difficulty of the development pathway, and a host of other factors. The best guide to the marketplace for similar licenses is the data made available by subscription services that track licensing agreements in that field. Those, however, can never be more than starting point. Remember, the goal of negotiation is to reach an agreement, not to extract every drop of blood from the other party.

While milestone payments can vary in amount for licenses in which the licensee is expected to develop the technology into a therapeutic approved for marketing in the U.S., the EU, or both, the timing of the milestones along the way are generally similar: filing of an IND for first human clinical trials or the beginning of the first human clinical trials, beginning the second phase of clinical trials, beginning the first pivotal or Phase III clinical trial, filing an NDA or request for marketing approval, and first approval for marketing in a market within the geographic territory covered by the license. The amount of the milestone payments generally increases with each milestone. In addition, the license may well provide that the milestone payments are an advance against royalties, with some percentage of the total milestone payments to be deducted from each year's royalties until the pre-commercialization payments have been wholly or partly recouped.

§3.5(E) The Bottom Line on University Licensing

Keep the nature of the technology and its role in the licensee's business plan foremost when negotiating the definition of licensed technology and field of use.

§3.6 Sponsored Research Agreements

Sponsored research agreements (SRAs) are commonly used by pharmaceutical companies to tap into the expertise of a particular university researcher and by biotechnology companies where the university scientist who developed a company's core technology may continue to participate in its development. In general, SRAs are simple agreements. The sponsor (the company) agrees to fund a specific dollar amount per year for some number of years to support a specified project or projects. The research funding provides for overhead to the university and direct funding to the laboratory or laboratories. What are the issues? Providing a description of the research to be performed, specifying a time period for review prior to publication and, the most difficult and significant issue, determining the sponsor's rights to inventions made during the course of the research. Each of these is briefly discussed in this subsection.

§3.6(A) The Description of the Research

The research that will be carried out needs to be described in sufficient detail to assure the sponsor that the university scientists are agreeing to the general plan but without such specificity as to limit the ability of the scientist to shape the research as time goes by and the scientist's knowledge and ideas about the technology evolve. For example, "developing *in vitro* and animal model assays for testing the efficacy of the X family of compounds in the treatment of brain cancer and optimizing the bioavailability and receptor-binding affinity of X compounds" is likely a sufficient description. This is certainly true where the scientist has a continuing relationship with the company or was a scientific founder.

§3.6(B) Time for Review before Publication

Of course, providing a time for review before data from the research is published assumes that the university is going to insist on the right to publish the data and that the only question is how long does the sponsor have to review the data prior to publication? That is a very good assumption: virtually every university with which a sponsor would want to collaborate would insist on such a right. That leaves the question of time period the only real issue. A major research institution frequently used the following clause in its sponsored research agreements:

> The Principal Investigator and other investigators of PRESTIGE U. (PU) have the right to publish or otherwise publicly disclose information gained in the course of the Research. In order to permit SPONSOR an opportunity to determine if patentable inventions are disclosed, the Principal Investigator will provide SPONSOR with copies of articles written by project personnel reporting on the Research prior to or coincidental with submission for publication. Whenever possible, efforts will be made by the Principal Investigator to provide drafts of intended articles as soon as they reach a stage suitable for distribution. SPONSOR shall inform PU and the author in sufficient time so as not to delay publication whether in its judgment the material contains information on which patent applications may or should be filed.

What needs to be done to this clause? The following language would better from the sponsor's point of view and may be acceptable to the university: "PU agrees that Sponsor may have up to 60 days to review the material and advise PU of its decision as to whether or not to ask for patent filings to be made." While the NIH Guidelines for sponsored research suggest 30 days as reasonable, it is a good idea to ask for sixty days and agree to try to speed the review even within a longer maximum time frame.

§3.6(C) Sponsor's Rights to Inventions Resulting from the Sponsored Research

There are really only two options possible here. The first is a limited period during which the sponsor has an exclusive right to review the technology and negotiate a license. The second option is to provide an option to license future inventions under specific terms, with the option providing a complete license agreement. The following language from Prestigious University's agreement is an example of the first option:

> PU agrees to cause patent applications to be filed and prosecuted in its name at SPONSOR's expense, prorated where more than one sponsor has an interest in the basic technology, on inventions or discoveries conceived and reduced to practice in the course of the Research. PU shall promptly notify SPONSOR and provide it a copy of any such patent application. From the date of notification, SPONSOR shall have a six (6) month period to negotiate the terms of a license agreement and PU agrees to negotiate these license terms in good faith. During this period PU will not offer a commercial license to any other party.
>
> In the event SPONSOR does not agree, within thirty (30) days after notification of PU's intent to file a patent application, to support said filing, PU may file at its own expense and SPONSOR shall have no further rights in that patent application.

These clauses give a sponsor a first option to negotiate a license. This is commonly done, but does leave the sponsor with some uncertainty as to what the price will be and some exposure to other companies taking the technology for a higher price than the sponsor was willing to pay during the negotiation period. A partial fix is to give the sponsor the further right to match any offer higher than its "best offer" but that leaves PU bargaining with 3rd parties with the significant handicap of the sponsor's right to match a 3rd party offer. A more complete fix is the second option, which is to negotiate a full license in advance, assuming that some sorts of inventions can be anticipated. If the goal is to generate an optimized compound, or to find an important receptor, those are the sorts of objectives for which the pain of negotiating a full license, and appending it to the sponsored research agreement as the terms of the option, may well be worth the reduction in uncertainty.

§3.7 Conclusion: University Technology and Biotechnology

Understand that universities must respect academic freedom, publish the results of their scientists' research, and play by the NIH's rules. The technology developed at major universities is usually worth living with those constraints. So far, it has generated an enormous industry in the United States and will likely continue to do so. Go get it.

CHAPTER 4

Intellectual Property (IP) Protection for Biotechnology

§4.0 Introduction: The Need for Patents and the Patentability of Biotechnology

From the beginning, the problem of patent protection has been one of the most significant issues facing the biotechnology industry. The many patent-centered disputes that have arisen over the years can be divided into two very broad basic queries. First, should biotechnology inventions be patentable at all or does the patenting of such inventions destroy the academic enterprise and unfairly reward those who discover basic scientific truths? Second, if biotechnology is patentable, how can the traditional patent doctrines developed for other fields be applied to this new and different field? This chapter examines these two questions, underscoring the importance of patent protection for biotechnology and providing an outline of intellectual property law and its application to biotechnology.

Developing a biotechnology product from a university research result to a product in the marketplace is an enormously costly endeavor. Particularly in the case of pharmaceutical products, the costs necessitated by FDA requirements for clinical trials and good manufacturing practices (see Chapter 8, Special Regulatory Issues: Human Health Care Products) can total more than $800 million for a single drug.[1] The necessity of intellectual property protection for biotechnology inventions begins with this premise: without the inducement of the

---

1. http://csdd.tufts.edu/NewsEvents/RecentNews.asp?newsid=6 (last visited May 5, 2006).

quasi-monopolistic prices a patent holder can command, no company could justify the expenditures necessary to research and develop biotechnology products. This tradeoff, of financial reward for investment in research and development, is the principal purpose behind the system of patent law and is even expressly recognized in the U.S. Constitution. U.S. Const. Art. I, §8, cl. 8 grants Congress broad power to legislate to "promote the Progress of Science and useful Arts, by securing for limited Times to Authors and Inventors the exclusive Right to their respective Writings and Discoveries." Thus the framers of the Constitution recognized that the reward of time-limited exclusive rights was a fundamental tool for promoting progress in science and technology.

In turn, Congress enacted the basic U.S. patent statutes, beginning with 35 U.S.C. §101:

> Whoever invents or discovers any new and useful process, machine, manufacture, or composition of matter, or any new and useful improvement thereof, may obtain a patent therefore, subject to the conditions and requirements of this title.

This brief statutory provision circumscribes the boundaries of what is patentable, or what patent lawyers refer to as "patentable subject matter" in the United States: that which is new, useful, and a process, machine, manufacture, or composition of matter, or "improvement thereof." The question as to whether or not the fruits of biotechnology research are patentable has taken two forms: first, whether such biotechnology products are "products of nature" and therefore not truly "new" and "invent[ed]" as required by §101; and, second, when such products include whole organisms, whether living things are patentable as items of manufacture or compositions of matter. The answer to the first form of the question actually long pre-dates the biotechnology era. It has been the case for many years that patent protection for a "natural substance" could be obtained by those who first devise a method for isolating and purifying a "natural" substance into a useful, isolated, and pure form not found in nature. An excellent and very early example was *Parke Davis & Co. v. H.K. Mulford & Co.* (196 F. 496 (2d Cir. 1912)), upholding a patent to an adrenal gland extract in language that still speaks, almost 90 years later, to the field of biotechnology:

> Before the patentee's appearance in the field, it had been discovered that the suprarenal glands of certain animals, when dried and powdered or in aqueous solution, possessed hemostatic, blood pressure raising and astringent properties which indicated that they might be utilized in medical, surgical, and other arts. They were so used, but for various reasons ... they were most unsatisfactory. It was impor-

tant, if possible, to ascertain what it was in these glands that possessed these physiological properties ... and ... to isolate it from its environment ... so as to obtain it in a stable, pure, and concentrated form, efficient and constant in action, and which could be used without danger of bringing deleterious and injurious bodies into contact with or introducing such bodies into the patient's system. This Takamine did, and by a process, which is the subject of another patent not here in suit, he produced a substance, a chemical compound—in the form both of a base and of a salt—possessing the physiological characteristics of the glands and in the practically pure and stable form which was desired. We are satisfied that his invention —this product—was a highly meritorious one and that it is covered by his patents. *Parke Davis* at p. 497.

Parke Davis was decided very early in the development of the modern pharmaceutical industry, and its principle was indispensable to the development of that industry. But, if the "product of nature" objection to the patenting of biotechnology was answered early in the twentieth century, the second variation of the objections to biotechnology inventions, the patentability of living organisms, could not be answered until the biotechnology industry made the genetic manipulation of those organisms possible. Thus, what *Parke Davis* represented for the young pharmaceutical industry was reaffirmed for the fledgling biotechnology industry in the decision of the United States Supreme Court in *Diamond v. Chakrabarty*, 447 US 303 (1980). According to *Chakrabarty*, the fact that a "manufacture or composition of matter" is in the form of an organism does not render it unpatentable, as long as that organism is a novel and nonobvious "product of human ingenuity." Thus the products of biotechnology are, in general patentable, whether those products are newly produced cDNAs, proteins, organisms, or processes of manufacture or treatment. There are many analysts of the biotechnology industry who felt that the fate of the biotechnology industry hung in the balance when the Supreme Court decided *Chakrabarty* and it certainly was a very important factor in the growth of the biotechnology industry that followed in the 1980s.

§4.1 Patents and Trade Secrets: Choosing a Method for Protection

The issuance of a United States patent entitles the holder to the exclusive right to make, use, or sell the patented invention in the United States for 20 years from the date of application.[2] In exchange, as discussed in §4.7 below, the patent application must disclose (and the issued patent will make public) the best method of practicing the invention known to the applicant at the time of the application, in such a manner as to enable a person of ordinary skill in the art to make and use the invention. In other words, the price of the patent and its exclusive rights is publication of the invention, so that when the patent period expires, the invention becomes part of the public domain. Furthermore, patent offices, including the United States Patent and Trademark Office (USPTO) publish patent applications 18 months after filing even though the patent may not issue until years later. Thus a decision to apply for patent protection is a decision to disclose to the competition what you are doing and how you are doing it, even before the patent has issued or patent protection has become a certainty. For this reason, it sometimes may be worthwhile to use the primary alternative form of intellectual property protection for biotechnology, trade secret, to protect an invention or discovery. The decision as to whether to protect a biotechnology invention by patent application or as trade secret involves balancing the advantages and disadvantages of each.

Trade secrets, unlike patents, are created by state law, usually by state common law (not by statute, but solely by a line of judicial decisions). Thus the details of trade secret law vary from state to state. Therefore this section can provide only a general view of the most important features of trade secret law as it is commonly provided in most states. Trade secrets are not issued by a government. Trade secrets are created by any business when it develops information which provides the company with a commercial advantage and which it makes an effort to keep confidential. As part of the effort required in order to maintain valuable information as trade secrets, every biotechnology business should require all employees to sign confidentiality agreements before commencing employment. The confidentiality agreement provides the employee's agreement not to disclose to anyone outside the company any of the company's confidential information. In general, such agreements provide an enforceable basis for trade secrets, as long as the company makes a reasonable effort to keep its information private and makes the necessary effort to enforce the agreements, particularly when employees leave the company. Trade secrets, unlike patents, are of potentially unlimited duration, ending only when the trade secret information is publicly disclosed. Should a com-

---

2. The 20-year term from the date of application, which was a change from the prior 17 years from the date of issue, was perhaps the most significant change in U.S. patent law that resulted from the Uruguay Round of the General Agreement on Tariffs and Trade (GATT 1994). *See* 19 U U.S.C.§3501 *et seq.* (1998) and 35 U.S.C. §154 (1998).

petitor steal a trade secret, either by actual theft or through the cooperation of a duplicitous employee, the wrong-doers are subject to damages and injunctions. However, again in contrast to patents, trade secrets do not provide any protection against competitors who duplicate the invention either through their own independent efforts or by "reverse engineering" a sample of a product legitimately acquired in the marketplace.

Because progress in biotechnology is so rapid, and competition is so fierce, most companies follow a practice of seeking patent protection for the core of each of their major areas of research and development, such as a new approach to drug discovery, the gene sequences coding for a sought-after protein, novel antibodies and their uses, and the formulations and methods of actually administering the product to a patient. If a patent ultimately issues, the company is protected against competitors who subsequently independently develop the same process or product. However, trade secrets are also ubiquitously employed in the biotech industry for protecting the secondary know-how that is developed in fine-tuning a product after the patent application is filed. For example, the determination that a particular change in cell-culture medium greatly enhances product yield is likely to be kept a trade secret (see §4.7 Written Description, Enablement and Best Mode). In that way, once the patent expires on the product itself, the original developer may retain a significant competitive advantage in the actual manufacture of the product. These trade-secret methods of manufacture may also provide significant obstacles to the rapid introduction of "generic" competition after the patent for the principal ingredient has expired (see §8.7 infra).

§4.2 Differences Between U.S. and Foreign Patent Law

In 1994, the United States Congress approved revisions to U.S. patent law as part of the adoption and implementation of the Uruguay Round of the General Agreement on Tariffs and Trade (the Uruguay Round Agreements Act, 19 U.S.C. §3501 et seq. and 35 U.S.C. §154, among other affected provisions of Title 35). Although the term of United States patents was brought into conformity with the patent terms of the major European and Asian nations, the United States still stands virtually alone in some of the key provisions of its basic patent law (although talks aimed at further harmonizing U.S., European Community, and Japanese patent law continue). While all the major patent regimes require the elements of novelty, utility, and nonobviousness (albeit phrased and construed a bit differently in the three jurisdictions), only the

U.S. awards a patent to the first applicant to invent, rather than the first inventor to apply. To clarify this distinction, let us assume that Company A and Company B both worked frantically to find the gene that codes for protein X, which will have extraordinary therapeutic value. Company A cloned the gene and isolated the purified protein on July 1st and filed for patent protection on August 1st. Company B cloned the gene and purified the protein on July 3 and filed for patent protection on July 31st. Under those facts, Company B would receive a patent in Europe and Japan as the first inventor to file a patent application. In the U.S. after receiving the application from Company A, the U.S. Patent and Trademark Office (PTO) would declare an "interference" between A and B, with B, as the senior party to file, gaining a presumption of prior inventorship. Assume, however, that in the course of the long, drawn-out, and expensive interference proceeding, A's carefully witnessed laboratory notebooks and other records document A's claim of prior inventorship. The U.S. PTO would ultimately award the U.S. patent to Company A, while B's first filing would allow it to maintain patent protection in most major markets outside the U.S.

A second major distinction between U.S. patent law and patent law in most of the rest of the world is the relationship between publication of an invention and the requirement of novelty. Novelty is a question of whether the invention is described in the prior art (which is measured by what was available to the public prior to the date of the application either by publication or nonconfidential disclosure). In the United States, a scientist who publishes an article disclosing her discovery may retain full U.S. patent protection so long as she files an application for a patent within a "grace period" of one year of the appearance of the publication or public disclosure of the invention (such as by poster at a scientific meeting). In most other countries, however, publication or public disclosure, even by the inventor, constitutes prior art and results in a complete loss of patent protection, with the published article placing the invention in the public domain.[3]

Despite these differences, and some differences between the U.S. and foreign countries in the interpretation of patentable subject matter and the requirement that a claimed invention not be obvious to the person of ordinary skill in the art (see §4.6 below), a certain amount of integration between U.S. and foreign patent systems exists by virtue of the Paris Convention, the Patent Cooperation Treaty (PCT), and the European Patent Convention and Euro-

---

3. Canada also has a 12-month grace period and Japan and Taiwan have 6-month grace periods.

pean Patent Office (EPC and EPO). The principal benefit of these various levels of international cooperation agreements is to allow for provisional applications to establish the priority date and to adopt the date of filing in any Paris Convention signatory jurisdiction (the U.S. is a signatory), as the original "priority" date for purposes of the international "first to file" standard.[4] Thus a U.S. company, by filing a provisional application with the USPTO, establishes its priority internationally as to the date of its application. The U.S. company can then maintain the possibility of protection in a large number of jurisdictions by filing with the PCT or under the Paris Convention within 12 months of its U.S. filing. PCT filing preserves (for eighteen months) the right to file for individual examination and protection in any of the 25 or so PCT member countries. In this way, the more substantial expense of national filings for examination and protection in individual jurisdictions can be deferred for approximately 30 months after the initial U.S. filing.[5]

§4.3 Patentable Subject Matter

From its inception, the biotechnology industry has been met with a series of challenges to the patentability of biotechnology inventions. The question of whether or not living organisms were patentable, answered affirmatively in *Chakrabarty* (see §4.0 supra), was hardly the end of inquiries into whether or not the claimed subject matter of a biotechnology patent application was patentable under 35 U.S.C. §101. Claims to genetically modified multi-cellular or higher organisms were the subject of a later ruling by the U.S. PTO in In Re Allen (2 U.S.P.Q.2d 1425, Bd. Pat. App. 1985) which held that non-

---

4. Provisional applications have two principal virtues that "regular" applications do not. First, they establish a priority date while postponing examination (and substantial additional filing costs) for up to one year. Second, the provisional applications can be amended during that year (allowing for additional data to be included in support of the original disclosure, but not the addition of matter which would be a new invention or indicate that the original invention was not complete at the time of the provisional filing). They should otherwise be written as a full final patent application in describing and claiming the invention.

5. In the European Union, it also possible to file a single application and have a single examination done that results in a patent that is binding in all of the European Union states that are designated in the application. The major limitation in such European Patent Office patents is that enforcement is still done separately in each country, according to each country's procedures and rules on damages, etc. *See* Joseph Greenwald and Charles Levy, *Protection of Intellectual Rights: Convention on the Grant of European Patents*, 1 Basic Docs. of Int'l Econ. L 983 (1994).

human multi-cellular organisms could be patented. The ultimate test of the patentability of higher organisms was presented in the claimed invention of a transgenic mouse expressing a human oncogene (a cancer-related variant of a gene involved in cell growth or replication), the so-called Harvard oncomouse. The PTO issued a patent allowing the claim to the mouse (U.S. Pat. 4,736,866) and shortly thereafter issued a patent for a method for making transgenic mammals (U.S. Pat. 4,873,191). Thus, in the United States, living organisms, including mammals, which have been genetically engineered, are considered compositions of matter and therefore patentable subject matter under §101.

The international position on animal patents is somewhat more complex. The European Union position is that it will allow patents on the methods of genetically modifying animals, as long as the method is not limited to one species. The EU will also allow claims to methods of using genetically modified animals, subject to the limitation that the methods must be applicable to more than one species. However, the EU will not allow patents on a "species," and permits examiners to deny claims to "processes for modifying the genetic identity of animals which are likely to cause them suffering without any substantial medical benefit to man or animal, and also animals resulting from such processes." Individual EU member states may also forbid the actual practice of particular animal inventions based on the host country's laws concerning the humane treatment of animals and other ethical concerns. (Directive 98/44/EC OF THE European Parliament and of the Council).[6] Thus the EU general ruling allowing animal patents is subject to rather ambiguous exceptions that may bar patenting in particular cases. For example, although the oncomouse patent was allowed (Eur. Pat. Off. V 0006-92) it was only after a particularized consideration of whether or not, on the whole, the suffering of the resultant cancer-prone mice could be justified in terms of the overall need for animal experimentation and the reduction in the actual number of mice needed when compared to conventional rodent cancer studies.

In Canada, the Canadian Supreme Court ultimately decided that genetically engineered animals, in particular the Harvard Oncomouse, were not patentable subject matter within the scope of the Canadian Patent Act.[7] The Harvard Oncomouse was again the test case. The Canadian Commissioner of Patents denied claims to the oncomouse on a number of grounds that are

---

6. http://europa.eu.int/eurlex/pri/en/oj/dat/1998/l_213/l_21319980730en00130021.pdf.

7. Harvard College v. Canada (Commissioner of Patents) 4 S.C.R. 45, 2002 SCC 76 (2002).

likely to affect most claims to genetically engineered mammals, particularly the fact that the invention was not reliably reproducible. Nevertheless, claims were allowed to various components and methods used in the creation of the oncomouse might allow the patent holder some protection against Canadian competition. The decision of the Commissioner to deny the claim to the actual mice was, after a lengthy battle in the Canadian Courts, finally upheld by the Canadian Supreme Court which, as the dissent noted, thus stands alone among the major developed western nations.

The problem of patentable subject matter took a different form in the challenges to biotechnology inventions asserting the "product of nature" exception. As previously noted, the "product of nature" objection was encountered by pharmaceutical companies long before the advent of biotechnology and is, in essence, an assertion that the subject claimed is not "novel" (i.e., as a product of nature it existed before the acts of the inventor) or that it was not "invented" (because as a product of nature its existence was obvious). Although the product of nature objection was one of those used to attack the claims to living organisms, *Chakrabarty* and its companion case, *In re Bergy*, involving an antibiotic producing microbial strain) are relatively easy to defend on that ground. The fact that an organism is genetically modified is strong evidence that the claimed invention is different from that produced in nature. The product of nature objection is also made to many other important products of biotechnology research, such as the genetic sequences for producing desired proteins or the recombinant proteins themselves. The earlier pharmaceutical inventions that raised the product of nature objection rebutted the objection by asserting that the form of the claimed invention (e.g. a pure isolate having a defined specific activity) was novel and that both the method for producing the purified, isolated form and the precise properties of the purified form were nonobvious. Thus far, the same approach has generally succeeded for biotechnology inventions. Although the gene sequence for human insulin certainly exists in nature, the human insulin gene, like virtually all human genes, does not exist separately, outside a cell, or even as a continuous sequence of "coding" nucleotides strung along the chromosome. Genes of all multi-cellular organisms consist of coding regions (exons) interspersed with non-coding regions (introns) that are deleted in the process of transcription into m-RNA (see §2.1(C)–§2.2 in Chapter 2, A Primer on the Basic Science of Biotechnology). Thus, the cDNA for human insulin (cDNAs are DNA sequences, free of introns, made by transcribing m-RNA back into DNA with reverse transcriptase) used to produce the recombinant insulin protein, does not exist in nature and a patent claim to "the isolated insulin gene" is in fact a claim to a DNA construct that never existed in a human cell.

Similarly, the "product of nature" objection to a recombinantly-produced protein, such as human erythropoietin (EPO), will fail if the inventor was the first to isolate and purify the protein in question. Yes, the claimed invention is produced in human cells and is found in human tissue, but a purified isolate of EPO did not exist until the gene was cloned and expressed in a recombinant host cell-line. The more difficult objection to the patenting of some recombinant proteins, such as EPO, is not that they are products of nature, but that the means for isolating and purifying them were obvious. The application of the nonobviousness requirement of §103 to biotechnology inventions is explained in more detail in §4.6, while some of the unique problems raised by biotechnology patents are discussed in §4.8 of this chapter.

§4.4 New and Useful: The Requirement of Utility

The §101 requirement of patentable subject matter has provided another fertile source of objections to biotechnology inventions. A patentable invention must be useful, referred to as the "utility" requirement, and the applicant must set forth a use of the invention, unless a use would be apparent to one of ordinary skill in the art. The use must be substantial and specific, as well as credible (USPTO's 1999 Revised Utility Guidelines 64 Fed. Reg.71440). Each of these three requirements requires further explanation.

First, "substantial" utility is a requirement interposed to make it clear that the inventor has a real-world use for the invention. While with traditional biomedical inventions (such as a vaccine or a blood-pressure medication) the real-world use is the entire point of the invention, the era of modern genomics, where the sequencing of genes is done on a mechanical (rather than target-driven) basis, brought the utility issue to the fore. This presented a dilemma for the privately-funded genomics efforts, such as that at Human Genome Sciences. Once the sequence of a gene was known, it required intellectual property protection or it would fall into the public domain once it was placed in a genome database by another sequencing effort. Yet the sequence itself provided only the barest hints as to function (each of the more than 100 G-protein receptors, for example, may serve just one of the many different purposes found in the group as a whole).

The currently indeterminate relationship between gene sequence and the gene product's function emphasizes the need for the utility to be substantial *and* "specific." The purposes of patent law ought to preclude claims to genes generated during mass sequencing, where the only utility is for mapping a chromosome (any specific stretch of DNA can do that) or assessing whether or not an individual has a variant form of the gene (not a substantial utility unless it can be associated with the diagnosis of a specific disease or is otherwise informative).

The resolution of this issue remains to be seen. The practical course taken by some sequencers was to file an application with as much of a specific function of the gene as could be determined from homology to other genes and the type of tissue expressing the gene (e.g. helper T-cells). An example of such a patent would appear to be the patent issued to Human Genome Sciences claiming the chemokine receptor CCR-5. The patent issued to Human Genome Sciences without any knowledge that the receptor was involved in the HIV disease process, although it was subsequently discovered by other scientists to be a necessary factor for HIV infection of T-cells. Instead, after identifying the receptor by general type (G-protein chemokine receptor) and describing how to screen compounds for activity against the receptor, the patent specification asserted a catch-all kitchen sink of possible uses:

> The compounds which bind to and activate the G-protein chemokine receptors of the present invention may be employed to stimulate haematopoiesis, wound healing, coagulation, angiogenesis, to treat solid tumors, chronic infections, leukemia, T-cell mediated auto-immune diseases, parasitic infections, psoriasis, and to stimulate growth factor activity.
>
> The compounds which bind to and inhibit the G-protein chemokine receptors of the present invention may be employed to treat allergy, atherogenesis, anaphylaxis, malignancy, chronic and acute inflammation, histamine and IgE-mediated allergic reactions, prostaglandin-independent fever, bone marrow failure, silicosis, sarcoidosis, rheumatoid arthritis, shock and hyper-eosinophilic syndrome.

Is this enough? The second requirement, that the utility be "specific" might suggest that unless the uses claimed by Human Genome Sciences are in fact applications of CCR-5 activating compounds, the inventor did NOT provide the specific utility of the gene. Whether patents such as this will be enforceable remains to be seen. In the interim, companies continue to file provisional applications with as much of a guess as to actual use as possible, while hoping that subsequent experimentation during the year allowed for amending the provisional application for examination will provide data to support the credibility of at least one of the claimed utilities.

This leads to a discussion of the third major component of the utility requirement, that the claimed utility be credible to one of ordinary skill in the art.

Again, the difficulty stems from the need to file a patent application as soon as possible, driven by the first-to-file system of determining priority among inventors that is used outside the United States. Because of this requirement, a patent application may be filed before much data has been produced that supports the claimed effect of the invention (and again, in the case of gene sequences, often before the function of the gene's protein is well understood). For example, a rational drug design company may have synthesized a compound that does bind, *in vitro*, to its target enzyme or receptor on a nerve cell. Cell culture experiments demonstrate that the compound promotes the survival of neurons that are challenged with beta amyloid protein (believed to be a key agent in Alzheimer's disease). A patent application is filed in which the compound is claimed and a claim is made to its use in neurodegenerative diseases generally and Alzheimer's disease in specific. The legal standard by which utility is measured is whether, based on the application, a person of ordinary skill in the art would find the claimed utility to be credible or would instead doubt that the invention performs as claimed (again see the USPTO's Revised Utility Guidelines, 64 Fed Reg. 71440 (1999)). An examiner might well question the reasonableness of whether or not such a limited experiment entitles the inventor to a claim the compound as a method of treating a human disease. However, because it may be possible to supplement an application with later experimental data and because the burden of supporting the objection rests with the examiner, the better practice for most biotechnology companies is to file for patent protection as soon as a particular invention has been fully conceptualized, even if further experimentation will be necessary to substantiate the probability of its commercial success.

§4.5 Novelty and Publication

In addition to the §101 requirement of patentable subject matter, the basic prerequisites for a patentable invention are §102 novelty, §103 nonobviousness, and §112 written description, enablement, and best mode. To turn to the first of these, §102 novelty is the product of a complex multi-part statute that as a practical matter has a relatively straightforward significance for most biotech inventions. An invention is novel if no one else has publicly described or displayed precisely the same invention (i.e., the inventor is the first to conceive of the invention).[8] Furthermore, an otherwise novel invention ceases to

---

8. The description of the invention will be a bar if it appears in any publication in any language anywhere in the world. For example, if a talk given to the Royal Chemical Society of Kuwait was published only in Arabic in the Proceedings of the Royal Chemical So-

be novel if the inventor waits too long to file for patent protection after (1) publishing a description of the invention, or (2) filing an application for patent in another country, or (3) putting the invention in public use, or (4) putting the invention on sale. In the United States, an inventor can obtain patent protection only if she files a patent application within one year of doing any of the listed actions. In most other countries, publication or public use or sale is an immediate bar to any patent rights, although there is increasing discussion of the benefits of a "grace period" in the EU and elsewhere.

While the problem of novelty is intertwined with the problem of competing claims of inventorship and the priority between competing inventors (which in the United States is resolved by interference proceedings, as briefly described in §4.2, the biggest issue of practical significance to most biotech companies is the loss of foreign rights that accompanies publication of the invention before a patent application claiming the invention is filed. Since the U.S. is only approximately one-third of the desired market for most biotech products, the one-year grace period provided by U.S. patent law for filing a patent application after publication is only a partial measure of grace. The importance of the foreign markets and the prevailing stringent foreign interpretation of novelty as barring any prior publication absolutely mandate that a patent application be filed prior to the publication of a report of the research or invention. Since the current international system does provide a period of one year for filing in additional jurisdictions after the first application is filed, it is not necessary to file world-wide before publication, but it is necessary to file in a country that is a signatory to one of the major patent conventions or treaties before publication. For inventions made in the U.S., a U.S. filing is generally necessary and will suffice to establish the international priority date.

Closely related to the issue of whether or not it is necessary to file a patent application before publishing the research results that disclose an invention is the issue of whether or not to publish those results after filing for patent. In the context of a university, where the goal is the dissemination of knowledge, there can be no question of keeping research results secret (although the slight delay that might be necessitated in order to prepare and file a patent application can be justified on the grounds that without patent protection the invention would never be developed for the betterment of hu-

---

ciety of Kuwait, with a circulation of 30, it would be an immediate bar to a U.S. inventor after one year and would be immediately non-novel virtually every else. More esoteric discussions of novelty center on manuscripts that exist only in one university library and scrutinize such facts as whether or not the manuscript appears in the card catalog for that library.

mankind). In the context of a biotech company, the decision is more complex, weighing considerations of building the value and reputation of the company through public knowledge of its discoveries against the competitive advantage of secrecy. That decision must take into account the now universal practice of publishing applications 18 months after filing, regardless of whether or not the patent has yet issued. Thus, although competitors can be kept in the dark for a while, the practice of publishing unissued applications does place an outer bound on the length of time such secrecy can be maintained.

§4.6 Nonobviousness

Closely related to the novelty requirement is the requirement, contained in 35 U.S.C. §103, that the invention not be obvious (that is, it must be nonobvious) to the person of ordinary skill in the art that is the subject matter of the invention. The link between §102 novelty and §103 nonobviousness is through the §103 concept of "prior art." While an invention is not novel if it is completely described in any single printed publication (again, anywhere in any language), it is novel (but possibly not nonobvious) if parts of the invention appear in different publications. The prior art consists of all of the publications in the world that bear upon the claimed invention and were available to the public before the patent application was filed. At that point, the relevant question is whether or not a person of ordinary skill in the art, in light of all of the prior art, would find the claimed invention to be obvious.

Of all of the requirements for a patentable invention, the requirement of nonobviousness is the most troublesome and vague. The difficulty is at least two-fold. First, in setting as a standard the expectations or understanding of the hypothetical "person of ordinary skill in the art," the standard clearly sets the stage for conflict as to the application of that hypothetical and subjective standard to any particular case. Second, beyond the general subjectivity of the "person of ordinary skill in the art" standard, because the state of the art in biotechnology is advancing so rapidly, a decision on one invention may be of little or no guidance in ascertaining the nonobviousness of a similar invention that claims a date six months or a year later.

The ambiguity of the nonobviousness standard in biotechnology was, at least for a time, over the core meaning of the standard. This problem took the form of a debate over whether the obviousness of a means of accomplishing a result made the end result obvious. One important context for this question in biotechnology concerned the gene sequences for known proteins. Once the process for going from pure protein to actual gene sequence became relatively routine, the USPTO took the position that the end result (the actual gene sequence for a particular protein) was obvious. The decisions of the top U.S. patent court, the Court of Appeals of the Federal Circuit (CAFC) in *In re Graeme I. Bell,* 991 F.2d 781 (Fed. Cir. 1993) and *In re Thomas F. Deuel,* 51 F.3d 1552 (Fed. Cir. 1995), provided a clear judicial rejection of the PTO's approach to this problem. The *Deuel* decision is particularly worth noting. Claims 5 and 7 of Deuel's application claimed the cDNA sequences for two new heparin-binding growth factors (HBGF), which were the highly homologous (163 out of 168 amino acids) bovine and human forms of the proteins. Claims 4 and 6 claimed all possible DNA sequences encoding the amino acid sequences of the two proteins (of course, the actual number of such sequences is staggeringly large). The PTO's rejection dealt primarily with the claims to the cDNA sequences. The PTO cited two specific examples of the prior art: Maniatis, the PTO's then-standard text on cloning and probing for genes; and Bohlen, (who had described almost identical proteins). The PTO's reasoning was simple: given the amino acid sequence disclosed by Bohlen and using the techniques disclosed by Maniatis, the person of ordinary skill in the art would have encountered little difficulty in fishing out and sequencing the precise cDNA sequences claimed by Deuel.

Judge Lourie's opinion for a unanimous panel reversed the PTO Board of Patent Appeals and held that the Deuel's claims to the bovine and human cDNA sequences were valid. Judge Lourie's opinion is particularly interesting in its attempt to set forth a general rationale for the obviousness of any gene sequence for which all or part of the protein's amino acid sequence is known. Drawing on *Bell* for support, Lourie declared:

> We today reaffirm the principle, stated in Bell, that the existence of a general method of isolating cDNA or DNA molecules is essentially irrelevant to the question whether the specific molecules themselves would have been obvious, in the absence of other prior art that suggests the claimed DNAs. (*Duel* at 1559.)

In other words, the fact that a method for finding an unknown gene sequence is obvious does not make any of the actual mammalian sequences obvious, where a large number of possible sequences exist.

Judge Lourie's sweeping pronouncement in *Deuel*, that an obvious method to produce an unpredictable compound does not, without other additional prior art, make the compound obvious, would seem to provide guidance for a wide range of biotech inventions. The same reasoning supports the conclusion that, given a known target and a standard method for screening, no particular "hit" is obvious. Similarly, given a known antigen and a routine method of making high-affinity antibodies, no particular antibody is obvious. Nevertheless, Judge Lourie's opinion opens another door of uncertainty under §112 just as wide as the obviousness door that was seemingly slammed shut. For while the Court of Appeals for the Federal Circuit reversed without remand the PTO's denial of Deuel's claims to the actual human and bovine cDNAs, it remanded to the PTO for further consideration his claims to all other gene sequences that would encode those proteins (claims 4 and 6). Since the specific cDNA sequence used by a cell is not obvious when there are 10^{30} or more possible sequences, Judge Lourie reasoned that "Because Deuel's patent application does not describe how to obtain any DNA except the disclosed cDNA molecules, claims 4 and 6 [to all codon sequences encoding the disclosed amino acid sequence] may be considered to be inadequately supported by the disclosure of the application" and remanded to the PTO for decision of this §112 issue. By this reasoning, providing a target and a method of screening or an antigen and a method of making antibodies would not necessarily support broad claims to the broad range of "hits" or antibodies not described in the specification, even though the description of the antigen would enable the making of a huge number of antibodies to the antigen. This problem is further discussed in §4.10 Broad Target Patents.

The uncertainties about claims to the range of DNA sequences that encode a protein with a particular amino acid sequence, and to the antibodies that recognize a disclosed antigen, were both resolved in the PTO's 2001 Guidelines for Written Description ("GWD") in favor of the adequacy of the written description.[9] The USPTO recognized that while the actual messenger RNA used by a particular species and its corresponding cDNA could not be predicted from the amino acid sequence, the first to discover and provide the amino acid sequence of a protein did both enable and describe the "genus" of DNA sequences that would produce the protein, simply by providing the codon substitutions possible for each amino acid (GWD example 11). However, the PTO in its GWD went even further, reasoning that the inventor who provided a novel antigen (antibody target) could properly claim all of the antibodies that were produced in response to that antigen by the standard methods of the art, even though the precise amino acid sequence and structure of those antibodies could not be simply generated from a table.

---

9. http://www.uspto.gov/web/menu/written.pdf.

What will this mean for biotech inventors seeking to protect inventions based on purified or recombinant proteins? Judge Lourie's opinions in *Deuel* and *Bell* shed some light but hardly answer all of a would-be inventor's concerns. If the invention consists, in part, of a newly identified or purified protein, which was not part of prior art, the inventor may have some assurance in the composition of matter claim to the protein. Nevertheless, under the reasoning of *Genentech v. Wellcome*,[10] there will not necessarily be protection against those who make substantial modifications to the protein that result in unpredictable differences in action, pharmacokinetics, etc. For patent applicants claiming gene sequences for previously known proteins the situation is even less clear. Although *Bell* and *Deuel* would seem to give the applicant enforceable claims to the actual cDNA sequences, provided that the sequences were not previously known,[11] those claims would not necessarily protect the inventor against others who sought to produce the protein by changing a significant number of the codons for the "degenerate" amino acids in the protein. Whether or not the doctrine of equivalents would provide protection against those using different gene sequences to produce the identical protein is far from clear (see the discussion of the Doctrine of Equivalents in §4.11. A strong argument can be made that if the only changes made to a patented gene sequence are to substitute codons for the same amino acids, then the resulting gene sequence performs the identical function (coding for an identical protein) in the same way (by translation through m-RNA) with an identical result (the protein itself).

While also relying on *Deuel*, inventors claiming gene sequences for known proteins should continue to provide the PTO with information that demonstrates that for the particular gene sequence claimed, particular steps in the sequence determination presented novel problems, which might well stymie the person of ordinary skill, using the prior art. For example, another, nar-

---

10. Genentech, Inc. v. Wellcome Found. Ltd., 29 F.3d 1555 (Fed. Cir. 1994). Genentech and Leuven sued Wellcome and Genetics Institute for infringement of patents covering purified natural human tissue plasminogen activator (t-PA) and the gene sequence that encodes it. The trial court verdict in favor of Genentech, based on the doctrine of equivalents, was reversed by the CAFC, which found that the substantially different half-life of the defendant's altered protein, as well as its much lower affinity for fibrin, distinguished the defendant's protein.

11. There is considerably less likelihood that any particular actual coding sequence for a known protein will be novel, given the publication of the complete human genome. However, for novel proteins, the actual coding may still be novel, since the "spliced" order of combination of coding regions would not necessarily have been obvious before the protein itself was isolated.

rower ground for upholding the claims in *Deuel* was that the cDNA for that rare protein was only derived by the use of a cell-type not previously known to produce the protein. While *Deuel* would seem to provide significant protection to the inventors of new proteins and their gene sequences, the last round in this battle remains to be fought. The requirements of §103 are a moving standard, particularly for a field like biotechnology where new methods and new techniques are being developed on an almost daily basis.

§4.6(A) Secondary Evidence of Nonobviousness

Despite the technical complexity of determining the obviousness of many claimed biotechnology inventions, one additional approach to the nonobviousness requirement is based not on science but on extrinsic or "secondary" evidence of nonobviousness. Thus, in the debate over whether the person of ordinary skill in the art would have considered a particular claimed invention one that was reasonably likely to succeed, the courts permit consideration of the "(1) commercial success, (2) long felt but unsolved needs, and (3) the failure of others to solve the problem", *Graham v. John Deere Co.*, 383 U.S. 1, 17–18 (1966). In other words, it is highly suggestive of the nonobviousness of the invention if it solves a well-understood problem for which the solution was likely to bring significant reward, thus more or less asking—if it was so obvious, why wasn't it invented before? *See also Hybritech Inc. v. Monoclonal Antibodies Inc.*, 802 F.2d 1367 (Fed. Cir. 1986).

While the publication of the Genome Project may mean that the Deuel issue of the obviousness of newly described sequences for previously known proteins will soon be only of historical interest, the broader issue of the obviousness of particular products in light of the "obviousness" of the process by which a desired product can be made or found will undoubtedly rise to the forefront of biotechnology patent issues again and again.

To summarize the requirements of §103, four aspects of nonobviousness require additional emphasis. First, an invention may not be nonobvious even though no single publication discloses the whole invention. Second, it may not be nonobvious even if all of the prior art together fails to disclose all of the elements of the claimed invention because the person of ordinary skill in the art would reasonably infer that which was not disclosed by the prior art. Third, a type of invention, nonobvious today, such as the function of a particular gene and its protein, may well be obvious next month or next year. Finally, the apparent or *prima facie* obviousness of a particular invention may well be overcome and a patent issued if the inventor can demonstrate that particular obstacles were encountered for which no solution was contained in the prior art or that the invention had significantly greater value (therapeutic effect, lack of adverse reactions, etc.) than would have been expected by one of ordinary skill in the art. The requirement of nonobviousness, in examining the hypothetical conclusions of one of ordinary skill in the art, is arguably the aspect of patent law that demands the greatest technical or scientific sophistication to apply to most claimed biotechnology inventions and is the requirement that creates the greatest uncertainty. Although this chapter is intended only to introduce the non-patent lawyer to the basics of biotechnology patents, the fundamental problem of extrapolating from all of the prior art, along with the use of extrinsic or secondary evidence of nonobviousness, provide the essence of the §103 requirement of nonobviousness.

§4.7 Written Description, Enablement, and Best Mode: The Requirements of §112

> The specification shall contain a written description of the invention, and of the manner and process of making and using it, in such full, clear, concise, and exact terms as to enable any person skilled in the art to which it pertains, or with which it is most nearly connected, to make and use the same, and shall set forth the best mode contemplated by the inventor of carrying out his invention.... 35 U.S.C. §112.

As discussed in the introduction to this chapter, the basic purpose of patent law is to reward investment in research with exclusive control of inventions for a limited time, which in the United States is twenty years after the application is filed, with the possibility of limited extensions under special circumstances.[12] As part of the basic quid-pro-quo of patent exclusivity for public good, the patent law requires that the inventor must teach the public the best way to make and use the invention (known to the inventor on the date of the application), so that the period of exclusivity will in fact end when the patent term expires. This requirement of teaching is embodied in §112 and, as can be readily seen from the statute itself, consists of three related requirements: written description, enablement, and best mode.

---

12. 35 U.S.C. §154(a)(2) (2006). This new term for patents was perhaps the greatest single change in U.S. patent law created by the December 8, 1994, legislation implementing the General Agreement on Tariffs and Trades (GATT).

Description: The first of these requirements, written description, is the primary means by which an inventor enables the public to reproduce or practice the invention after the expiration of the patent term. Although the written description is in this way intimately bound up with the enablement requirement, it has a life of its own and can also be used to show that the inventor had possession of the invention (in other words, had completely worked out all essential elements of the invention) as of the time of application. At this time, the decision of the Court of Appeals for the Federal Circuit (the primary federal appellate court for intellectual property disputes) in *Enzo Biochem v. GenProbe*, 323 F.3d 956 (Fed. Cir. 2002) (Enzo II), has reiterated that the written description is an independent requirement, but left somewhat unclear its precise scope.

Although in 2005 the Court of Appeals finally held that Enzo's patent was invalid by virtue of having been offered for sale more than one year prior to the patent's filing (Enzo III), in the 2002 Enzo II opinion, which dealt with the adequacy of the written description, Enzo's patent was held to have adequately provided a written description of the claimed gene sequences. Enzo's patent claimed nucleic acid probes that would hybridize to N. gonorrhea in preference to the closely related, but non-pathogenic, N. meningitidis. The patent did not provide the actual sequences that were specific to N. gonorrhea but instead referred to the deposit of a nucleic acid with the American Type Culture Collection[13] and claimed DNA sequences which hybridized to the deposited nucleic acids. The CAFC explicitly endorsed the PTO's Written Description Guidelines and upheld the sufficiency of the written description by classifying it as a description by "functional characteristics ... coupled with a known or disclosed correlation between function and structure" (323 F.3d 964) quoting from the Guidelines (at 66 Fed. Reg. at 1106).

For inventions involving DNA, the ease of sequencing DNA has increased exponentially since Enzo made their deposit and filed their patent application, and the better practice is to include the sequence of nucleotides comprising the claimed sequence in the written description. Patent attorneys do broaden sequence claims a bit by specifying the complementary sequence and claiming all sequences that would hybridize to the complementary sequence under stringent conditions. This practice relies on the reasoning of the PTO in the GWD (and endorsed by the CAFC in Enzo II) that specifying such conditions for hybridization effectively limits the number of variants that are described and that

---

13. ATCC, a commonly used depository for biological materials referred to in patents and scientific articles.

minor codon variations would literally infringe such a claim and not necessitate use of the Doctrine of Equivalents to protect would-be competitors.

Enablement: The enablement requirement, in the terms of the statute, is a written description of the invention that would permit the person of ordinary skill in the art (the same "objective" standard used for purposes of nonobviousness) "to make and use it, without undue experimentation" using the description in the patent application and the knowledge in that field (referred to in patent law as "art").[14] It is, to use a simple analogy, the complete recipe, with all the ingredients, measures, and instructions, so as to enable the reasonable chef to produce a dish that is essentially indistinguishable from that produced by the inventor.

The enablement requirement is often somewhat more burdensome for the inventor in biotechnology than in many other areas of science and technology, because biological systems may be somewhat less predictable than mechanical or electrical devices. For example, one can be reasonably assured that creating a particular electrical circuit used to convert an analog signal to a digital signal, according the diagram contained in a patent, will produce precisely the same result as the electrical circuit produced by the inventor. There may be a more significant variation where the claimed invention is a process for making human Factor VIII-C in Chinese hamster ovary cells. A person of ordinary skill in the art who follows the written description of the process for inserting the gene into the cells may nevertheless obtain a resulting cell-line that is significantly less productive of pure protein than the original inventor or, if other conditions are sufficiently varied (temperature, medium, etc.), may even produce a protein that is folded or glycosylated differently and has little or no biological activity.

The variability inherent in biological systems has caused courts to enunciate the doctrine that for such less predictable processes as are involved in biotechnology, "the scope of enablement obviously varies inversely with the degree of unpredictability of the factors involved."[15] In other words, the greater the potential variability in outcome, the more extensive are the requirements of enablement and best mode.

One solution to the problem of enablement in biotechnology has been the practice of depositing microorganisms and cell cultures with depositories such as the American Type Culture Collection (ATCC), as a supplement to the writ-

---

14. U.S. v. Telectronics, Inc. 857 F.2d 778, (Fed. Cir. 1988).
15. Amgen, Inc. v. Chugai Pharm. Co., Ltd., 13 USPQ2d 1737, 1776–7 (D. Mass. 1989). Judgment *aff'd* in part, *vacated* in part by 927 F.2d 1200 (Fed. Cir.), *cert. denied,* 502 U.S. 856 (1991).

ten description (See Enzo II, above §4.7). These depositories maintain the biological materials deposited and may make them available to the public after the issuance of a patent. In that way, an inventor can more easily demonstrate to the patent office that the applicant has enabled others to reproduce the invention.

Best Mode: The requirement of best mode is an extension of the enablement requirement. The patent application must not only enable the invention to be duplicated by the person of ordinary skill in the art, but must also reveal the best mode of carrying it out known to the inventor at the time of filing the patent application. For example, if a particular patent application claims a process for producing a recombinant protein in a yeast system and, at the time that the application was filed, the inventor believed a particular known strain of yeast to be the most efficient producer of the recombinant protein, then the application must disclose that strain. Once again, deposit may be necessary for the enablement of others to practice the best mode, where one of ordinary skill in the art would be unable to replicate the inventor's best mode (i.e., cell-line, host organism, etc.) without considerable experimentation and trial and error.

For both the §112 requirements, enablement and best mode, as in the case of the ordinary skill in the art for §103 purposes, the rapid progress of the state of the art in biotechnology leads to changes in the nature of the enablement requirement. For example, in only the past few years, the ability to produce and select monoclonal antibodies has increased dramatically. For that reason, where some years ago the deposit of the antibody-producing hybridoma cell-line was considered essential to enabling the practice of the best mode of an antibody invention, it may no longer be so. For example, it was not necessary for enablement in *Johns Hopkins v. Cellpro*, 152 F.3d 1342 (Fed. Cir. 1998).

The best mode requirement has been the subject of rather cynical commentary that asserts that it is common for companies to deposit something other than their preferred biological materials. For example this criticism is found in Cooper, Biotechnology and the Law §5.04[2] (1992). Part of the problem lies in the way that the failure to disclose the best mode is raised as part of a defense of invalidity in an infringement action. In such a case, the burden falls on the defendant to prove that the patent holder concealed the best mode known to the inventor at the time of application. As stated in §4.1, it is quite common and even preferable to file for a patent before the best mode for scale-up and manufacture has been determined, with subsequent refinements in methods kept as trade secrets. In such a case, it is difficult or impossible for an accused infringer to show that the inventor concealed the best mode known to him at the time of filing.

In *Amgen v. Chugai* (*see* note 48), part of the controversy over enablement and best mode turned on the fact that although Amgen had deposited E. coli and yeast strains of the type it had originally used to make recombinant human erythropoietin (EPO), Amgen's preferred method for production used Chinese hamster ovary (CHO) cells which were not deposited. Amgen had specifically claimed transformed CHO cells as well as other host cell types and had described a method for making EPO in the strain of CHO cells it preferred. The District Court held that Amgen had enabled its best mode and the Court of Appeals agreed. If at the time of application the inventor contemplates a particular mode as best, "the inquiry is whether his disclosure is adequate to enable one skilled in the art to practice the best mode or, in other words, whether the best mode has been concealed from the public."[16] Thus, a written description that permits one of ordinary skill in the art to replicate the invention and best mode, without undue experimentation, will suffice even if several modes are described and the application does not point out which of the listed methods is preferred by the inventor and even where the preferred mode is described but deposit is made of cells used in other described modes not preferred by the inventor.

§4.8 OTHER SPECIAL PROBLEMS FOR BIOTECHNOLOGY PATENTS

In addition to the general problems of applying the requirements of utility, novelty, nonobviousness, enablement, and best mode to biological inventions, discussed above, there are several patent-related problems that are particularly difficult for biotechnology. The equitable ownership of products derived from human tissue, the problem of inventions that draw on the traditional knowledge of indigenous peoples, the patenting of animals (particularly farm animals), and broad claims to "disease pathways" are significant and controversial issues for biotechnology patent protection. This section discusses each of those problems.

§4.8(A) Protection from Foreign Competition

Many biotechnology companies have directed their efforts toward the production of previously known substances by use of the increasingly powerful

---

16. Amgen v. Chugai, 927 F.2d 1200 *at* 1209 (1991).

tools of biotechnology. One example is the attempt to produce a known but rare or difficult to purify protein by determining the gene sequence that encodes the protein and then producing commercial quantities of the protein using genetic engineering. This was the means by which Amgen produced the most successful biotech product to date, erythropoietin. In such a case the gene sequence might be patentable. But what if the end product was already known (for instance purified erythropoietin or purified Factor VIII-C) and the gene sequence, which makes commercial production possible, is the true invention? Could competitors use that gene sequence to produce the previously known protein in a country in which the gene sequence is not patented and import the resulting product into the United States?

This problem was among the issues involved in the Amgen litigation against Genetics Institute and Chugai (*see* notes 48–49). While the battle in the District Court and CAFC raged over the validity of Amgen's gene-based EPO patent and Chugai's purified-EPO-from-nature patent, Chugai was making recombinant EPO abroad and bringing it into the U.S. Amgen claimed that Chugai's importation of recombinant EPO violated a section of the Tariff Act of 1930 that prohibited the importation into the U.S. of

> articles that—(ii) are made, produced, processed, or mined under, or **by means of, a process** covered by the claims of a valid and enforceable United States patent. (Tariff Act of 1930, as amended in 1988, §1337(a)(1)(B)(ii)) (emphasis added). *Amgen* at 1534.

Amgen had received a patent on the gene sequence for making recombinant erythropoietin, but had not then been granted a patent on the process of using the sequence to make the protein. Nevertheless, Amgen argued that its patent claims to various materials, including gene sequences, vectors, and host cells were equivalent to a "process of making" EPO and that Amgen was therefore entitled to protection by the Tariff Act against the importation of recombinant EPO.

The International Trade Commission rejected Amgen's request for protection on jurisdictional grounds and the CAFC affirmed the result of the ITC's decision ruling that Amgen's claims to various compositions of matter were not equivalent to process claims and the statute only protected holders of patents with process claims. The result was that inventors of recombinant versions of previously known substances would not be protected by U.S. patent or tariff law against foreign competitors so long as the process by which the recombinant protein was made, rather than the materials used in the process, was not itself new, nonobvious, and therefore itself a patented process.

Congress ultimately amended the patent laws to provide explicit protection to such biotechnology inventions:

(b)(1) ... a biotechnological process using or resulting in a composition of matter that is novel under section 102 and nonobvious under subsection (a) of this section shall be considered nonobvious if— (A) claims to the process and the composition of matter are contained in either the same application for patent or in separate applications having the same effective filing date.... 35 U.S.C. §103.

The result of the addition of subsection (b)(1) to §103 is to allow process claims for biotechnology inventions that use standard methods to express novel gene sequences or to produce proteins that had not previously been isolated or purified. This gives U.S. patent holders effective protection under the Tariff Act of 1930 against importation of a previously known protein produced abroad with the U.S. patented process. However, an analogous problem and perhaps even more serious problem that is not addressed by the statute can easily arise in the context of patents on newly identified targets for drug discovery and their use in screening techniques to identify new drugs. While such a screen may be patented in many countries, it may be practiced anywhere that it is not patented and the resulting drugs or drug candidates can be brought into the U.S. for further development or sale without infringing the screening patent. See §4.10 Broad Target Patents.

§4.8(B) Biotechnology Products Derived from Human Tissue

The question of the rights to products derived from human tissue was most famously raised in the case of *Moore v. Regents of University of California*, 51 Cal. 3d 120, 793 P.2d 479 (Ca. 1990). In *Moore*, the plaintiff Moore's treatment for leukemia by physicians on the faculty of UCLA included the removal of his spleen. Golde, Moore's physician, along with his co-researchers found that tissue from Moore's spleen was richly productive of a number of cytokines. Research on Moore's tissue ultimately enabled Golde to find the gene sequences for a number of important cytokine products that were included within the claims of U.S. Patent 4,438,032 (March 20, 1984), assigned to defendants Genetics Institute and Sandoz. Moore claimed that he had not been apprised of the potential commercial use of his tissue, that he had not truly consented to such use, and that he was entitled to a share of the profits from the products developed from research that used his tissue.

Although the California Supreme Court allowed Moore's claim for violation of his right to give informed consent to procedures to which he was subjected, the court rejected Moore's claim to a right to share in the profits from products developed from his tissue. The court, in a 5-2 decision, ruled that a person had no property interest in tissue removed from his body, thus effectively ending any claim of patients to share in profits from biotechnology products under California law. The court's reasoning has been subjected to considerable criticism, however, and it is not binding on the courts of other states. Thus the issue that was raised in the *Moore* case should still be approached with caution by companies deriving products from human tissue or using human tissues in research. The greatest caution ought to be exercised in ascertaining the source of such tissues and, where possible, to ensuring that the person from whom tissue is to be obtained fully understands that his or her tissue is to be the subject of research that might lead to a commercially valuable product for use in the treatment of disease.

§4.8(C) Patent Protection for Stem Cell Research and Development

Research and development of stem cells has spawned its own generation of patent issues and battles. In general, the United States Patent Office has treated stem cell patent applications like other applications for cell-based and biological products, applying the basic framework and requirements of patent law. A number of such patents have been issued. Perhaps the most significant such patent was issued to James A. Thomson of the University of Wisconsin, assigned to the Wisconsin Alumni Research Foundation and licensed, in part, to Geron Corporation. That patent, U.S. Patent No. 6,200,806 (March 13, 2001), covers methods of producing quantities of purified pluripotent human embryonic stem cells and preparations of such cells which meet specific criteria for viability and stability. It has been the subject of scholarly attack[17] but has not yet been the subject of any substantive court ruling on the validity of its broad claims. In the EU, the situation has been significantly complicated by specific provisions of the European Patent Conventions dealing with embryos and human beings. As a result, such broad patents have generally not been issued.[18]

---

17. *See, e.g.,* Peter Yun-hyoung Lee, *Inverting the Logic of Scientific Discovery, Applying Common Law Patentable Subject Matter Doctrine to Constrain Patents on Biotechnology Research Tools*, 19 Harv. J. Law & Tech. 79 (2005); Amy Ligler, *Egregious Error or Admirable Advance: The Memorandum of Understanding That Enables Federally Funded Basic Human Embryonic Stem Cell Research*, 2001 Duke L. & Tech. Rev. 37 (2001); Arti K. Rai and Rebecca S. Eisenberg, *The Public Domain: Bayh Dole Reform and the Progress of Biomedicine*, 66 Law & Contemp. Prob. 289 (2003).

18. Heike Vogelsang-Wenke, *Patenting of Stem Cells and Processes Involving Stem Cells According to the Rules of the European Patent Convention*, 23 Biotechnology L. Rep. 155–167 (2004).

§4.9 Biotechnology Patents: Moral and Ethical Issues

While there have been numerous objections to the morality or ethics of biotechnology patents in general, two areas deserve special mention. The first involves the problem of the ethical and fair way for biotechnology companies and their scientists, searching for potentially useful valuable knowledge of plants and other natural resources, to interact with persons from traditional societies.[19] A second area in which unique moral and ethical issues has been raised is in the patenting of whole animals, particularly in the EU where patent law specifically permits the denial of patents to inventions that contravene morality. This problem is discussed in §4.9(C).

To provide a context for the first problem, concerning interactions with indigenous peoples, it generally arises in situations when employees of biotechnology or pharmaceutical companies visit with traditional healers in less-developed regions of the world in order to learn what plants those healers use to treat various illnesses, such as diabetes, arthritis, or infection (when described symptomatically to the traditional healers). Samples of the various plants used by the local peoples would then be taken back to laboratories in the U.S. or Europe and analyzed for the chemical constituents that might be the active ingredients responsible for any reported therapeutic activity. This process is often referred to as "bioprospecting." The controversy has two distinct and rather different dimensions. First, what should be the rights of the indigenous peoples who contributed to the effort that produced the invention? Second, should the active ingredient of a traditional botanical or herbal medicine be patentable?

§4.9(A) The Rights of Indigenous Peoples

Questions raised concerning the rights of indigenous peoples to value created from their traditional knowledge have not been completely resolved. The problem is complicated by the inherent potential for conflict between any recognition of the rights of the indigenous peoples and the rights of the sov-

---

19. *See* Marc Lacey, *An Age-Old Salt Lake May Yield a Washday Miracle*, N.Y. TIMES, February 21, 2006. The article recounts criticism of Genencor for profiting from sales of an enzyme derived from microbes found in African Salt Lakes and commercially used in "stonewashing" jeans. Though not a pharmaceutical product, the industrial enzyme engendered much the same debate about biodiversity and biopiracy.

ereign state in which the indigenous people live and from which the botanical source of the remedy is obtained. The short answer to the question "What are the rights of the indigenous people?" is that there are no international legal conventions that grant any property rights of this type to the indigenous people beyond that which is given to them by their national laws.

The United Nations Convention on Biodiversity, to which the United States is a signatory (United Nations Conference on Environment and Development, June 5, 1992, S. Treaty Doc. No. 20 (1993)), declares that the "genetic resources" contained within a sovereign state are owned by that state and use of those genetic resources requires fair compensation. However, it is not at all clear that the identification of a small molecule active ingredient of an herbal remedy, which is then reproduced synthetically, requires any particular compensation under the treaty even to the sovereign state in which the original plant was found. Although theoretically a traditional healer could obtain some rights by contractual bargaining, as a practical matter there is no real possibility of true negotiation between a traditional healer and a major pharmaceutical company, unless the sovereign state in which the indigenous people reside decides to intervene on their behalf, rather than bargain for some return to the sovereign state. Of course the sovereign state also has the power to place whatever conditions it desires on the entry of research scientists or their removal of biological samples on their departure.

The absence of formal international legal protection does not mean that there is not a widespread recognition that the knowledge of the indigenous healers has real value and an ethical, if not legal, claim for compensation. This situation has led to two different models for companies wishing to use traditional knowledge of botanical remedies as a component of drug discovery. Under the Shaman Pharmaceutical model, the company pledged a part of its profits to assist indigenous peoples. Under the Merck model, a conservancy organization of the host country is the contractual partner, with compensation to the host country then being used to promote conservation and biodiversity.

As the reference to the U.N. Convention on Biodiversity demonstrates, the problem of indigenous knowledge and "bioprospecting" has been further complicated by the potential connection between the effort to find useful pharmaceutical compounds in the botanical resources of lesser developed countries and the need to protect the biodiversity of the biologically rich environments of the least developed portions of those countries, such as the Amazon rain forest in South America. Clearly, if a small molecule that can be synthesized is identified from a plant source, the process of bioprospecting does not contribute to the loss of biodiversity. Nevertheless, the position of the U.N. Convention is that bioprospecting should be made to contribute to the solution of the problem of the loss of biodiversity, in part because pharmaceutical companies stand to be among the principal commercial beneficiaries of the rich chemical portfolio contained within biologically diverse environments. In conclusion, the moral claims of indigenous persons to compensation for the use of their knowledge are widely viewed as strong and legitimate moral concerns, but the recognition of those claims has, to date, been voluntary.

§4.9(B) Patent Protection for Drugs Derived from Traditional Remedies

The second aspect of the intellectual property issues surrounding bioprospecting is whether or not drugs with active ingredients derived from a traditional botanical remedy can be patented. The answer to that question is an unequivocal "that depends." If the active ingredient has not previously been known or described in the chemical literature, then it is novel and useful. The question as to whether or not it is nonobvious is the source of the uncertainty. Nonobviousness in turn would hinge on the difficulty that a person skilled in the art would anticipate encountering in isolating, identifying, and characterizing the active ingredient. In most cases, given the state of that general art, a researcher would not necessarily believe that success is likely without considerable "non-routine" effort. Only where success in characterizing the active ingredient is likely to be relatively routine would the identity of the source make the end result obvious. Of course, the more that is known about the particular traditional remedy, the more the prior art might make the end product obvious. Nevertheless, a pill containing a standardized dose of a synthetic active ingredient is clearly a different product than a leaf to be chewed or brewed, and is therefore a potentially patentable invention.

§4.9(C) Patenting Animals: Beyond Patentable Subject Matter to Agricultural Policy and Animal Rights

There appear to be primarily two concerns of those who oppose the issuance of patents on higher organisms. First, there are those for whom the issue is primarily an ethical one. For many of those opposed, the patenting of animals is wrong because it essentially gives an inventor dominion over an entire species and rewards the manipulation of species, both of which have somewhat God-like overtones. For these opponents of animal patents, owning the rights to an entire species of animal is an ethical wrong unrelated to the right to claim property rights in individual animals. Of course for animal rights advocates, such patents are simply a gross extension of the unethical property interests generally permitted in animals.

A second group of opponents of animal patents are farmers who are concerned about the implications of such patents for the economics of farming. As the allowed patent claim quoted above indicates, the rights of the patent holder extend not only to the first generation of animals produced or sold, but also to all subsequent generations of such animals during the patent term. Thus future sales of patented animals could be different from the long-established pattern for sales of traditionally-bred animals. When one sells a traditionally-bred cow that is extremely productive, the seller generally derives no additional benefit from her offspring even though they are likely to inherit the desirable productivity of their mother. With patented animals, the seller might well demand and receive a payment not only for the animal sold but also a royalty on, and control over, offspring of the patented animal.

A number of bills have been introduced in the U.S. to deal with the two basic objections raised to animal patents although none have passed. The two basic solutions posed are a moratorium on such patents, to allow for additional reflection upon the ethical, environmental, and economic issues such patents raise, and a "farmers exemption" to permit farmers who purchased patent animals to breed the animals and use the offspring without being liable to the patent holder for infringement. Although the U.S. has not enacted such legislation, the European Union Directive on biotechnology inventions does grant the farmer a limited right, in derogation of any right of the patent holder, to breed patented animals for farm purposes other than commercial sale. O.J. Eur. Comm. NO. C 110, P. 0017 (Feb. 26, 1998).

§4.10 Broad Patent Claims to Drug Targets

The most recent "special" controversy to affect the field of biotechnology patents is the issuance of patents with claims to any pharmaceutical treatment of a particular disease (or in some cases broad areas of disease such as "autoimmune") that operate by acting on a newly identified target. Such claims are often referred to as "reach-through" claims, as they allow the inventor of the target to "reach through" to claim rights to subsequently developed drugs. A particularly noteworthy example of such a patent was issued to a group of researchers, including Nobel laureate David Baltimore, who first identified the NF-1B signaling pathway. The patent was assigned to the Massachusetts Institute of Technology, the Whitehead Institute for Biomedical Research, and Harvard University and licensed to a biotechnology company. The first claim (of 203 separate claims!) of U.S. Patent 6,410,516 (June 25, 2002) reads as follows:

> 1. A method for inhibiting expression, in a eukaryotic cell, of a gene whose transcription is regulated by NF-.kappa.KB, the method comprising reducing NF-.kappa.B activity in the cell such that expression of said gene is inhibited.

The significance of that claim lies chiefly in the extraordinary potential for therapeutics that operate through the claimed method, that is by reducing the activity of NF-1B, which includes an enormous range of diseases involving inflammation, viral infection, and autoimmune disease, among others. While scientists are now working to develop therapeutics that regulate NF-1B activity, the Baltimore group which had first identified the molecule as a key cellular activator of inflammatory genes had filed their original patent application in 1986. The patent that issued in 2002 was one of a number of "offspring" divisional and continuation applications of that 1986 patent and is now a potentially dominant patent over virtually all of the ongoing work on the NF-1B field.[20]

While an extensive analysis of the Baltimore patent or others of that type is beyond the scope of this chapter, it is possible to use the Baltimore patent to illustrate the issues that are likely to arise when the holders of such patents seek to enforce them against the developers of drugs that operate on the target described within the broad method claims. First, the claim is a method claim that on its face would cover virtually any chemical composition that might be developed to reduce the production of NF-1B. Methods are subject to all the other requirements of patentability, so the further issues are whether the method is novel, nonobvious, adequately described, and enabled. Since Baltimore discovered NF-1B and its biological function, it was clearly new and almost certainly nonobvious. The more difficult question is whether the inventors have provided sufficient guidance in their application to permit others to do what is claimed, i.e. reduce the expression of NF-1B. The requirements that are most difficult in such cases will almost always be whether or not the written description is enabling and provides the best mode of practicing the invention then known to the inventor. In such cases, the test is whether the written description would enable the person of ordinary skill in

---

20. Ariad Pharmaceuticals, Inc. is the licensee of the patent and is currently in litigation with Eli Lilly and others to enforce their claims. *See, e.g., Ariad Pharms., Inc. v. Eli Lilly & Co.,* 2005 U.S. Dist. LEXIS 10941 (D. Mass., 2005), N.Y.Times May 6, 2006 at C1-2, *Lilly Loses Patent Case to Ariad* (reporting a jury verdict for Ariad in the District Court).

the art to practice the invention with out "undue" experimentation. While the patent office ultimately came to the conclusion that only "routine" experimentation would be required to use the claimed method, that conclusion is certainly one that will be vigorously contested in the case of Baltimore's patent and others like it.

§4.11 Patent Infringement and the Doctrine of Equivalents

In patent law, when the owner of a patent (patentee) wishes to enforce her exclusive patent rights against a would-be competitor, the patentee would bring an action in infringement.[21] When the accused infringing product is clearly identical to the patentee's invention, the case can be considered one of literal infringement. Literal infringement occurs when every limitation in the claim[22] is found in the accused product or process. To avoid literal infringement, the developer of a second, competitive product might make changes in the product. When an alleged infringer has simply made minor, easy, or relatively obvious changes in the product, the courts have developed the equitable doctrine of equivalents to prevent "what is in essence a pirating of the patentee's invention."[23] As traditionally stated, an accused product or process infringes a patent if it performs substantially the same function in substantially the same way to give substantially the same results,[24] because, in patent law, "if two devices do the same work in substantially the same way, and accomplish substantially the same result, they are the same, even though they differ in name, form, or shape."[25]

---

21. 35 U.S.C. §271.
22. Jurgens v. McKasy, 927 F.2d 1552, 1560 (Fed.Cir. 1991). A patent claim is an assertion of what the invention purports to accomplish and claims (found at the end of a patent) define the invention and the extent of the grant. Any feature of an invention not stated in the claims is beyond the scope of patent protection. See 35 U.S.C. §112 (2006), second paragraph.
23. Loctite Corp. v. Ultraseal Ltd., 781 F.2d at 870 (1985); see also Hormone Research Foundation, Inc. v. Genentech, 904 F.2d 1558, 1564 (Fed.Cir. 1990), cert. dismissed, 499 U.S. 955 (1991).
24. *Graver Tank*, 339 U.S. 605, 608 (1950).
25. *Id., see also* Sanitary Refrigerator, 280 U.S. at 41–42 (1863): "There is a substantial identity, constituting infringement," where a device copies a claimed device, without variations or with such variations 'as are consistent with its being in substance the same thing.'" *quoting* Burr v. Duryee, 68 U.S (1 Wall.) 531, 573 (1863).

The Federal Circuit and the Supreme Court have re-examined the test for infringement under the doctrine of equivalents twice in recent years.[26] In the Warner Jenkinson case the Supreme Court upheld the viability of the doctrine while ruling that the determination of equivalence should be applied as an objective inquiry on an element-by-element basis.[27] The Supreme Court said:

> the particular linguistic framework used to determine 'equivalence,' whether the so-called "triple identity" test or the 'insubstantial differences' test, is less important than whether the test is probative of the essential inquiry: Does the accused product or process contain elements identical or equivalent to each claimed element of the patented invention?[28]

Thus the major impact of the Supreme Court's review of the doctrine of equivalents in Warner Jenkinson was this requirement of element-by-element analysis. In the subsequent *Festo* decision[29] the Supreme Court was forced to clarify the impact on the doctrine of equivalents of a patent's patent office history ("prosecution history"). A hypothetical involving human versus porcine insulin may help illustrate the ruling in *Festo*. For a very long time prior to the advent of biotechnology, diabetics were treated with purified porcine insulin, which differs only slightly from human insulin, and which was derived from pig pancreases. Let us assume that a patent applicant originally claimed the gene encoding "mammalian" insulin, but in interaction with the Patent Office agreed to limit the claim to the gene encoding "human" insulin. May the patent holder then successfully bring an infringement action against a company, which produces recombinant porcine insulin, alleging infringement under the doctrine of equivalents? In such a case the fact that the patent holder amended his application to narrow the particular limitation in question (mammalian to human) would likely prevent a finding of infringement under the doctrine of equivalents, even though the production of recombinant porcine insulin may well have been facilitated by the purification and publication of the human insulin gene. While the Supreme Court's Festo decision is sure to be the subject of extensive future litigation and analysis, for pur-

---

26. Warner-Jenkinson Co., Inc. v. Hilton Davis Chem. Co., 520 U.S. 17 (1997); Festo Corp. v. Shoketsu Kinzoku Kogyo Kabushiki Co. 535 U.S. 722 (2002); Hilton Davis Chem. Co. v. Warner Jenkinson Co, Inc. , 62 F.3d 1512 (Fed. Cir. 1995).
27. Warner-Jenkinson Co., Inc. v. Hilton Davis Chem. Co., 520 U.S. 17 (1997).
28. *Id.* at 40.
29. *Festo Corp.*, *supra* note 59.

poses of this chapter it may suffice to say that while the doctrine of equivalents is still alive, prosecution history will frequently limit its application and the best course of action will always be to get the broadest literal claims that can be supported.

§4.11(A) Reverse Equivalents

Although the doctrine of equivalents was created to provide a patentee with equitable protection against competitors who avoid literal infringement by making obvious or easy changes to the patentee's invention, the concept of reverse equivalents was judicially created for the reciprocal problem. The reverse doctrine of equivalents protects second inventors when their product might be found to literally infringe the claims in a prior patent, yet works in a substantially different way or produces a substantially different result.[30] In these situations, the concept of reverse equivalents serves the same equitable purpose as the doctrine of equivalents by allowing the patented invention to be construed in such a way that the form of the claim does not triumph over the substance of the patented invention.[31]

The purpose of "reverse equivalents" is to prevent unwarranted extension of the claims beyond a fair construction of the first inventor's contribution.[32] The contribution of an inventor is used here to refer to that which the inventor's patent application has truly enabled others to do.[33] A patentee must provide a sufficient description of her invention to enable others to practice the invention and is not entitled to a patent that claims more than what is enabled. Thus, reverse equivalents protects a second innovator who can truly assert that her claimed invention was not made possible (enabled) by what was described

---

30. An accused article may avoid infringement, even if it is within the literal words of the claim, if it is "so far changed in principle from a patented article that it performs the same or a similar function in a substantially different way." *Graver Tank,* 339 U.S. at 608–09.

31. See Scripps Clinic & Research Found. v. Genentech Inc., 927 F.2d 1561, 1565 (Fed. Cir. 1991); Pennwalt Corp. v. Durand-Wayland Inc., 833 F.2d 931 (Fed. Cir. 1987), *cert. den.,* 485 U.S. 961 (1988); and *In re* Hogan, 559 F.2d 595, 607 (C.C.P.A. 1977) ("Like the judicially-developed doctrine of equivalents, designed to protect the patentee with respect to later-developed variations of the claimed invention, the judicially-developed 'reverse doctrine of equivalents,' requiring interpretation of claims in light of the specification, may be safely relied upon to preclude improper enforcement against later developers.")

32. See Scripps Clinic *supra* note 64.

33. *See generally* 35 U.S.C. §112, first paragraph; Edward T. Lent, *Adequacy of Disclosures of Biotechnology Inventions,* AIPLA Q. J. 315 (1989).

by the prior inventor and bars enforcement in such cases against later true innovators.[34]

§4.11(B) Unexpected Results

One additional concept of patent law that is closely related to reverse equivalents is the rule that a second compound will be patentable (in particular with respect to nonobviousness), despite its structural similarity to a prior compound, if the second applicant can provide evidence that the second compound produced results that are desirable and would not have been predicted.[35] Under this rule, once the Patent and Trademark Office demonstrates that there is a structural similarity between the prior art and the second compound and that the prior art would suggest a reason to make the second compound, it has established the *prima facie* obviousness of the second compound.[36] The burden in some cases then shifts to the applicant to rebut by "test data that the claimed compositions possess unexpectedly improved properties ... that the prior art does not have."[37] In such cases the doctrine of unexpected results can be thought of as a close sibling of reverse equivalents, for the unexpected results are a basis for arguing that the second compound, despite its similarity to the first, performs its function in a substantially different way. Reverse equivalents is used when the second compound is accused of infringing; and its sibling, the doctrine of unexpected results is used when the second compound is denied a patent on the grounds that a prior compound rendered it obvious.

§4.12 Conclusion: Future Directions for Biotechnology Patent Law

As in other areas of biotechnology law, biotechnology patent law seems likely to evolve at a rapid pace during the next ten years of the biotechnology industry. Major changes may well come from efforts to harmonize the U.S. patent system with the other major patent systems of the world; from pressures to make life-saving pharmaceuticals available at a reasonable cost in developing countries; from the continuing rapid pace of scientific development

---

34. *In re* Hogan, 559 F.2d 595, 194 USPQ 527 (CCPA 1977).
35. *In re* Mayne, 104 F.3d 1339, 1342 (Fed. Cir. 1997) (*"Mayne"*).
36. *Id.*
37. *Id.* at 1342 (*quoting In re* Dillon, 919 F.2d 688 at 692 (Fed. Cir. 1990)).

(particularly from the Human Genome Initiative); and, from the maturation of the still young field of biotechnology patent law.

§4.12(A) International Patent Issues

In the international arena, negotiations aimed at harmonizing the patent systems of the U.S., European Community, and Japan have been ongoing for years. Three major changes have already been made to U.S. patent law as a result: inventors may establish their priority with lower-cost-of-filing provisional applications for up to one year before full filing; the twenty-year term of patent protection in the U.S. now conforms to the prevailing international rule; and, U.S. patent applications are published eighteen months after filing. Among the major issues which remain are the difference between the foreign "first-inventor-to-file" vs. the U.S.'s "first to invent" standard for determining inventorship and the U.S. provision for a one-year grace period after publication without compromising novelty. Among other issues where the differences between countries seem to be narrowing are the scope of patentable subject matter and the general approach to obviousness or inventiveness.

Although it is difficult to predict the final resolution of international negotiations to harmonize U.S., European Union, and Japanese patent laws, some predictions seem warranted. First, the U.S. may well abandon its first-to-file system and interference practice to adopt the world standard of first-inventor-to-file as it has now conformed to the general practice of publishing patent applications eighteen months after filing. In return, it is possible that the Europeans will adopt a grace period for filing after publication that more closely resembles the current U.S. practice.

§4.12(B) Life-Saving Pharmaceuticals in Developing Countries

One of the greatest controversies over biotechnology patents in the past decade has been over the need to make life-saving pharmaceuticals, particularly for HIV, available to the low-income populations of developing countries.

Whatever the final resolution, international patent issues will continue to be of great significance to the biotech industry and continuing change in that field is a certainty.

§4.12(C) Scientific Progress, the Human Genome Initiative, and Changes in Patent Law

As noted in §4.6, the basic patentability requirement of nonobviousness necessarily changes with the rapid changes in the ordinary skill of the art. One of the forces driving the rapid pace of advance in the ordinary skill in the art is the Human Genome Initiative, the multi-national, long-term effort to map and sequence the entire human genome. The impact of the Human Genome Initiative (HGI) on patent law comes in two ways. First, the Genome project produced much more powerful techniques for gene sequencing, which led to an enormous increase in the ability of scientists in universities and the private sector to produce gene sequences independently of the HGI. A great number of patents were filed by private companies using these newer techniques in the period prior to the publication of the completed draft sequence in June of 2000.[38] Whatever the resolution of the issues surrounding the patenting of those gene sequences, such as Human Genome Sciences patent on CCR-5 (see §4.4), it seems very likely that the era in which the cloning of a newly-discovered gene is patentable will soon end. Virtually every naturally occurring human gene sequence (at least in its most common form) will very likely have been placed in the public domain or been the subject of a patent application before the end of this decade.

To conclude, the biotechnology industry has developed, at least in large part, because of the availability of patent protection for the fruits of biotechnology research. Although there are many unresolved issues for biotechnology patents, such as the application of enablement and written description requirements to the reach-through claims of drug target patents and the future harmonization of international patent law, a great deal has already been clarified. The basic pre-existing framework of general patent law, patentable subject matter, novelty, nonobviousness, and enablement has been more or less successfully adapted to the needs of the new biotechnology. The changes in biotechnology patent law that are likely to occur over the next several years will generally be in the nature of fine-tuning and adjustment than radical change and revolution.

---

38. *See* Nicholas Wade, *Reading the Book of Life*, N.Y. TIMES, June 27, 2000, at A1.

CHAPTER 5

Biotechnology Business Strategy

§5.0 Introduction (Including a Brief History of Biotech Business Models)

This chapter is about planning for the future of a biotechnology company, particularly a company that has a technology in the general field of human health. This introduction provides a brief history of the trends in biotech start-up business plans. §5.1 of this chapter examines the most basic decisions to be made in the early phases of a biotech start-up with a reasonably broad technology platform. §5.1(A) is a brief analysis of the market issues in the biotech business plan, while §5.1(B) examines the risk factors to be considered in the early planning stages. §5.2 of this chapter integrates the analysis of market issues with regulatory strategy. §5.3 provides a synopsis of the financing issues involved in planning for a new biopharmaceutical in the context of a small, dedicated biotechnology company[1] as well as the interplay between financial and other strategic issues. §5.4 is a brief summary of the planning issues and

---

1. Dedicated biotechnology company, or DBC, is a term used to refer to small and emerging companies that have been founded with the exclusive intention of developing and applying biotechnology to produce a new product or products. Since biotechnology's techniques are now used throughout the pharmaceutical industry as well as other industries, such as agriculture, the term DBC is particularly useful in distinguishing small, exclusively biotechnology-focused companies from larger, well-established companies, such as large pharmaceutical companies, that have incorporated biotechnology into their research and development. Business planning for a DBC involved in developing a new biopharmaceutical is, necessarily, very unlike that of a large pharmaceutical company such as Merck or Pfizer. Such large companies are easily able to finance development of new products from existing revenues, whereas DBCs generally are financed piecemeal, over time, with the status of the planned next funding always dependent on the success of the prior stage of de-

the interrelationship among the financial, regulatory and intellectual property components of a biotech business plan.

The Brief History: Putting the problem in perspective requires a brief account of the history of fads in biotechnology venture capital funding. When the first generation of biotechnology companies (such as Genentech, Chiron, and Amgen) were founded in the late 1970s and early 1980s, the naively simple intention of their scientific founders was to use some or all of the tools of the new biotechnology to develop great therapeutics and become biotech versions of the pharmaceutical giants. For the most part, that did not prove to be a realistic plan.

Thus the second generation of companies, founded in the mid to late 80s, was founded on two different strategies. Some of these companies were founded on relatively narrow, albeit seemingly important discoveries, such as Telios' founding on a cell surface target for wound healing and metastasis, CYTEL's founding on a group of carbohydrates involved in inflammation, or Ligand's founding on an important family of intracellular receptors.[2] The other second-generation biotech companies were aimed at technological niches, such as gene therapy, drug delivery, antisense, various antibody therapeutic strategies, and so forth. With few exceptions, both of these kinds of second-generation companies were still founded with the idea of developing great therapeutics and becoming smaller biotech versions of the pharmaceutical giants. For the most part, that did not prove to be a realistic plan either. The problem was that the risk of any particular project failing was high, the costs of developing even the first candidate to the later stages of development were very high, and the capital markets both too uncertain and too quick to pull the plug at the first sign of failure.

This led to a change in biotech start-up strategy that persisted through the 1990s: the demand by venture capitalists for a "technology platform." A technology platform is, in general, a technology that has the ability to provide or enable a great number of further discoveries or leads across a fairly wide-range of diseases. For example, in the early 1990s a number of companies

---

velopment. It is impossible to overstate the significance of these funding constraints for the DBCs, particularly for strategic planning.

2. Ligand actually shifted gears a few years after its founding, deciding that it finance much of its research by offering its receptors to pharmaceutical company partners in different therapeutic areas. This provided the kind of transition to a technology-platform company that other companies followed. *See, e.g., Progenx Changes Name to Ligand Pharmaceuticals to Reflect Research and Development Activities*, Bus.Wire, November 9, 1989 (available on Lexis).

were founded in on "genomics" platforms, to use different approaches to the growing power of gene sequencing and genomics to produce valuable goods and services. Later in the 1990s several companies were founded on proteomics platforms, with equivalent intentions. Other, more recent "platforms" include protein re-engineering, high-throughput structural determination, families of related molecules (e.g. analogs of neuropeptides), and focus areas such as proprietary methods for bacterial target screening or the genetics of aging.

There were basically two reasons for the mass shift of venture capital to broader platform technologies. First, the broader the platform the less critical would be a disappointment from any one effort based on the platform. Perhaps even more important was the second reason, which was that once the venture capital markets had finally realized that biotechnology was far too expensive to rely on venture funding to complete product development, the answer appeared to be to use technology platforms to bring in pharmaceutical company money at a very early stage. This appeared to allow shifting much of the capital requirements from the venture funds to the pharmaceutical industry. The broad-platform companies were not generally founded with the idea that they would ultimately make great therapeutics and become small pharmaceutical companies. Rather, they were primarily founded to provide research tools (some were called "toolbox" companies) to the pharmaceutical industry in order to profit initially from the sale of services and then later from the royalties from the drugs pharmaceutical companies discovered using their tools.

This model ultimately proved frequently disappointing as well. This time the problem was not that the vision was too grandiose, but rather that the profits from the toolbox side were generally small (the pharmaceutical companies quickly learned to drive hard bargains) and the profits from any royalties were too far off for most investment horizons (a decade from discovery might be a good average timeframe). In addition, the royalty stream was likely to be calculated at single-digit percentages, meaning that the lions' share of the profits went to the lions. Naturally that made investment in lions much more attractive and investment in the lions' food chain much less so.

That brings this history to the present time, when venture capital is once again reinventing its investment philosophy in light of the most recent lessons. It is much harder to put a succinct label on the latest generation of start-ups. Those that are "fundable" are likely to have either or both of two characteristics: first, an enabling technology that can provide multiple opportunities (some kind of "platform"); and second, the potential to bring the lead products (which are increasingly in-licensed from other companies) to later stage development (i.e. into or through Phase II clinical trials) in a reasonably short time frame (i.e. 3–4 years).[3] The best business plans are likely to generate immediate revenue from using their technology platform for pharmaceutical partners in selected areas, while proceeding as rapidly as possible to develop their own products in the area fenced off from pharmaceutical partnerships.

The idea is that the partnering activities protect against the capital market problems while the development of lead products to later stages means that they can be licensed at much higher royalty rates. A drug with a successful Phase II clinical trial might command a royalty well into double digits, meaning real revenues for the biotech company just a few years after licensing. The current vogue is for hybrid companies with both partnering and development plans, and sometimes with plans to in-license leads candidates that can be brought to the middle stages of development in the very near-term. In theory, the lead compounds provide a fairly firm intellectual property basis for going forward even if the technology platform is quickly outmoded. Of course, there is still the risk that the compounds will fail in the clinic and the platform will be outdated, but that is why these early investors are called ("ad")venture capitalists.

§5.1 Strategic Planning for the New "Hybrid" Biotechnology

The central problem for most recent biotechnology companies is simple to state but hard to solve. How can a company's technology generate revenue from large pharmaceutical companies while the technology is also used to develop products that will be the company's own, at least until some later stage of development? For example, if a hypothetical biotechnology company "ProteoMaxx" has a proteomics technology that enables it to very precisely identify novel targets for drugs in a wide range of diseases, from cancer to neurodegenerative diseases, what should guide the company's decisions and how can the financing model be made to fit the business plan?[4] Of course, what-

---

3. Increasingly biotech companies are licensing in candidate drugs that have already been through preclinical toxicology and perhaps even a Phase I study, but that for whatever reason do not fit into a large pharmaceutical company's plans.

4. §8.5(a) discusses the strategic planning problem primarily from an FDA perspective and with respect to choosing an indication for a particular drug with potential uses in several diseases. The discussion in §8.5(a) touches on some of the same issues and can usefully be read in parallel with this section.

ever the direction, the intellectual property protection issues must be dealt with: both the partnered and retained areas of development absolutely depend on a reasonably good IP position.

§5.1(A) The Role of Markets in Strategic Planning

ProteoMaxx has very difficult decisions to make from the beginning. For example, its proteomics technology can produce targets for drug development in neurodegenerative diseases, autoimmune diseases, and cancer. The following analysis will be limited to those three, although there may well be applications to cardiovascular disease and just about everything else as well. Analyzing every other possible therapeutic area in this section and the sections immediately following would unnecessarily complicate the discussion. For most of the areas open to the company, there will be an effort to generate partnerships and research collaborations; for one or two, the company will retain rights and attempt the early phases of development on its own. How can the choices be made?

For most business plans, whether for a cookie franchise or an Internet software venture, the process begins with an effort to predict the potential market for the potential product. The role of the potential market variable in strategic planning for a new biotech product differs from market considerations for a consumer product. There are two major differences: predictability and importance. While the market for a drug is more predictable, market size is less important. For some kinds of consumer products predicting the market size can be very difficult. For most biotechnology-based health care companies the task of determining the size of the potential market is actually somewhat easier than predicting how many people will buy a new flavor of ice cream because the market for these biotechnology products is determined by the disease or health-related condition for which the product might be used. In most cases, reasonable data as to the number of patients per year with a particular disease is readily available, whether for the U.S. market, the European Union member states, or Japan. In the United States, the Center for Disease Control is a major source of such epidemiological data, although there are also patient advocacy groups that have fairly good data on the numbers of sufferers for many diseases.

The market for a biopharmaceutical or diagnostic is more or less restricted to the precise indication on its product label (more precisely, its prescribing information as it appears reproduced on the manufacturer's package insert) as it was approved by the FDA. Thus a drug's label might state, for example, that it is indicated for use in the treatment of relapsing/remitting multiple sclerosis, or it might state that it is for use in the treatment of all forms of relapsing multiple sclerosis (a more inclusive patient population than the former indication). Although a cookie baker may try to entice absolutely anyone to buy his wares, whether to eat as a snack or to break up and use as a pie crust, the FDA has narrowly circumscribed the right of a manufacturer of a new drug to promote the use of its product for any indication not approved for its labeling by the FDA.[5]

Nevertheless, it is perfectly proper (and commonly done) for a physician to prescribe a drug for a condition for which it has not been approved, for example by prescribing human growth hormone for an older person whose levels of growth hormone have declined normally with age. Such uses are referred to as "off-label" and in some cases may increase the market for a drug significantly beyond the total patient population for which it has been approved. Despite the potential for significant off-label use of some drugs, in most cases the market for a drug will be some percentage of the patient population diagnosed with the indication for which FDA approval is obtained.[6] Big pharmaceutical companies may try to use their marketing budgets and prowess to increase their share against competitive therapeutics; biotech companies need to address serious, unmet medical needs for which market share is driven more by data than by beating the drum.

If market predictions are the basis of most companies' business plans, then it would seem to follow that the relatively predictable market size for biotechnology therapeutics would make good strategic biotech business decisions easy. That conclusion, however apparently logical, would be 100 percent wrong. Market sizes may dominate pharmaceutical company decision making; but pharmaceutical companies live in a different universe. Pharmaceutical companies might have 20 early-stage development projects proceeding in parallel, with no need for external funding for any internal R&D. In the world of biotechnology start-ups a company may have only one chance to fail. There is unrelenting pressure to meet milestones in order to receive the necessary funding for the company to continue operations.

---

5. James O'Reilly and Amy Dalal, *Off-Label or Out of Bounds? Prescriber and Marketer Liability for Unapproved Uses of FDA-Approved Drugs*, 12 ANN. HEALTH L. 295 (2003).

6. The ultimate share of patient population that a drug achieves will obviously depend on its advantages over other therapeutics and its price, as well as other factors. For a breakthrough drug that offers significant efficacy for a serious disease for which there currently is no effective therapy (which is the goal that drives most biotechnology companies) the likely actual market for the drug may reasonably reach 60 percent or more of the patient population some few years after product launch.

If market size should not determine a biotech start-up's business plan, what should? Maximizing the probability of success is the principal goal. That requires minimizing the risk of failure, while achieving the greatest possible progress in the shortest period of time with the minimum necessary resources. Faster and more certain progress in product development is far more important than the size of the ultimate market for that first product, as long as that market is sufficiently large to bring in a reasonable revenue stream.[7] Pharmaceutical companies need multi-billion dollar drugs to satisfy the capital markets. Even today a biotech company can be quite content with $300 to $500 hundred million in annual revenues from its first product.

§5.1(B) Minimizing Risk in Clinical Development: Animal Models, Clinical Trial Costs (Sizes and Duration), Clear Pathways, and Unmet Medical Needs

How can the hypothetical proteomics company, ProteoMaxx, make decisions about risk in the context of choosing between entire fields of research, such as neurodegenerative diseases, cancer, autoimmune diseases, and so on? Are there characteristics of these entire fields that actually result in different relative levels of risk? Surprisingly, the answer is yes, at least for some of those broad categories, whether the problem is determining priorities for a platform or for evaluating licensing opportunities. This section looks at a number of different variables that affect the risk of a drug development program and vary to some degree from disease field to disease field.

Perhaps the most important single factor in determining the risk of a drug development program is the predictive value of animal models for the disease target. A good animal model is one that is clearly very close to the human disease in as many dimensions as possible: causation, genetics, biochemistry, phenotypic manifestation, rate of progression, and response to already-developed agents.

Can these broad disease categories be distinguished on the basis of the quality of available animal models?[8] Looking, for example, at ProteoMaxx's as-

---

7. Companies that have spent far more to develop a product than its first several years of gross sales have certainly been punished in the marketplace. A minimum market might be one that brings in net annual revenues that are several times the total direct development costs (direct development costs are the sum of the costs of preclinical testing, clinical trials, and regulatory submissions). The return on investment in that case is actually one that any capital market ought to appreciate.

8. This discussion of risk factors such as the availability of animal models is intended only to illustrate the decision-making framework and represents only the author's personal

sumed choice between cancer, autoimmune disease, and neurodegenerative diseases, the answer is yes. The animal models in cancer have been the least predictive and generally the least close to the human in causation. Implanted human tumors in the nude mouse are the most extreme example of differing causation.[9] A casual familiarity with the past few decades of news stories is sufficient to recognize that a thousand cures for cancer have been found in rats or mice and failed in human trials. The reasons for that would make for an interesting dissertation topic, but the high level risk in Phase II human cancer trials is clear. Biotech companies with powerful platforms are often attracted to cancer targets for a variety of good reasons. Many have been excited enough by their laboratory data to proceed to clinical trials, where far more have failed than succeeded. Pharmaceutical companies are increasingly interested in cancer drugs and we can all be thankful for that. Biotech companies might do well to be more wary.

The animal model criteria would seem to be a strike against keeping cancer for ProteoMaxx's own development activities. What about autoimmune diseases and neurodegenerative diseases? Autoimmune diseases present an interesting problem. Although the animal models are not perfect, there are a number of them and they are getting better. More and more animal models are being created using genetically engineered mice.[10] The characteristics of autoimmune disease in an experimental animal are generally clinically and pathologically similar to the human disease, even though artificially generated by in-

---

familiarity with some of the scientific literature in these diverse fields; other sources may differ. The proposition that good animal models are of vital importance to the risk of a drug development program is not in issue. In designing a startup business plan, consult a variety of experts to determine in which areas the core technology can be tested against the most predictive animal models.

9. *See, e.g.,* P. J. McLaughlin et al., *Opioid growth factor inhibition of a human squamous cell carcinoma of the head and neck in nude mice: Dependency on the route of administration.* 24 INT. J ONCOL. 227–232 (2004).

10. *See, e.g.,* T.T. Glant, A. Finnegan, K.Mikecz, *Proteoglycan-induced arthritis: immune regulation, cellular mechanisms, and genetics,* 23 CRIT. REV. IMMUNOL. 199–250 (2003), abstract:

> There are numerous rodent models that simulate some or many of the clinical, immunological, or histopathological features of the disease. Recently, it has become a strong working hypothesis that MHC and non-MHC genetic components share loci that are common in various autoimmune diseases, and in corresponding animal models. The most relevant animal models of rheumatoid arthritis appear to be those induced by cartilage matrix components such as type II collagen or proteoglycan aggrecan. (PubMed abstract).

jecting the putative auto-antigen (the molecule at which the immune system is misdirected, which differs in the various autoimmune diseases). Therefore, on the basis of animal models, autoimmune diseases are a viable program area. The story is similar for neurodegenerative diseases (some of which have a strong autoimmune component). While there are not perfect animal models across the range of diseases in either field, there are reasonably good models being developed in both areas. So on the first criteria, both autoimmune disease and neurodegenerative diseases seem to come out ahead of cancer.

The second dimension of risk in platform development business strategy is clinical trial cost. It is a function of the length of time required for completing clinical trials and the size of those trials. The length varies with the time required to measure efficacy, and the size of the trials depends varies with several factors, which can include the expected placebo effect rate, the severity of the disease, and the predictability and speed with which it progresses. The basic rule is that diseases that generally progress slowly or manifest their symptoms at unpredictable intervals (e.g. multiple sclerosis) will require more patients and longer trials to show efficacy, absent some widely agreed on "surrogate" endpoint.[11] The longer the trial, the more expensive it is, both in direct and indirect costs. By the time ProteoMaxx would reach the end of a Phase II trial that requires 12 to 18 months for its lead drug, its overhead would be very substantial. Similarly, the more varied the course of a disease, the more patients required to show efficacy, because in small groups the normal variation can mask the drug's effect. Finally, the higher the anticipated placebo-effect rate, the more patients that will be required.[12] At the same time, increasing the number of subjects also proportionately increases the direct costs. If a trial's direct cost is $15,000 per patient, then a 400 patient trial would cost $6 million whereas a 100 patient trial would cost only $1.5 million. What about the duration and size of clinical trials in cancer? The story is complex, of course, with cancer trials designed to show survival, duration of response, or reduction in tumor volume all having different time frames and being acceptable in different cancers at different stages. On these grounds, cancer trials actually may have an advantage. Trials in serious later-stage cancers, in which the expected progression of the disease is rapid, would require fewer patients for shorter periods of time.

---

11. *See* §8.5(a).
12. This factor seems to vary within disease areas as well as between disease areas. The placebo effect in cancer survival data is generally quite low, but in other disease areas it can vary widely from indication to indication. A review of the clinical trial literature in the prospective disease areas quickly gives a basis for estimating the anticipated placebo effect.

This criterion makes most neurodegenerative diseases costly targets. Parkinson's, MS, and Alzheimer's are all diseases for which clinical trials are generally longer and may require more patients.[13] Stroke is about the only CNS target for which short-term trials are feasible, however even there the patient-to-patient variability in outcomes may require more subjects. On the question of the duration of clinical trials, however, the autoimmune diseases would be viable, since Phase II and some Phase III studies can certainly be done in six months. For example, two of the three trials relied upon for the approval of Enbrel for rheumatoid arthritis used six months as the duration of treatment and enrolled 89 and 234 patients respectively.[14] Remicade was approved in Crohn's disease on the basis of one short, small trial and one longer, medium size trial. The shorter trial enrolled 108 patients and produced a very strong signal with statistically significant evidence of efficacy in just four weeks. The Alzheimer's market is much larger, but clinical trials in Alzheimer's are generally enroll several hundred to more than a thousand patients.

So, at the end of two rounds, considering animal models and the cost of clinical trials, cancer and neurodegenerative diseases each have one strike, while autoimmune diseases appear to be in the lead with no strikes. Of course, ProteoMaxx may feel so confident that its newly-discovered cancer targets are so important that the risks of failure are low, or it may feel that the offer from big pharma for their autoimmune opportunities is too good to pass up. The questions of whether the disease indication is a relatively serious unmet medical need and the difficulties presented by the anticipated placebo effect remain to be considered, but are not be discussed here.[15] Nevertheless, the foregoing discussion provides an example of at least a part of a rational framework for decision making.

---

13. In MS, Avonex's clinical trials enrolled 300 and 382 patients respectively, with a median time per patient of around 18 months. This is a moderate size clinical trial, but a long one. In Alzheimer's the current drugs are of modest effect, but the trials are a bit larger and shorter than in MS. One of the two pivotal studies for Aricept, the best-selling Alzheimer's drugs, enrolled 473 patients in a 30-week study, while Exelon's 26-week study enrolled 700 patients. Both drugs had longer-term studies that continued after approval to follow the duration of the drugs' benefits.

14. See Enbrel prescribing information at http://www.enbrel.com/pdf/enbrel_pi.pdf. The third trial enrolled 634 patients for approximately a year.

15. The placebo effect can generally be determined from the response rate of the control group in previous placebo-controlled trials for the same indication, while the question of whether or not the indication represents a serious unmet medical need is a relatively easy one for any physician specializing in that disease area to answer.

§5.2 Integrating Regulatory Strategy and Market Issues

After ProteoMaxx has decided which fields to partner with big pharmaceutical companies and which initial field to pursue for its own products, it needs to pick an indication within that field, based on a variety of *in vitro* and animal model experiments. For biopharmaceuticals different markets can require substantially different costs, both in development time and in actual dollars expended on direct development costs. Most biotechnology companies are in the research and development phase, burning significant amounts of money without having a product on the market, let alone earning a profit. It may therefore make more sense to embark on a four-year, $25 million dollar effort to win approval to enter a $100 million market than to work on a six-year to eight-year, $120 million effort to enter a $500 million dollar market. These kinds of time and money differences can exist even when the same compound could be developed for two or three different indications (Remicade, for example was approved first for Crohn's disease and later for arthritis). Chapter 8 explains the process for optimizing targets within a field, while the remainder of this chapter addresses the problem of how to match the money to the business plan.

If for example, we estimate that the time from initial animal tests to FDA approval for a first product will be eight years, with clinical trials on seven hundred patients for all of phases II and III combined, with a per patient cost of $15,000,[16] then the initial market entry will carry with it direct clinical trial costs of $10.5 million dollars plus another million or more for the formal animal toxicology and, depending on the size of the Phase I trial, perhaps total costs (direct costs exclusive of overhead) of $12 million. In these days of 20 and 30 million dollar venture capital rounds, these price tags do not seem too high, until the overhead costs are added in. By the time the company is nine

---

16. This discussion assumes that which is generally true, i.e. that longer clinical trials not only are more expensive by virtue of the continuing indirect costs (rent, depreciation, patent life amortization, general payroll, etc.) that are consumed during the trial, but also in their direct costs. This is because the longer trials generally require a proportionately greater number of patient observations, laboratory analyses, and physician examinations. If, for example, the protocol requires blood to be drawn from each patient every month for complete analysis, a two-year trial requires 18 more such tests than a six-month trial. The same logic affects each of the major components of direct clinical costs. Additionally, the more data, the more difficult and expensive will be the process of data analysis and FDA filing.

years old, there are likely to be from 200 to 500 or more employees and a corporate burn rate of approximately $25 million (including the cost of clinical trials)![17] Where is that kind of money going to come from, before there is a single dollar from product sales?[18] The choice of the initial target then helps define not only the product development pathway, but it also goes a long way towards defining the financing pathway.

§5.3 Financing Biopharmaceutical Development: Considerations and Strategy

The development pathway and initial target provide the three key parameters that drive financing: the anticipated development costs, the anticipated revenues, and the timing of development milestones. The first two of these, the anticipated development costs and the anticipated revenues, together with the estimated per unit price and rough cost, give an annual figure for gross profits. Gross profits are the key variable in determining the rate of return on varying levels of prior investment. Thus, the business plan should demonstrate to the venture capitalist providing early funding, or to the investment bankers providing later-stage financing, that the well-thought-out first indication will translate into an adequate return (risk adjusted for the 90 percent chance of failure at the earliest stage down to the approximately 15 percent chance of failure after Phase II) on the investments required to pay for the anticipated development costs.[19]

---

17. As a historical case study, IDEC Pharmaceuticals was spending almost that much in 1994 and 1995 dollars, when the company was 9 and 10 years old, finishing a successful Phase II on its lead product in 1994 and enrolling the patients in Phase III during 1995. Information is available in IDEC Pharmaceutical's annual reports from EDGAR on-line.

18. The Tufts Center for the Study of Drug Development estimated the cost of bringing a new drug to market to be approximately $800 million, a figure which included the costs of all failed programs as well as the time cost or opportunity cost of the money invested in R&D, http://csdd.tufts.edu/NewsEvents/RecentNews.asp?newsid=6. While that figure is controversial, it is clear that most biotech companies have been in business for ten or more years and spent more than $250 million before reaching the approval of their first major drug (*see* IDEC history in the preceding note). This total expenditure is exclusive of the time cost of money, which would likely double that total over the time period involved.

19. Of course, this part of the business plan helps sell the investment, but what venture capitalists really need to see are milestones in product development that translate into "exit" opportunities, either through an IPO or through an acquisition by a larger company, where the exit provides a return of at least 5-10 times their investment within six years.

For the third parameter, the timing of development milestones, correlating projected milestones with funding opportunities is complicated by the fact that criteria for initial public offerings or major corporate partnerships change over time, varying with the condition of the capital markets. Nevertheless, for planning purposes, the milestones that are penciled in for the first indication can be prospectively equated with financing benchmarks, and in some cases, opportunities for the earlier investors to cash out or exit. A plan that is premised on beginning Phase II trials in 36 months might assume that a major venture round or major financing commitment from a corporate partner could be closed prior to Phase II trials if no adverse results are produced in phase I or a phase I/II. Similarly, positive Phase II data certainly should be enough to bring a strong possibility of an IPO in all but the bleakest capital markets when a potentially less rewarding, from an investor return perspective, sale of the company to a larger company may be necessary. The corollary of all this is that given the estimated expense of development as it is projected forward in time, each **preceding** investment or investments must be sufficient to carry the company to the next development phase and its accompanying opportunity for new money.

Thus the lawyer or business executive who is involved in the negotiation of a venture capital financing or a corporate partnering/licensing agreement for a biotechnology company needs to understand the milestones provided for in the particular agreement and also that round of financing's place in the overall structure of product development. Although plans are inevitably subject to change due to a variety of unpredictable circumstances, management, with the assistance of legal counsel, needs to develop and frequently revise an overall plan for financing the company through to profitability. One of the greatest challenges for the emerging biotechnology company's executives and its legal counsel is to integrate its intellectual property and product development strategy with its financial plan. An example of such a plan for ProteoMaxx, as it might look at the very beginning of development, appears in Figure 1. Please note that it would be extremely unusual for any actual biotechnology company to progress as far as ProteoMaxx while spending as little as the amounts shown in Figure 1. The main purpose of Figure 1 is to portray the way in which both development and financing proceeds in stages and it is imperative to plan each stage of development to fit within the corollary stages of financing.

Figure 1
ProteoMaxx Initial Plan

Initial Product (PM123)[20]: Targeting Ulcerative Colitis
(partnered in Europe after Phase I, partnered in Japan and Asia after Phase II).

Income $	Source	Total Inc./ Year $
6,000,000	Series A $1.50* 4,000,00	6,000,000
1,000,000	1st Partner upfront payment	
300,000	1st Partner Yr. 2 R&D funding	7,300,000
600,000	1st Partner Yr. 3 R&D funding	
15,000,000	Series B 2.50* 6,000,00	
250,000	1st Partner accepts lead	
1,000,000	Sign second Partner	
150,000	2nd Partner R&D funding	24,300,000
480,000	1st Partner R&D funding	
360,000	2nd Partner R&D funding	25,140,000
1,000,000	Partner PM123 in EU	
15,000,000	Series C 3.00*5,000,000	
480,000	1st Partner R&D funding	
500,000	1st Partner Phase 1 filing milestone	
360,000	2nd Partner R&D funding	
250,000	2nd Partner accepts lead	42,730,000
0	1st Partner terminates	
	Phase II requires Phase 2B	
360,000	2nd Partner R&D funding	
500,000	2nd Partner Phase 1 filing milestone	12,000,000
15,000,000	Mezzanine round	
5,000,000	License N.A. rights to PM 123	
15,000,000	R&D funding for PM123 P3	78,590,000
10,000,000	Rights to PM 123 in EU	
50,000,000	IPO	
		143,590,000
5,000,000	Rights to PM 123 in Asia	
10,000,000	Complete Phase III milestone	
		153,590,000

20. It is very important to give your lead compounds at least three digit numbers. Lower numbers signal the reader or potential investor that you have not screened enough compounds.

5 · BIOTECHNOLOGY BUSINESS STRATEGY

Second Product (PM 609): Targeting Rheumatoid Arthritis
Partnered Field Products (one in Neurodegenerative Disease, one in Cancer).
(PM 175) Adenocarcinoma (colon, head and neck)
(PM 325) Parkinson's

Expenses $	Description of Expenses	Total Exp./ Year $	Year End
2,000,000	G&A year 1	2,000,000	1
3,500,000	G&A year 2		
280,000	1st P'r R&D cost	5,780,000	2
560,000	1st P'r R&D cost		
1,000,000	Tox. on PM123		
6,000,000	G&A year 3		
130,000	2nd P'r R&D cost	13,470,000	3
3,000,000	File/do PM123 Phase 1		
7,000,000	R&D + G&A year 4	23,470,000	4
3,000,000	Begin Phase 2 PM123		
1,000,000	Tox. on PM609		
10,000,000	R&D + G&A year 5	37,470,000	5
3,000,000	Complete P. II on PM123		
R&D + G&A year 6			6
6,000,000	Phase IIb, plan Phase 3		
3,000,000	Phase I on PM609		
12,000,000	R&D + G&A year 7	73,470,000	7
5,000,000	Begin Phase II PM609		
12,000,000	R&D + G&A year 8	90,470,000	8
5,000,000	Phase II on PM609		
15,000,000	R&D + G&A year 9	110,470,000	9
15,000,000	Cost of Phase III on PM123		
500,000	File NDA		
15,000,000	R&D +G&A year 10		
		140,970,000	10

This effort at long-range planning is complicated both by the fact that the fundraising value of any planned product development milestone, in either investment capital or in the form for a pharmaceutical join venture of particular development milestones, will vary with the cycles of the pharmaceutical companies' need for product and the general stock market conditions for biotechnology. In addition any plan needs constant and sometimes major adjustments as product development provides both new opportunities and disappointments.

For example, let us assume that ProteoMaxx Biotech, the company whose financial and development history is outlined in Figure 1, is developing a technology based on targets and leads discovered by its proteomics platform. The technology provides the ability to identify novel targets and, at the same time, develop basic structural information about small molecule ligands on which to build combinatorial libraries for screening and target validation. Based upon this brief synopsis of the technology, proprietary rights might be sought for ProteoMaxx's proteomics method, the targets discovered, the small molecules identified (in a form that allows for variant "Markush" forms)[21] the use of the targets as a screen for small molecules that might be agents in the relevant disease areas, antibodies against the targets and their use in diagnosis and therapy as well as particular antibodies that have been identified as useful because of their affinity and specificity.

The strategic issues of relating those proprietary rights to regulatory strategy and financing goals can be appreciated from a brief look at the initial technology and its possible development paths. Assume that the first target that is discovered in ProteoMaxx's autoimmune program is a transcription factor active in T-cell activation, which could prove to be a valuable target in one or more autoimmune diseases. The first program would be directed to synthesizing or screening for small molecules that also bind to and inhibit the transcription factor, or making antibody fragments that bind to the target and inhibit the binding of the target to the gene promoter that it regulates.

---

21. Markush claims are claims to families of chemicals. The basis of a Markush claim is a central structure, which is precisely described (for example, a five carbon ring) with one or more side groups labeled as R1, R2, etc. The claim then goes on to list the various possible allowable substitutions claimed for each R group (e.g. "where R1 is either a methyl or butyl group and R2 is a hydroxyl or carboxyl group." In practice, the list of possible substitutions at each position can be rather large, although as the total number of theoretically claimed molecules grows, the rules for the prosecution and examination of the patent become complicated. *See* Hypertext MPEP: 803.02 Restriction—Markush Claims, *available at* http://patents.ame.nd.edu/mpep/8/803.02.html. The MPEP is the official U.S. Patent Office Manual.

While all these possibilities are being considered and juggled, with some development opportunities selected for initial exploration, much will eventually also be learned about the particular T-cell driven autoimmune diseases, the level of the transcription factor's activity in each disease type at different stages, and the effect of using the first "hits" from the screening program, in various animal models of autoimmune disease. As the second or third generation compounds are developed, they will generate additional proprietary rights that need to be pursued. If the company is successful in generating sufficient venture capital to take the best of its early "leads" ("hits" that look good in animal models and preliminary toxicology studies) through a relatively successful Phase I/II trial, the model in Figure 1 suggests that milestone ought to be sufficient for another round, with more cash and development funding. This could well be partly accomplished by licensing foreign rights to the new lead program while retaining domestic rights until much later in development. It also would be important in generating partnering interest in any other areas in which the platform could be used.

It is at the stage of initial corporate partner negotiation that the strategic issues of proprietary rights, financing, and regulatory strategy become urgent, while at the same time revealing their complexity. How can the (hopefully ever-) increasing bundle of proprietary rights be allocated between ProteoMaxx and its first partner? What can ProteoMaxx keep for itself and what should its own development priority be? What problems might arise if the line between *meum* and *teum* (ProteoMaxx's and partner's) is not sharp enough to avoid conflicts of interest and ultimately competition?[22] There are no hard and fast rules that can guide the strategy and the negotiation, but it is clearly essential to have a good grasp of the technology, its development costs for different kinds of compounds and indications (particularly with respect to obtaining FDA approval), and the scope of available intellectual property protection.

§5.4 Summary: The Multiple Strands of a Biotech Business Plan

ProteoMaxx's development and financing scenario, as sketched in the previous section, raises several issues. First, the company spent between $110

---

22. *See, e.g.,* Ortho Pharmaceutical Corp. v. Amgen, Inc., 887 F.2d 460 (3rd Cir.1989), where Ortho had licensed the rights to market erythropoietin in all but the end-stage renal dialysis market.

and $120 million by the time it was ready to begin a Phase III trial on its first product. That is a somewhat conservative estimate of what a company might spend IF just about everything went perfectly, as it did for the hypothetical ProteoMaxx. The initial compound made it through preclinical toxicology and Phase I without any difficulty; and, while two Phase II trials were done to iron out dosage and patient criteria, efficacy was also not difficult to establish. However, even in this rosy scenario, the company could not have been able to begin planning Phase III in year 7 without a significant IPO and three rounds of venture capital. If the development of PM123 had not gone so smoothly, or if the capital markets had been very tight at any of the key junctures, then ProteoMaxx would have faced very difficult circumstances. As many biotech companies have been forced to do, ProteoMaxx would have been forced to slow down its development program, or downsize, or sell the North American rights to its autoimmune program, or all three. Of course, in really bad times with even worse technology failures, biotech companies have occasionally been forced to sell out for pennies on the dollar, or close the doors, or file for liquidation. It is a good thing for biotech executives that fluoxetine hydrochloride ("Prozac") is now generic; Zocor is generic as well, and big pharmaceutical companies are looking to fill their pipelines.

This look at biotechnology business planning underscores the importance of patent predictability as well as FDA predictability in any biotechnology business strategy. In terms of biopharmaceutical development strategy, the argument made in this chapter is that it is clearly important to get into the marketplace sooner, with a greater patent term remaining, even to the extent of influencing, if not dictating, the choice of initial target indication for a biotechnology company's lead compound.[23] It is also better for an early stage company to pick an area with less risk and a smaller market than a target with more risk and a bigger potential market. At the same time, any examination of the issues in planning for a biotechnology company also reemphasizes not only how important it is for the FDA to act swiftly, but also how important it is for the FDA to provide companies with enough information about the re-

---

23. The patent term extension provisions of 35 U.S.C. §155a *et seq.* partially mitigate the effect of the period of time necessary to test and obtain approval for new drugs, as only one-half of the testing time can be credited and the total credit (one-half testing time and all time from NDA filing to approval) is capped at a maximum of five years. Thus every year of additional time in the clinic can extend the term of a patent by no more than six months.

quirements of prospective clinical trials to enable companies to make appropriate decisions about alternative development pathways.[24]

---

24. It is, of course, always possible that new information or understanding about a disease will necessitate changes in regulatory requirements for clinical trial data. For example, as continuing research into AIDS changed our understanding of the dynamics of the disease, it was both natural and appropriate that the FDA require that clinical trials measure viral load as well as CD4 counts. However, absent such significant changes in the basic understanding of a disease, once the company and an appropriate FDA official have agreed that in a Parkinson's trial the principal measurement will be clinical measurements of muscle strength, flexibility and function, for example, it is simply devastating to have the New Drug Application rejected because the reviewer felt that NMR data on neuronal deterioration should have been supplied.

CHAPTER 6

Financing the Process of Development

§6.0 Introduction

Chapter 5 is about business planning for a biotechnology company and discusses development times, milestones and costs. The amount of money necessary to enable a biotech company to go from start-up to major therapeutic launch is in the hundreds of millions of dollars over a period of eight or more years, through a combination of venture capital, corporate partnerships with large pharmaceutical companies, and public stock offerings. This chapter provides an overview of these various financing vehicles, describes their structures, and discusses each in the context of the stages of development for which it may be best suited. This chapter also examines some related issues, such as stock incentives for employees, which are of critical importance (financial and otherwise) to biotech companies. §6.1 explains some of the preliminary choices in finance and the basic concepts of the corporation and of common and preferred stock. §6.2 discusses the basic legal framework that governs the sale of shares in the U.S., particularly the requirements that must be met in order for a private offering to be exempt from the requirements of registration of the offering with the Securities Exchange Commission (SEC). §6.3 is about "Angel" investments for very early stage financing. Venture capital financing is covered in §6.4. Corporate partnering and strategic alliances are discussed in §6.5 and going public is discussed in §6.6.

§6.1 A Brief Introduction to Corporations and Securities: (The MBAs and Lawyers May Wish to Skip This Section)

A corporation is created by filing articles of incorporation, which then becomes its principal governing document, more or less analogous to its constitution. A corporation is a separate legal entity, created by the law of the state in which the business is incorporated and the articles of incorporation are filed. As a "legal entity" the corporation can buy, own, and sell property and enter into contracts of all sorts. The important consequences of this concept of "separate legal entity" are: first, it establishes "limited liability" so that the corporation, rather than the individuals who manage it or work for it, is primarily responsible for most of the consequences of the business's operations (such as a traffic accident involving a company delivery driver, or the payment of money owed for office supplies); and, second, it enables the ownership of the corporation to be divided up among many individuals, who may have little or nothing to do with the day-to-day business of the company.

While there are numerous alternative forms of limited liability business entities, such as limited liability corporations (LLCs), Subchapter S corporations, limited partnerships, and so on, the vast majority of biotechnology companies are formed as basic state corporations. Delaware is commonly chosen as a state of incorporation because of the expertise of the Delaware courts in the legal issues involving corporations and finance and because of the generally "pro-business" orientation of the state's corporation law, although other states also have favorable laws and can be equally practical choices. Many start-ups are incorporated in the state in which the business is principally located and then reincorporated in Delaware prior to an initial public offering. Any good business lawyer can advise would-be biotech entrepreneurs on the pros and cons of various states for incorporation.

The second important consequence of forming a corporation, the possibility of distributing ownership among many individuals (some of which may also be corporations), is achieved through the sale of stock. Stock, or shares representing fractional ownership interest in the company, can be in one of two classes, common or preferred. There is also a hybrid class known as *convertible* preferred, which are preferred shares that can be converted into common shares upon some event or upon the election of the holder. Convertible preferred shares are generally what venture capitalists acquire for their investments. Preferred shares derive their name from the various preferences that can be assigned to them by the articles of incorporation or bylaws. In the context of a biotechnology start-up, the most important preference generally given to preferred shares is the liquidation preference. In general, the liquidation preference is only of significance in the generally catastrophic event that the company can no longer continue in business and the assets are to be sold. In liquidation, or the distribution of the assets, preferred shareholders are to be reimbursed for their preferred price, plus any accumulated, unpaid preferred dividend that is provided for in the articles or bylaws, before the common shareholders can receive anything. Of course, even the preferred shareholders are only entitled to the assets that remain after the companies' creditors are paid.

While these preferences are, of course, desired by the venture capitalists who invest in the company, they actually work to the benefit of common shareholders as well. Because the preferences make preferred stock more valuable, they enable common shares to be bought by the founders and employees at a significantly lower price. Moreover, they allow the granting of options to founders and employees at an option price that can be a small fraction of the price that the venture capitalists paid for theirs, without the options grant being taxable income at the time the grants are made or the options exercised.[1] So the division of the company's shares into preferred and common stock allows the venture capitalists to receive additional protection for their investment and the founders and employees to be given options to buy shares at an affordable price and without immediate tax consequences.

While the governance of a corporation can vary in a great many details, there is an almost universal basic structure: the shareholders elect the members of the Board of Directors and the Board oversees the general management of the corporation; appointing the officers and making the major business decisions, from approving budgets to approving major acquisitions or dispositions of assets. In a typical start-up, the VC investors will be guaranteed, by the terms of the preferred stock offering, a certain percentage of the Board seats, for example two of five, or 40 percent, which is generally more than sufficient under the bylaws to block major decisions without their consent. For example, if the bylaws require that two-thirds of the Board approve any merger, or the issuance of new shares, then the two VC representatives

---

1. This applies to "non-qualified" stock options, in which case the options are not taxable income when issued or exercised, but the gain, if any, made when the shares are sold is income. Check with your tax advisor. *See also* Gilson and Schizer, *Understanding Venture Capital Structure: A Tax Explanation for Convertible Preferred Stock,* 116 HARV. L. REV. 874 (2003).

can, by their votes alone, block any such action. Issues involved in the terms of a venture capital investment are discussed below, in §6.4.

§6.2 Selling Securities: Caveat Vendor

Before beginning the discussion of raising money, every entrepreneur needs to keep in mind the most basic requirements of securities law. The sale of stock and other "securities" has been a highly regulated activity in the United States since the crash of 1929 and the resulting enactment of the Securities Act of 1933[2] and the Securities Exchange Act of 1934.[3] The significance of this or entrepreneurs who wish to raise money is that in order to sell stock in a company-to-be the stock offering must be registered with the SEC (an expensive and impractical alternative for a start-up) or the offering must fall within one of the exemptions from the registration requirements.

The principal exemption upon which most biotech companies rely for their early-stage financing is the "private placement" exemption created statutorily by the 15 U.S.C. §77d(2) of the Securities Act of 1933 and further defined in the regulations at 17 C.F.R. §§230.501–508. The essential requirement of such a private placement is that the securities may be sold only to a limited number of "accredited" investors. The regulation, 17 C.F.R. §230.501, defines accredited investors in detail and includes a variety of institutional investors, including most venture capital funds, as well as high-net-worth individuals (such as the "angels" discussed in the next subsection). Despite the exemption from the registration requirements of the 1933 Act, the seller of securities in a private placement is still subject to the anti-fraud provisions of the 1934 Act.

§6.3 In the Beginning: Is There an Angel in the House?

While venture capital firms have long been the primary source of funding for biotechnology companies in their early stages, so-called "angel" investors are playing an increasingly important role in the very first round of biotechnology company start-up financing. Angel investors differ from venture capital investors in that they invest their own money, generally in smaller amounts

---

2. 15 U.S.C. §§77a *et seq.* (2006).
3. 15 U.S.C. §§78a *et seq.* (2006).

and only for one "seed" round. Angel investors can play an important role in providing the preliminary financing often necessary to enable a very early-stage company to reach the point at which venture capital is willing to step in.

For example, suppose that ProteoMaxx (the hypothetical company discussed in Chapter 5) believed it could quantify the changes in expression of all the active kinases in any two cell samples (for example a T-cell from a rheumatoid arthritis patient and one from a non-arthritis patient). A patent has been filed on this core, "protein profiling" technology by Illiana University, which would like to license it to a start-up company. However, venture capitalists might not be interested until work has been done on a number of human tumor, autoimmune disease, and normal cell lines. That work may take up to six months and cost $250,000 to $500,000. If the founding scientist does that work in her university lab, with NIH funding, the rights to all of those targets then become the university's, which could then be licensed to the start-up as well, but which would complicate the IP and future revenue structure of ProteoMaxx. That is when angels may step in where venture capitalists fear to tread.

Who are these "angels"? They are all very wealthy, sophisticated investors and predominantly veterans of successful high-tech and biotech companies themselves. Having succeeded in building their own fortunes in the industry, they become angel investors to keep their hands in the industry, provide some guidance to new companies, and possibly increase their net worth with successful investments as well. Most geographic areas with thriving biotech industries will have angel investor "clubs" to which would-be entrepreneurs can pitch their initial business plans and link up with potential angel investors.[4]

In most respects, the legal and business considerations surrounding angel investment rounds are similar to those in a venture capital financing, with two significant exceptions. First, while good venture capital investors are prepared to invest in subsequent rounds and may, at least informally, commit to doing so provided the company meets certain milestones, angels generally are in only for the initial funding. While angels may participate in future rounds, they do not expect to carry the company financially beyond the "seed" round. The second concern, which is not a problem with sophisticated angel investors, is making sure that the initial capitalization does not entail issuing too many

---

4. For example, in San Diego, one such group calls itself "Tech Coast Angels" and provides a website with substantial guidance for would-be entrepreneurs and meets regularly to hear entrepreneurs pitch their business plans and opportunities. *See* http://www.techcoastangels.com/. Most experienced business lawyers in a community would be familiar with the local groups.

shares, thus creating an unworkable capital structure upon which to build future financings. This would result from selling too many cheap shares to less-sophisticated investors, thus making it difficult for any venture capitalist to come in later and acquire a reasonable percentage of the company for an appropriate per-share price. Most venture capitalists want to acquire a 40 percent to 51 percent interest in the first round of venture financing and will do so at a price between $1 and $2 per share. So if the first venture round is going to be $6 million and will be valued at $1.50 a share, the VCs would take four million shares. If the company has already issued more than four million shares to angels, founders, and the university, it becomes very difficult to structure the initial VC round, unless the existing shareholders agree to a reverse split to decrease the number of shares outstanding. The issues of preparing for due diligence, valuation, capital structure, and other private financing concerns apply to both angel investors and venture capital investors and are discussed in the next section on venture capital financing.

§6.4 Venture Capital Financing Issues

In the past twenty years, venture capital has become an enormous factor in the U.S. economy, planting the initial seeds for the growth of Silicon Valley, the computer industry, the Internet, and biotechnology. Venture funds invested over $100 billion in 2000 before greatly diminishing in the wake of that year's stock market collapse, but even 2005's total of $21.7 billion still represents a great deal of private investment.[5] Moreover, since the dotcom bubble burst, the "life sciences," which include biotech and medical devices, now receive more venture capital than any other single sector.[6] So venture capital still remains the primary source of capital for early stage biotech companies. In what may be an apocryphal tale, Willie Sutton was famously quoted as responding that he robbed banks "because that's where the money is."[7] Modern day entrepreneurs are similarly motivated in their choice of funding sources.

What are venture capital funds? There are two general types: private venture capital funds and corporate venture capital funds. Private venture capi-

---

5. *See* http://www.ventureeconomics.com/vec/news_ve/2005VEpress/VEpress01_25_06.pdf.

6. *Id.*

7. Steve Cocheo, executive editor of the American Bankers Association's *Banking Online*, provides an interesting short account of the "real" history of the Sutton quote *available at:* http://www.banking.com/aba/profile_0397.htm (last visited on May 1, 2006).

tal funds are typically structured as limited partnerships. The general partners of the firm invest and manage a portfolio on behalf of the limited partners, which consist of both institutional and individual investors. Corporate venture capital groups are usually the captive venture capital arms of a principal, such as a pharmaceutical company or investment banking firm. The general partners of the venture fund make the final decision on whether or not to invest, although they may rely substantially on the work of their "associates" in finding and appraising potential investments. The general partners are persons with substantial business and investment experience and often have extensive experience in the particular industries in which they generally invest.

§6.4(A) Due Diligence

Venture capital financings are generally exempt from the extensive registration and disclosure requirements of public sales of securities as private offerings (see §6.2) to sophisticated or qualified investors. In lieu of a full prospectus and other public offering documentation, venture capitalists rely on their own painstaking investigation of a potential investment opportunity, in the elaborate process known as "due diligence." Since getting a passing grade in the VC's due diligence is critical to obtaining financing, the entrepreneur should be well prepared.

A venture capitalist's interest is sparked by a business plan; however, the majority of venture capitalists believe that the most important ingredient of a successful investment is the people who will be involved. While business planning is covered in Chapter 5 and lining up the right team is beyond the scope of this book, the next step in the due diligence process discussed here is examining the intellectual property rights.

Patent applications are a key part of the initial package a venture capitalist evaluates after his or her first contact with the start-up biotechnology company. Very few scientists or their institutions are inclined to allow third parties to review a scientific discovery prior to filing patent applications, even pursuant to a confidentiality agreement, because of the risk that someone else working in a similar area or on a similar approach may use the trade secret or know-how to beat the inventor to the patent office. Biotechnology is characterized by a high degree of parallel innovation. It is not uncommon for more than one scientist to "invent" virtually the same thing independently.

Once a patent application has been filed, the inventor has established a priority date, after which it would be difficult to successfully pursue a competing, later-filed patent prosecution. Ownership of the patent, as opposed to inventorship, is much more difficult to discern, and failure to identify the owner correctly can be very costly. Many biotechnology discoveries and patents come from scientists who have conducted research at more than one academic institution or company. At each of these places the scientist would ordinarily have agreed to assign all inventions made during that period of employment to the employer. Therefore, it is critical at the start-up stage to determine who really owns the core technology—the intellectual property that forms the basis for the investment.

The first question to ask is where the inventor (or more complicated yet, inventors) has been employed or affiliated in the past. After conducting a thorough background search on the inventor's past affiliations, the potential investor must examine past employment agreements, as well as the policies and procedures of prior employers that relate to ownership of their employees' intellectual property.

In determining patent ownership, it is very important to fix the date of conception of the invention, as opposed to the date of filing of the patent application. For example, assume an inventor associated with the newly-formed ProteoMaxx, was a professor at New University for Biotechnology ("NUB") prior to November 30, 2004, moved to Prodigious University for Biotechnology ("PUB") on January 1, 2004, and filed his patent application on August 1, 2004. Although it may appear that PUB "owns" the patent application, if it relates to an invention that was conceived by the inventor in September 2003, NUB is the probable owner under the policies to which its professors and graduate students agree as a condition of employment or enrollment.[8]

The investor must also beware of a very common provision in many biotechnology companies' employment contracts relating to rights to inventions conceived within a certain period after an employee's separation from the previous employer. Though very narrowly construed by courts in most states in favor of the employee-inventor, if an invention is close enough in time and subject matter to research conducted for the previous employer, the previous employer may still have rights to the invention, even though the date of conception and date of filing of a patent application may both have occurred later than the date of separation.

If the start-up company claims to own a technology, the investor will look for the license to the company from the inventor or the inventor's place of employ at the time of invention. The license for the core technology would, in most cases, be an exclusive worldwide license that effectively conveys to the company the rights to develop all applications of the technology. The license

---

8. *See* §3.2, Assignment of Rights.

will usually obligate the licensee to future consideration in the form of minimum royalties, progress payments, and royalties on product sales and may have already resulted in the issuance of common stock to the licensor.[9] It should be noted, however, that a license does not divest an earlier employer of any right it may have; it only conveys the inventor's or other licensor's rights, which may turn out to be subject to a prior right in a prior employer.

The patent and licensing issues are critical in the early due diligence. When funding a start-up biotechnology company, no matter how exciting the technology. If the start-up does not own it, or its ownership is questionable or tainted, it will invariably increase risk and diminish potential return to the investor. Any real possibility of a fight over ownership that arises in due diligence almost certainly will preclude a green light on funding. Patent applications and licenses can establish ownership of the technology, but they do not establish that the science is sound and will be able to be developed in the direction contemplated by the business plan. An important element of most venture capitalists' due diligence is the independent appraisal of the science by their own scientific consultants, who are often top scientists in the particular field.

§6.4(B) Valuation of the Investment (and the Company)

Pre-money and post-money valuation are key terms for the business novice to understand. For example, assume ProteoMaxx had issued one million shares to its founder and her university. It then raised $1 million from angel investors, who received two million shares at $.50 per share. The "post-money" valuation of that round is computed by the number of shares outstanding after financing (three million) times the per share price of the last round ($.50) or $1.5 million. The original million shares outstanding had a "pre-money" value of $500,000, based on the $.50 per share price paid for the new angel shares. So if the initial venture capitalists acquire 3 million shares at $2.00 per share, the number of shares outstanding increases to 5 million and the post-money valuation increases to $10 million ($2.00 times 5 million). The pre-money valuation immediately preceding the financing was $2.00 times the then-issued 3 million shares, which is $6 million. The $4.5 million increase from the post-money valuation of $1.5 million immediately after the angel round to the pre-money valuation of $6 million prior to the first venture round is attributable to the additional value (in terms of progress and

---

9. University licensing issues are discussed in §3.5.

risk-reduction) accomplished with the proceeds of the angel investors' money. This however, merely defines the terms commonly employed in financings. How are those $1.5 million, $6 million, and $10 million values actually established? How does a venture capitalist decide that an investment in ProteoMaxx is worth $6 million dollars for a 50 percent stake?

The intrinsic "value" of most start-up biotechnology companies lies in their core technology. Although investors are increasingly betting on the "jockey" (the management team) as much as they bet on the "horse" (the scientists and technology), jockeys alone cannot enter, much less win, races. The valuation of the initial investment by a venture capitalist (or sophisticated angel) is usually the first objective assessment that the core management team of a biotechnology start-up receives. In general that assessment the outside investors is based on one of two methods of valuation.

The first valuation method is based on the valuations placed on comparable companies at the same stage of development and with similar business models. ProteoMaxx can again serve as an example. After developing its technology with the Angel investment, ProteoMaxx would positioned for potential investors as a platform company with a strong platform, several proprietary targets for therapeutic development, and a number of compounds in early preclinical development. So, if Venture Capital firm 1 paid $5 million for a 40 percent stake in ProteoCure as a start-up, and ProteoCure was at a similar stage in developing a different approach to discovering pharmaceutical targets and leads (see §4.2 for a discussion of patent issues related to target-discovery technology) this would be a good starting place for Venture Capital firm 2's valuation of ProteoMaxx. Other factors that may affect this starting point for valuation would be the ProteoMaxx's management team's level of experience and association with previous biotechnology successes or with large pharmaceutical companies, business plan, competition (ProteoCure may have the edge by being earlier), and stage of technological development.

Even when there are closely comparable companies or technologies, market conditions will have a significant impact. The market for venture capital investments itself is subject to the laws of supply and demand. If the capital markets are very active and there is active competition among venture capitalists for good investments, valuations of start-ups will increase because more money will be chasing fewer deals. All other things being equal, ProteoMaxx may have a potentially lower valuation if, at the time it seeks funding, the venture capital market is in a "down" cycle relative to when ProteoCure was funded. If ProteoMaxx had the misfortune of needing venture capital in April of 2002, it would certainly have received a lower valuation than ProteoCure received in February of 2000. In biotechnology, as much as in stand-up comedy, timing can be all-important.

The more traditional and theoretically sounder way to value the start-up is to rely on the net present value of the future stream of earnings from the commercialization of the technology, adjusted appropriately for risk. If the start-up's business plan truly reflects the underlying commercial value of the technology, the *pro forma* operations statement projecting earnings for at least five years after the predicted date of commercialization of the technology should provide the data from which net present value and current valuation can be calculated. For a more complete introduction to valuing future earnings, *see* http://www.investopedia.com/articles/stocks/06/BiotechValuation.asp (last visited June 1, 2006). While this method provides greater apparent objectivity to value, it certainly rests on a base of enormous uncertainty, for the journey from business plan to revenue stream is a long and perilous one. The comparable market valuations may provide greater short-term security and, in reality, a much more solid starting point.

Another very important factor that affects the start-up's valuation is the estimated time to the next important milestone. If ProteoMaxx is only months from beginning a Phase I trial of a drug, as opposed to two years, a venture capital investor may be more inclined to affix a premium to the valuation. The comparable valuations are investments in companies at a similar stage of development, even if those companies were funded a few years back or have a very different technological base. Often, the venture capitalist will also require rights of first refusal to provide future funding or to participate in future financings on a pro-rata basis. This raises issues of capital structure, discussed in the next subsection.

§6.4(C) Capital Structure: Authorized Shares

An important part of the capitalization and financing plan is determining the number of shares authorized for issuance, which must be divided into common shares and preferred shares. Shares issued from those previously authorized do not require further shareholder approval. Increasing the number of authorized shares is more cumbersome, so it is better to anticipate the need for future shares and authorize them in advance. In determining the number of common shares to initially authorize, the company should take into account not only the shares to be issued to founders and employees, but also those that will be issued to future employees, stock option grants, and the right of preferred shareholders to convert their shares into common at a later date. The preferred shares authorized must be planned in anticipation of the stock issued to the initial venture capital investors and take into account future financings; including secondary and tertiary venture capital financing, corporate strategic alliances, the initial public offering ("IPO"), secondary public offerings, and stock splits. Ten million common shares and twenty million authorized shares may be enough, although more recently, with larger and more frequent public offerings, double that amount is not uncommon.

Increasing the number of authorized shares at an early stage has the advantage of avoiding the necessity to seek shareholder approval every time the number of authorized shares needs to be increased. On the other hand, Delaware and a number of other states impose franchise taxes using formulas that are based in whole or in part on the number of shares issued or authorized for issuance, so the convenience and practical advantages of authorizing sufficient shares at an early stage can be expensive.

When the number of shares authorized for issuance is increased, it is advisable to eliminate or waive preemptive rights that may still be in effect under the articles of incorporation, the by-laws, or the initial shareholders' agreements. "Preemptive rights" are provisions found in many "standard-form" corporate documents, which give initial or early investors rights to purchase newly authorized or issued shares at a certain price or under certain specified terms to protect them against the dilution of their stake in the company. Although they resemble anti-dilution provisions,[10] they differ in that they require an additional investment to maintain the right-holder's stake.

§6.4(D) Stock Options Plans

Generally, the shareholders need to approve increases in the number of stock options authorized under a company's stock option plan. Since it may become more difficult to obtain proxy majorities for increases in a company's stock option plan later in a company's development, particularly after a company has gone public, it is advisable to provide well into the future for employee option grants authorizing a large enough number of option grants at as early a stage as possible.[11] This is essential for the start-up biotechnology company because stock options are the greatest incentive for attracting top entrepreneurial and scientific talent.

The stock option plan should be designed so that any options granted vest over a period of time, such as four to five years, and are forfeitable when an employee leaves the company for any reason. This gives good employees an

---

10. See §6.4(F).

11. There may be a significant impact for accounting purposes when the options are actually granted, however that is beyond the scope of this book. See http://www.nysscpa.org/cpajournal/2005/805/essentials/p30.htm.

incentive to stay with the company, since a presumably favorable, low exercise price is locked in at an early stage, even though the actual vesting occurs over time and remains "at risk" until the vesting period has expired. Options that are granted but are later forfeited (e.g., because an employee has been separated from the company) can be returned to the stock option plan for issuance at a later time if the plan is properly drafted. As an additional safeguard for the most valuable original employees, many venture capital investment agreements will tie up founders' stock in escrow or provide for retirement of founders' shares to treasury stock if milestones are not met or if founders "jump ship" and leave the company.

Finally, in connection with a start-up's financing, the "founders" group of scientists and businessmen responsible for setting up the company must give up what appears to be a huge slice of the pie to a third party, the VCs. However, it is important for the entrepreneur and scientist to remember that any percent of something is greater than 100 percent of nothing, which is what they may end up with if the venture capitalists decline to invest!

§6.4(E) Stages of Venture Capital Investment

It is common for venture investors to plan on investing increasing amounts in at least three venture rounds. Some of the large funds plan on investing a total of $20 million or more in their portfolio companies, beginning with a $5–$6 million dollar first round and then, together with additional companies, similar or slightly larger investments in subsequent rounds of perhaps $20 million or even $30 to $40 million in total. However, the subsequent funding depends on reaching the milestones set out in the previous round.

This structure presents some serious issues. On the one hand, holding back installments until a certain development milestone has been achieved gives the biotechnology company an incentive to achieve development and commercialization consistent with the timelines in the business plan and the *pro forma* operating statement (which purportedly supports the net present value, which in turn supports the valuation of the company and the price paid by the venture capitalist). On the other hand, a start-up biotechnology company that experiences delays in its development of a technology generally has no other sources of funding draw upon to achieve the delayed milestone.

The compromise that can bridge this gap is for the investors to provide the subsequent rounds even if the company is behind schedule, but at a valuation that is lower than would otherwise have been the case, thus further diluting the insiders' share, but proportionately increasing the investors' share. However, at some point, the investors may be unwilling to provide further funds, the markets may be generally poor for financing, and an outside investor may only be willing to provide funding to "save" the company at a price that is significantly less than the previous round. In some cases, the offer requires a reverse stock-split of the existing shares (one share for two to four shares or more) prior to issuing new shares in exchange for the new funding. These financings, sometimes referred to as "cram down" for obvious reasons, greatly dilute the value of the ownership interest of the existing shareholders and trigger any anti-dilution rights granted to earlier investors unless they agree to waive those rights.

§6.4(F) Anti-Dilution

Anti-dilution provisions can take several forms. The principal purpose of most common anti-dilution provisions is to provide some degree of protection to investors against subsequent rounds of financing at a lower, per-share price ("down" rounds). Such anti-dilution clauses protect the absolute value of existing investors' share of the company by giving them a bigger piece of the company if the company's value declines. Anti-dilution provisions can give full protection or partial protection (permitting some dilution as a result of later investment, or increasing the investors' stake to a level that does not fully restore the original value of their investment).

§6.5 Beyond the First Round: Licensing and Corporate Strategic Alliances

When does a biotechnology start-up stop being a start-up? The rite of passage from start-up to mid-stage is often associated with a company's first major licensing deal or corporate strategic alliance. Typically, these agreements involve the performance of research and development activities by the biotechnology company in return for receiving funding from and licensing technology rights to a larger company. It can actually be an oversimplification to characterize what is really a "research and development" arrangement between a very small and a very large company as a "strategic alliance," since with very few exceptions it is the biotechnology company that conducts the basic research and transfers technology to the larger company. It is more appropriate to view these "technology development ventures" as a continuum ranging from a basic license to an acquisition of the biotech company by the larger company.

The most popular type of arrangement or "strategic alliance" has elements of a license agreement,[12] a research and development support agreement, and a marketing and distribution agreement. Very often, these arrangements also include an equity investment in the biotechnology company by the larger company.

Why do biotechnology companies and their venture capital investors seek this type of financing, which requires the company to give up substantial rights to future revenues from the commercialization of a significant portion the company's core technology?[13] Most small biotechnology companies either start with the knowledge, or come to the realization very quickly, that the infrastructure necessary to conduct a full-scale pharmaceutical development program is far more expensive than their venture capitalists' capacity to continue funding the company.

Small biotechnology companies also come to realize that if there have been few significant developmental breakthroughs in their first few years of operation, their ability to raise money from other investment sources, including the public markets, is limited. These potential "tertiary" investors usually look for signs of credibility and legitimacy, such as an alliance with an established pharmaceutical company, presumably expert in evaluating the kind of science involved. Not to be overlooked, however, are the strategic partner's proven capabilities in clinical and regulatory affairs, marketing, sales, and distribution of products similar to those being developed by the biotechnology company. In this way the loss in value attributable to the rights that are licensed can be more than offset by the credibility and value in the marketplace that the pharmaceutical company deal brings with it.

Finally, biotechnology companies also come to realize that as the technology becomes more mature, the level of resources needed to develop it to the next stage are not only staggeringly expensive, but may be much more prudently "rented" from a corporate partner than acquired in-house. Multi-center clinical studies, pilot manufacturing facilities, and market focus groups, for example, all involve not only exponentially increasing expenditures, but also skills and facilities that are not readily or easily available.

The corporate strategic alliance addresses these needs. First, the strategic alliance with a larger company adds value by legitimizing and lending credi-

---

12. Many of the basic licensing issues are discussed in the context of university-company licensing in §3.5.
13. §6.5(A) discusses strategic issues involved in deciding what areas of a company's technology platform can be partnered out while keeping other portions longer for internal development.

bility to the biotechnology company's technology. Second, it provides necessary medium-range funding to continue the company's activities to a much more advanced stage. Finally, the larger company can instantly provide access to resources that it would take the small biotechnology company many more years and many more dollars to develop and, even then, at substantial risk of doing so unsuccessfully.

In short, the strategic alliance is a very effective way of leveraging the small biotechnology company's position by accelerating its development programs and diminishing the risks associated with capital depletion and organizational growth.

§6.5(A) The Framework for a Strategic Alliance

As in the case of start-up financing, the strategic alliance can take multiple forms. The following model examines a strategic alliance containing many more elements than usual, in order to examine all of the potential of this financing vehicle.

The essential element in virtually all strategic alliances is a license of technology from a small biotechnology company to a larger a pharmaceutical or chemical company. The license can vary in scope, depending on the overriding purpose of the alliance; R & D support, marketing in a particular territory, or products for a certain application or field of use. Licenses have five basic dimensions: a definition of the licensed technology; a level of exclusivity; geographical limits; a specified field of use; and. the consideration, for example license fees, royalties, transfer prices for manufactured products, and milestone payments.[14]

In addition to the license terms, the agreement provides a research and development arrangement, in which the large company undertakes to fund or perform various parts of the project, such as further research, preclinical and clinical studies, and process scale-up. The payments are usually staged based on installments over time or based on specific milestones. The research and development prong of the alliance should involve enough funding to assure that the technology will be commercialized for the particular products covered by the license or development agreement. The R&D agreement can really provide a sort of partnership, in which each party agrees to undertake the tasks for which it is best suited; for example technology de-

---

14. Chapter 3 discusses the key terms of a license in the context of university technology transfer; however, the issues are very similar in the biotech to pharma context. See §3.5 for a more detailed discussion of licensing terms.

velopment at the biotech company, planning for Phase II or III clinical trials, regulatory filings, and marketing primarily by the pharmaceutical company. Funding for the research performed by the biotechnology company often is based on a formula that assigns a cost per person for example $300,000 per year for each full-time employee ("FTE") and a number of persons to be assigned to the work.

Strategic alliances frequently involve a private placement of stock by the biotechnology company to the large company at a higher price which reflects a premium over previous valuations, along with warrants to purchase more stock over a period of time usually five to ten years. In this connection, the corporate partner's interest is much the same as the venture capitalist's and, as a consequence, many of the same features, such as anti-dilution clauses, are found in corporate partnership investment agreements as are found in venture capital deals.

§6.5(B) Timing of Alliances and the Impact on Valuation

There is a direct relationship between the stage of development of a technology and the relative bargaining positions of the biotechnology company and the larger company. All other things being equal, for example, a technology that has been developed to the Phase II clinical trial stage is going to command a much richer deal than one in preclinical stages. Similarly, a drug ready for Phase III can be licensed at terms that provide a very significant share of the profits to the biotech company, while the pharmaceutical company picks up the costs of the Phase III trial and assists with the process of filing the NDA.

More importantly, the very process of development can result in the biotechnology company's better defining itself, its own strategy, and the range of products and applications for which its own technology is best suited. For example, if ProteoMaxx has a basic technology that it believes has application for the development of anti-cancer drugs, but after initial research ProteoMaxx learns that the technology also gives rise to diagnostic applications, ProteoMaxx may adopt a much different approach in establishing its strategic alliances. ProteoMaxx may decide to approach two different types of companies, one for the two kinds of applications, with different structures and pricing models for each.

One of the most difficult challenges of the young biotechnology company contemplating a strategic alliance is to retain as much of its core technology as possible for as long as possible. Although most biotechnology companies entering into licensing agreements or strategic alliances try to hold on to as much as possible of the value that can be generated by their core technology, in most cases biotechnology companies must license away significant opportunities provided by their core technology in order to buy the security of a long-range, comprehensive partnership with a large drug or chemical company.

The key to success for the company is in holding on to its core technology, in ensuring that any strategic alliance is structured in such a way as to provide not only funding but enough room for biotech company's future growth and other partnering opportunities. The first strategic alliance should not lock up all of the value of the technology with one partner in all territories, for all fields, for the life of the patents covering the technology. In this way, the biotechnology company can procure the ongoing development funding, manufacturing, marketing and distribution support it needs maximize the value of its core technology in other applications.

Sufficient limits on the first strategic alliance can be achieved by manipulating a number of variables. The first is time. For example, a marketing agreement can extend exclusive rights to a corporate partner to sell products based on a technology for a certain number of years after relevant regulatory approvals, with the biotechnology company reserving co-marketing rights after the initial exclusive period. The number of years can be roughly calculated by estimating the earnings generated by the technology after it is commercialized, and then estimating the number of years that are required for the corporate partner to generate sufficient sales revenues and earnings to achieve a predetermined, acceptable, risk-adjusted, return on its investment.

Similarly, if some geographic territory, such as Japan and Asia, can be excluded from the licensing agreement, then the value of those rights will increase while the partner helps fund development for the U.S. and E.U. Another strategy is simply to provide a sliding royalty scale which increases upwards as sales exceed specified benchmarks. While an upward sliding scale is not common, it actually makes great basic business sense as each successive benchmark represents a substantial increase in marginal profits. The return on the first $100 to $200 million in sales can actually be relatively low, given the high initial costs of approval, manufacturing and marketing. Once a drug reaches blockbuster status, however, that "second" billion in sales has huge marginal profits and justifies a sharing of those increased rewards with the biotech company whose technology was instrumental in achieving those returns.

All agreements to license to or co-develop technology with a large pharmaceutical company must impose objective milestones upon the corporate partner, particularly if the partner is going to be significantly involved in development, testing, pre-marketing, marketing, distribution or manufacturing. For a platform technology company, such as the hypothetical Proteo-

Maxx, that slices off "leads" to partners in different disease areas, partner progress in development is not only significant for the funding it provides, but also because the partner's development progress adds credibility and value for ProteoMaxx and its investors. The biotechnology company can also use noncompete provisions, so that if the partner develops or acquires technology or sells products that compete with the licensed technology, then certain rights would revert back to the biotechnology company.

The pharmaceutical company on the other side of the table seeks to get the greatest value from its deal with the biotech company by acquiring rights to the most valuable opportunities the biotech company can generate. ProteoMaxx may find that its initial partner wants rights to all applications for cancer, or autoimmune disease while ProteoMaxx would prefer a product-by-product approach, with separate agreements for different products in the same area, or even different indications in an area such as cancer. However, there are pitfalls to this approach in that each succeeding slice of the pie can be harder and harder to sell off. Larger companies are disinclined to make big, risky investments for niche products based on technologies in which other companies, some perhaps their competitors, are also participating. The dangers increase markedly when different products enter the marketplace as largely interchangeable commodities. For example, there is little risk in licensing the same antibody separately for diagnostic purposes and therapeutic purposes. However, licensing the same protein, for example erythropoietin, to different companies for different indications, such as end-stage renal disease to one partner with all other severe anemias licensed to a separate company, is likely to produce difficulty and litigation (See note 3, §3.5(c)).

The biotechnology company may reap short-term benefits by concluding many deals; but unless a certain critical mass is established in a strategic alliance, by definition it cannot be very "strategic" or critical to the large or small biotechnology company's business development. Serving many masters can also become a problem, particularly when some projects look more promising than others, or when scarce resources are stretched to the limit.

The successful strategic alliance will be one that is a "win-win" situation for both the biotechnology company and the industry partner. The relationship must be something more than a financing vehicle for the biotechnology company. It must be viewed as the best way to develop its business at a given time early in its evolution, providing credibility and legitimacy for its technology, added value, maturity, depth, and direction to its still evolving business plan. For some biotechnology companies, the establishment of a strategic alliance represents the first time the company thinks and acts like a business instead of a research and development department.

§6.5(C) Foreign Partnerships

For an American biotechnology company, a license or strategic alliance with a foreign company can represent an excellent opportunity to augment company resources, assure rapid and competent commercialization of its technology in foreign markets, and preserve all its rights in its most important market, the U.S. Drugs and health care are transnational industries in the broadest sense, and what is determinative of the "foreign" character of a strategic alliance is not the domicile of the partner, but rather the territory covered by the arrangement. Companies like Novartis, GlaxoSmithKline, Roche, and Aventis may be based in Europe, but their business is global.

It is somewhat less commonly the case, however, that a Japanese pharmaceutical company is directly involved in many markets outside Japan, Taiwan or South Korea. Health care is one of the few high-value added industries in which the Japanese presence is secondary, and the Japanese have traditionally been substantial net importers.[15] As a consequence of these differences between Europe and Japan, a large European drug or chemical company is more likely to seek an alliance which extends beyond Europe, due to its global manufacturing, marketing and distribution presence. A Japanese partner, on the other hand, is less likely to insist on obtaining rights outside of the Far East. Even if it does, it is less likely to make a convincing case that it can deliver the support (particularly regulatory filings and marketing) necessary to achieve rapid commercialization outside its local territory.

A viable strategy for many biotechnology companies has been to seek separate strategic alliances for Europe and Japan with companies that have a major presence in each respective market, while seeking to retain substantial rights in the U.S. market, whether in the form of a co-marketing agreement or other division of U.S. sales for some period of time. In this way, the biotechnology company may fund its product development through early clinical development before sharing some or all of the U.S. rights. Retaining the U.S. rights until a successful Phase II clinical trial should eventually enable the biotechnology company to access the public markets a significantly higher val-

---

15. In 1998 Japanese pharmaceutical imports were almost twice exports, while in the U.S. imports exceeded exports by a bit less than 20 percent. The biggest net exporters of pharmaceuticals were Germany, Switzerland, the UK, Ireland, France, and Sweden. Data from Glaxo is *available at* http://www.schoolscience.co.uk/content/4/biology/glaxo/pmb1trade1.html (last visited May 6, 2006).

uation, or to conclude a strategic alliance for the U.S. rights in exchange for a much bigger piece of the pie.

§6.6 Introduction to the Public Offering

The phrase "going public" evokes images of fast-track executives jetting from coast to coast, lauding the virtues of investing in their emerging companies to underwriters eager for a piece of the action. Indeed, a few initial public offerings do conform to this glamorous image. Most, however, do not. If a company is not meeting investor expectations, as is the case of many biotechnology companies today, public disclosure and scrutiny of its activities can be like living in a fish bowl.

Glamorous or not, the initial public offering ("IPO") is certainly the most significant of financing vehicles and is a uniquely important event in a company's evolution, with far-reaching consequences. Viewing the IPO from the insider's point of view, one should ask a number of basic questions: How does a company's management decide to go public? What are some of the mechanics involved in preparing the company for such a significant step? What can be expected from other participants in the public offering process and what do these other participants expect of the company? And finally, what can a company expect in the aftermath of going public?

§6.6(A) The Decision to Go Public

Some biotech sages would advise that the decision to go public is an externally driven one: take the money when the public financing window is open and an offering can fly. Such a cynical view is dangerous. Going public makes subsequent financing more difficult and going public too soon for too little can be disastrous. How then does a company's management decide that it should raise money by going public? The first step is to compile a "needs profile." First, does the company need the money, and if so, how much? Second, for what is the money needed? Third, is the additional capital needed for ongoing operations or for research and development? Fourth, does the company need the money to expand its manufacturing operations or its marketing activities; or is the company in the development stage and operating in the red? Finally, how much money will be needed for it to become a commercial enterprise, and therefore profitable? Once a public offering is made, subsequent financings, whether secondary offerings or private investments in public equity (PIPEs), are priced by the market. The company that has gone public and suffered development setbacks will be punished by the financial markets and find it difficult to raise money or find the terms on which money is offered very onerous.

Once the company has decided that it needs additional capital, it should next determine whether or not the public offering is the best vehicle for obtaining this capital. A public offering can be a very expensive way of obtaining additional capital, since commissions and expenses can often exceed 15 percent. However, a public offering generally brings needed capital into a company with much less dilution than would be the case with most of the private alternatives. Furthermore, unlike venture capital, there are few, if any, strings attached to the source of financing. Public shareholders own stock in the company, and the obligation to shareholders is generally defined by the company's articles of incorporation and bylaws, not by an escrow agreement with an institutional investor or the specified milestones in a venture capital agreement.

The next step is to examine the current capitalization of the company. Who are the current major shareholders—top management, or venture capitalists? The shareholder mix can make a big difference. Venture capitalists are motivated primarily by their investors, who generally desire rapid liquidity. Venture capitalists may demand that the public offering be done at a price significantly higher than the price at which they invested, and their recommendation to go forward or to postpone the current opportunity to go public is likely to be based on their own needs, not the company's. Venture capitalists will almost always favor an early exit and the public offering over private sources of funding. Finally, venture capitalists involved in the company will invariably have piggyback registration rights so that when the company goes public and registers its shares publicly, it must also register the venture capitalist's shares. Such an action might depress the price of the stock once the company has gone public if too many of these shares are sold (which will likely be precluded for some period by the underwriting agreement).

The crucial decision on whether or not to go public should involve a thorough examination of the corporate structural changes that must be made. Should "shark-repellent" or "poison pill" amendments to the articles of incorporation or bylaws be promulgated in order to prevent a hostile takeover attempt? Does the company currently have cumulative voting rights for director elections? If it does, management may wish to change the bylaws in order to maintain firmer control over the election of directors. The articles or bylaws may also have to be changed to eliminate preemptive rights of shareholders, which is the right of a shareholder to maintain a proportionate ownership in the company. The lawyers for the issuing company and for the underwriting investment banks will undoubtedly have a long list of items to address to assure that the offering itself and the post-offering public company comply with the securities laws, including the requirements of the securities reforms (particularly Sarbanes-Oxley)[16] enacted in the wake of the stock market crash and scandals of 2001–2002.

§6.6(B) Preparing the Public Offering: The Letter of Intent

The letter of intent deals with the major terms and aspects of the public offering. Essentially, it is an agreement to agree. Generally the underwriter offers the letter of intent to the company. Except for provisions regarding payment or reimbursement of the underwriter's costs, the letter of intent is normally not binding. The price is usually not set in the letter of intent. Although the underwriter recommends a tentative price to the company, the actual price is not set until the effective date. This gives the underwriter the flexibility to recommend changes in the price should market conditions change significantly between the date the letter of intent is signed and the effective date. The letter of intent also covers the issue of the percentage of equity to be sold by the company, that is, the dilution factor. For example, the company and the underwriter may decide that the amount of shares sold to the public should be 20 percent of the total shares outstanding after the public offering.

Even though the letter of intent is normally not a binding contract, it is nevertheless a very important document because it sets the tone for how the public offering is to be conducted and also defines the most important aspects of the relationship between the company and the underwriter. The utmost care should be taken in drafting and negotiating the letter of intent, and the input of all relevant parties should be obtained before it is signed.

§6.6(C) Going Public: The Team

A public offering, particularly an initial public offering, involves a great deal of work and preparation. Usually, management of companies with an IPO underestimates the amount of distraction arising from the process of going public. Presumably top management is already fully occupied with the operations of the company. Thus, during the registration period and often far

---

16. Sarbanes-Oxley Act of 2002, 15 U.S.C. §7201 et.seq. (2006).

beyond it, key officers will find themselves overextended. It is extremely important to put together a team that can make the process of going public function as smoothly as possible. Inevitably the team needs to include the CEO, the COO and the chief financial and legal officers. Outside counsel can provide the necessary securities law expertise and the company's outside auditors, presumably from a firm with considerable expertise in representing emerging public companies, can also provide considerable service and support in the offering process.

The company's venture capital investors and investment bankers, if different from the underwriter should be involved in the negotiations with the underwriter, especially on such issues as price, dilution, costs, and commissions. They can be instrumental in acting as buffers between the company's management and the underwriter in negotiating the terms of the IPO.

Finally, if the company has engaged an outside public relations firm, it should be one with investor relations experience and should be involved as well. A pattern of a consistent flow of information to the public should be established as soon as possible.

It is extremely important for all team members to know each other and function well together. Furthermore, they should be able to function with their counterparts representing the underwriter. To the extent that these people have worked together in the past, the going will be that much easier.

§6.6(D) Going Public: Due Diligence

Due diligence is the underwriter's investigation of the company and confirmation of all the material elements of the business. If the underwriter is not the company's regular investment banker, the due diligence process will necessarily involve more time and will likely delay the process of going public. Essentially, the underwriter is seeking a level of comfort with respect to the business of the company. The due diligence process primarily serves to verify the information on which the underwriter originally based the decision to underwrite the sale of the company's stock. The underwriters will seek to verify important licenses, contracts, leases, backlogs of orders, development of new products, and occasionally something as fundamental as the existence of the facilities and background of the employees.

In particular, the underwriters will look for problems that on disclosure to the public could prove embarrassing or could detract from the value of the company, like excessive loans to insiders, "sweetheart" deals with related parties such as consulting agreements with relatives of insiders, and any existing debt or equity instruments that may be convertible to common stock, thus having a significantly dilutive effect. In the case of biotechnology companies, scientific and technical due diligence is routinely performed to validate objectively the commercial potential of the company's technology base. All correspondence with the FDA about any products in development will also need to be disclosed.

The most important thing to remember about due diligence is that it is a dress rehearsal for going public. In exchange for the right to sell shares to anonymous individuals in the public market, your company must make publicly available all information that is important to an investor's decision to buy, sell, or hold shares. Thus, the company must be able to back up every public statement it makes. The underwriters will want to examine copies of all significant documents, including material agreements, licenses, back orders, claims, and lawsuits. Every significant matter involving the company or its business operations are important to the due diligence process and should be appropriately documented, and available for the due diligence and public disclosure. If an important item is not already in writing, the underwriter will very likely demand that it be put in writing before the registration process continues.

§6.6(E) Going Public: Filing the Registration Statement

When the registration statement has been completed by the combined efforts of the underwriters, management, and counsel for all parties, it is filed with the SEC. Once the statement has gone to the SEC, the process is out of the company's hands, and the time frame will then depend on factors beyond its direct control, such as the backlog of other registration statements being reviewed at the SEC.

Once the SEC has reviewed the registration statement, it usually comments on it. If the comments are not very significant or lengthy, the company can respond in a short time. However, if the comments relate to significant matters or are very extensive, the process will naturally be lengthened. It is not unusual for the SEC to make two, three, or even more sets of comments. Each time the comments are made, the company must respond to them to the satisfaction of the SEC, further delaying the effective date on which sale of the stock to the public can begin. Again, the emphasis is on good preparation. Depending on the backlog at the SEC, the first set of comments is likely to be received within three or four weeks.

A word about terminology: Frequently-used terms include "registration statement," "prospectus," and "red herring." These terms are all interrelated. The registration statement filed with the SEC includes a prospectus, which is the document used to offer the stock to the public. All disclosures material to a decision to invest in the company should be made in the prospectus. However, before the company actually goes public, that is before the SEC officially clears its stock for sale to the public, a preliminary prospectus is issued that allows the underwriter to preliminarily market the issue prior to the effective date. This preliminary prospectus is called a "red herring" because of the disclaimer printed in red on the cover informing the public that the document is only a preliminary prospectus and not an offer to sell the securities. Of course, the primary use of the prospectus is to conditionally offer to sell the securities, because an actual offer to sell can only be made through the final prospectus that has been distributed after SEC.

§6.6(F) Going Public: The Effective Date and the Closing Date

Approval by the SEC of an IPO is usually a cause for celebration! The registration statement has been filed with the SEC, the Commission has commented on it, and the company has reviewed the comments and responded to them. When the SEC has notified the company that it is satisfied with the registration statement, it will ask for an effective date. If the market is perceived to be at a crest, the underwriter will usually suggest a date soon after SEC clearance. However, if the market has become depressed since the letter of intent was signed, the underwriter may wish to postpone the effective date until market conditions are better. If the underwriter does not feel that market conditions are going to improve soon, the alternative is to renegotiate the price or lower the number of shares offered. The lowering of the price or number of shares will often result in far less capital being raised than originally anticipated by the company. Of course, whether to shelve an offering or lower the price is a decision that can only be made in light of the company's financial needs and other available sources of capital.

On and after the effective date, most underwriters will ask the company to hold information meetings at one or more locations where the underwriter expects to market the stock or, if more than one underwriter is involved, where other underwriters of the issue are located. These meetings, fondly referred to as "dog-and-pony" or "road" shows, can be important marketing tools for the company's public stock offering. At the meetings, key management people usually make presentations and answer questions from an audience that is largely composed of brokers and institutional investors. No written material other than the prospectus may be disseminated at these meetings, nor can the company distribute DVDs, CD-ROMs, or any other tool or medium for containing and displaying information. At these meetings it is very important that the company representatives do not say anything materially different from what is included in the company's registration statement. Also, the discussion should steer clear of forecasts and projections. Before any presentations, the company's stock should be properly "Blue-skyed," which means state-registered,[17] in those states where the meetings will take place. In addition to federal registration, state registration is required in any state where the stock is to be offered for sale, so that stock purchased by residents of that state can be traded in that state.

Once the SEC has given the company an effective date, the stock can be sold publicly and, more importantly, the company can receive the net proceeds. The underwriting agreement specifies the closing date, the date on which the funds are actually transferred to the company's coffers.

§6.6(G) Going Public: The Quiet Period

After the SEC has approved a registration statement and an effective date has been set, the company must be careful not to divulge any information that differs materially from the prospectus. This is the so-called quiet period, a period of 90 days after the effective date during which the prospectus, and not any other representations made by management, must serve as the primary selling vehicle for the company's stock.

How does a company handle the quiet period? Should management go to the extreme of placing a gag order on all employees? Probably not. The first thing management should do, however, is make all management—and, if it is a small company, perhaps all employees as well, aware of the quiet period and what it means. This does not mean that no one can talk about the company at all. It simply means that no one can tell the public anything that differs materially from what is in the prospectus. The idea is that if the information is important enough to tell a potential investor, then it should be disclosed in the prospectus. Basically, let the prospectus be your guide during the quiet period. If new information arises that is material, then the prospectus should be amended to include it. On the other hand, a presentation at a scientific meeting by a company scientist *might* be okay, if it concerns research that was already made public and no new data is disclosed.

---

17. Blue Sky laws are the generic term for state securities registration requirements. The term is purported to derive from the desire of state legislators to prevent unscrupulous promoters from selling securities consisting of or backed only by the pure blue sky above.

§6.6(H) Going Public: Lock-Up Periods and the PSLRA "Safe Harbor"

Underwriting agreements generally provide a "lock-up" period during which management, the board of directors, and other significant pre-IPO investors, are precluded from selling any significant amount of their stock. A lock-up period of 90 to 180 days is quite typical. This is to prevent insiders from dumping their stock into the newly made market and driving the price down from the IPO price. The underwriting agreement may also limit the percentage of any such pre-IPO investors shares that can be sold in any one quarter. All "insiders," including all senior management, must also disclose to the SEC their sales of shares even after the lock-up period.

Management must also have a comprehensive plan for keeping the public and investors informed about the company's progress following the quiet period. It is important to have periodic broker meetings for the company's market makers, institutions, and investors in large retail markets. If public relations counsel has been retained, it can assist in coordinating such investor relations activities.

The company should also set up a schedule of regular shareholder communications, such as regular quarterly and annual reports, news releases, and other communications that may be of importance to shareholders to comply with the laws, rules, and regulations of the SEC and of the states in which the stock is registered, as well as with the rules of the NASD or the exchange on which the stock is traded. Such disclosures include the preparation of quarterly reports to the SEC on Form 10-Q, periodic reports on the use of the proceeds of the offering on Form SR, and an annual report on Form 10-K. In addition, if the company is registered under the Securities and Exchange Act of 1934, it is required to clear its proxy statements through the SEC.

The obligation to comply with all of the securities laws is often pointed to as being one of the most serious drawbacks to going public. This need not be the case. A well-managed public information and regulatory compliance program can furnish required disclosures to the public in compliance with securities laws without causing a great deal of extra cost to the company or distraction from its operations. The challenge is to integrate compliance with the company's business activities and its regular flow of information. If this is done, preparing documents and reports for the SEC and providing information to shareholders, brokers, and potential investors can be easy (and when things are going well, enjoyable) tasks.

A company's profile can be as high or low as desired. Many companies report a "fish bowl" phenomenon brokers, shareholders, and the media calling, all interested in the company's activities. Only a few people should be appointed as company spokespersons for handling these inquiries, for example, primarily an investor relations manager and, w hen necessary, the CEO. Spokespersons must be consistent in their answers and prepared to refuse comment on matters that should not be openly discussed.

In this post-Enron, post-ImClone era, biotech executives must be acutely aware of their potential for liability for misleading the public concerning the condition of their company. Any public statements, oral or written, about anything from financial returns to progress in product development, should be accompanied by a "safe harbor" disclaimer that protects "forward looking statements" from being actionable under the securities law.[18] However, no "safe

---

18. The Private Securities Litigation Reform Act of 1995, 15 U.S.C. 78u-5 (c)1 *et seq.* (2006) provides:
    (c) Safe harbor(1) In general
    Except as provided in subsection (b) of this section, in any private action arising under this chapter that is based on an untrue statement of a material fact or omission of a material fact necessary to make the statement not misleading, a person referred to in subsection (a) of this section shall not be liable with respect to any forward-looking statement, whether written or oral, if and to the extent that—
    (A) the forward-looking statement is—
    (i) identified as a forward-looking statement, and is accompanied by meaningful cautionary statements identifying important factors that could cause actual results to differ materially from those in the forward-looking statement; or
    (ii) immaterial; or
    (B) the plaintiff fails to prove that the forward-looking statement—
    (i) if made by a natural person, was made with actual knowledge by that person that the statement was false or misleading; or
    (ii) if made by a business entity; was—
    (I) made by or with the approval of an executive officer of that entity; and
    (II) made or approved by such officer with actual knowledge by that officer that the statement was false or misleading.
    (2) Oral forward-looking statements
    In the case of an oral forward-looking statement made by an issuer that is subject to the reporting requirements of section 78m(a) of this title or section 78o(d) of this title, or by a person acting on behalf of such issuer, the requirement set forth in paragraph (1)(A) shall be deemed to be satisfied—
    (A) if the oral forward-looking statement is accompanied by a cautionary statement—
    (i) that the particular oral statement is a forward-looking statement; and
    (ii) that the actual results might differ materially from those projected in the forward-looking statement; and
    (B) if—
    (i) the oral forward-looking statement is accompanied by an oral statement that additional information concerning factors that could cause actual results to ma-

harbor" shields statements that should be known to be false or misleading. There is considerable uncertainty about whether or not a prior statement, for example about progress towards FDA approval, must be "updated" if later developments change the scenario.[19] Once public, the company must have an active and ongoing SEC compliance program. Furthermore, Sarbanes-Oxley[20] requires public companies to have an audit committee composed entirely of outside directors, with at least one member of that committee being an expert in finance or accounting. In addition, it requires management's certification that the audited results are true and provides criminal penalties for intentionally misleading the auditors in order to produce a misleading financial statement.

"Going public" is an important event in the evolution of a company and should be a memorable, rewarding undertaking. It can open new doors for a biotechnology company and can help it make its dreams come true. It is, however, merely a means to an end. A company must continue to conduct its business and work to achieve its goals of growth and profits. If this is kept in mind, and if management is careful and skillful in following all the necessary steps, the IPO should result in a maximum of benefit and a minimum of disruption.

§6.7 Conclusion: Biotechnology Finance as a Four-Dimensional Rubik's Cube

This chapter has provided a brief overview of the three primary means by which biotechnology companies raise money: venture capital, corporate partnerships and licensing, and public offerings of securities. The complexities of each individual type of transaction, whether it is a venture capital offering or a strategic alliance, pale in comparison with the complexity of the overall venture. Biotechnology companies are faced with the extraordinary challenge of

---

terially differ from those in the forward-looking statement is contained in a readily available written document, or portion thereof;

(ii) the accompanying oral statement referred to in clause (i) identifies the document, or portion thereof, that contains the additional information about those factors relating to the forward-looking statement; and

(iii) the information contained in that written document is a cautionary statement that satisfies the standard established in paragraph (1)(A).

19. For a thorough discussion of this issue, see Steven E. Bochner and Samir Bukhari, *The Duty to Update and Disclosure Reform: The Impact of Regulation FD and Current Disclosure Initiatives*, 7 STAN. J.L. BUS. & FIN. 225 (2002).

20. Pub. L. No. 107-204 (2002), 15 U.S.C. §7201 et.seq. (2006).

financing and developing cutting edge science, preferably with proprietary technology, over an enormously complicated seven to ten year average period of preclinical and clinical data collection, aimed at succeeding in one of the business world's greatest challenges, obtaining FDA approval of a new drug, biologic, or device.

To succeed requires not only great science but a business plan that is sufficiently focused and yet somewhat flexible, managers who can deal with both the internal demands of product development and the external demands of selling their technology and their company to investors of all types, and a great deal of money. Moreover, that money comes in at different times, in different amounts, through different types of transactions, and it is never easily obtained. In other words, the plan to produce the financial life-blood to sustain the company from start to finish must, over time, take into account the extraordinary complexity of the business itself, with its Rubik's cube facets of science, personnel, competition, and regulatory strategy. It is no wonder that failure among biotechnology companies is more likely to be financial than scientific: most companies that fail will run out of money long before they run out of scientific promise.

The art of planning and managing the finances of a biotech company is an indispensable element in success. In each stage, the "burn rate" or monthly expense of operation must be carefully gauged against the projected time until the next expected financing, with a major allowance for the possibility that the next expected financing could be delayed by a year or more by market conditions or technological difficulties. Since the "survival index," which calculates the number of months of life left to a company based on current cash and burn rate, is a major concern for everyone in biotechnology, when financing times are difficult it may be necessary to make downward adjustments in spending every six months. Even with that strategy, no early to mid-phase company can be assured of more than two years continued existence without additional capital inflows. Thus product development goals, particularly those milestones most likely to bring in either corporate partnership financing or the opportunity for additional public equity, must be realistically obtainable within the budgetary constraints of a cash-conserving financial strategy. If this sounds like a description of trying to tap dance and play first-violin at the same time, that may just be the closest analogy to the demands of financing biotechnology. That also makes a versatile transactional lawyer and a top "business development" executive, both with excellent negotiating and drafting skills, indispensable parts of a growing biotechnology company.

While there is a direct relationship between development progress and valuation (because, other things being equal, the risk diminishes with the achievement of each product development milestone), there is also a direct relationship between development progress and the absolute amount of investment needed. For example, although a company may command a greater price for its stock after it has obtained permission from the FDA to begin Phase III clinical trials in humans, it will need to raise a great deal more money in order to conduct those trials. In the end, the rewards of bringing a biotech company from start-up to success make this risky, complex process perhaps the most exciting business challenge there is.

CHAPTER 7

An Overview of the Regulation of Biotechnology

§7.0 Introduction

Regulation plays a major role in the development of biotechnology, with layers of regulatory scrutiny accompanying the development of any biotechnology product, from inception in the research laboratory until final approval for distribution in the marketplace. When biotechnology first emerged from university laboratory research into large-scale commercial development, a debate began as to whether a new federal regulatory scheme was necessary to control this powerful new technology, or if the public interest could be adequately served by the application of existing statutes by established agencies. In 1984, during the Reagan Administration, an interagency working group resolved the debate in favor of the existing statutory framework. The conclusions of the working group were that existing statutes were adequate and that the products of biotechnology could be reviewed by agencies accustomed to dealing with analogous products created by traditional means. While the statutes themselves have been changed during the past twenty years, the basic decision to regulate biotechnology within a framework of general statutes and existing agencies has remained unchanged.

The interagency working group on biotechnology regulation's efforts resulted in the publication of the Coordinated Framework for the Regulation of Biotechnology by the Office of Science and Technology Policy on June 26th, 1986 (51 Fed. Reg. 23301). In the Coordinated Framework, each of the major agencies involved (NIH, FDA, EPA, USDA, and OSHA) set forth their initial position on the regulation of those applications of biotechnology that fell within their respective jurisdiction. This "application-derived" regulatory framework confronts the development of biotechnology with a diverse array of statutes, agencies, and regulatory standards. One of the chief advantages of this approach is its implicit recognition that the risks of different applications should be treated differently and require different kinds of expertise. For example, the agricultural biologists at the USDA are likely to be better able to review a genetically engineered microbial pesticide, while the FDA would be better equipped to evaluate a DNA probe for the detection of Hepatitis B virus. The most significant principle established by the inter-agency working group was that the issues presented by biotechnology should be regulated in the same general way as risks from other technologies, a principle that is still debated and not always strictly applied.

The advantages gained by entrusting the regulation of biotechnology to existing statutes and agencies are certainly partially offset by the problems created by that approach, particularly that of overlapping jurisdiction among the agencies. In some cases, such as microbial pesticides, reliance on existing statutes also results in the problem of judging the risks of living organisms within a regulatory framework designed for inert chemicals.

The basic structure set forth in the Coordinated Framework is relatively straightforward. The National Institutes of Health Recombinant-DNA Advisory Committee (NIH-RAC) Guidelines control all research at institutions receiving any NIH support (effectively all not-for-profit biomedical research institutes and universities in the U.S.). The FDA, under the Food, Drug and Cosmetic Act (FDCA), regulates the development of biotechnology products for human health care, new animal drugs, food additives, and some aspects of other food products produced using biotechnology. The USDA regulates the development of new animal pharmaceuticals classified as biologics under the Virus, Serum, Toxin Act, plus experimentation with recombinant and transgenic plant and microbial species under the Plant Protection Act. The EPA regulates genetically engineered pesticides and other plant regulators (a term of art meaning a substance used to affect plants, not the bureaucrats who promulgate regulations about plants) under the Federal Fungicide, Insecticide and Rodenticide Act (FIFRA), as well as any major use of recombinant organisms or products not regulated under other agencies and statutes under the Toxic Substances Act (TSCA).

This simple description of the role of each agency under the Coordinated Framework can be deceiving. In some cases, two or more agencies will have the power to review the same experiment at the same time, such as when the USDA reviews, for potential plant pest characteristics, the proposed testing of plants engineered to produce a biological pesticide while the EPA reviews the same testing proposals for the environmental and health implications of the pesticide and its use. Similarly, the development of a single product may be subjected to scrutiny by different agencies, using different statutes and different standards, at different stages of development. For example, a transgenic food plant may be regulated under the NIH-RAC guidelines while being developed in a university, reviewed by the USDA before it can be experimentally grown outdoors by a commercial venture, and then require additional review by the FDA before it can be sold as a food product in the United States.

This chapter surveys the regulation of biotechnology, with a brief history of the controversies surrounding the regulation of genetic engineering, continuing with a general overview of the regulatory matrix of agencies, statutes, and rules currently in place, synopsizing each of the major statutes now being used to regulate biotechnology. There is an extended analysis of the regulation of food products produced from genetically engineered crops in §7.7.

§7.1 HISTORY OF THE REGULATION OF GENETIC ENGINEERING

The regulation of biotechnology has been made difficult by the technology's complexity, many scientists' distrust for lawyers and government officials, and public misperceptions about the nature of genetic engineering and its potential dangers. It is not surprising that since the first days of recombinant DNA experimentation, the regulation of biotechnology in the United States has provided both controversy and confusion. The popularity of such works as *The Andromeda Strain*, combined with popular press debates over the possible creation of world-ending viruses and the *Brave New World*-potential for "cloning" human beings, led to substantial public concern over genetic engineering experiments done by molecular biologists. For their part, molecular biologists were unaccustomed to such public attention and ill-prepared to deal with the problems of public and media relations.

The era of controversy over regulating the new biotechnology has its origins in the 1971 proposal of Stanford scientist Paul Berg to use a phage (a virus which infects bacteria) to transfer DNA from a virus which caused cancer in monkeys (SV-40 tumor virus) into a strain of E-coli (a bacteria which is quite common and a number of varieties of which are frequently found in the human intestinal tract).[1] Berg's desire to study the function of the tumor-virus

---

1. This subsection owes much to Swazey, Sorenson and Wong, *Risks and Benefits, Rights, and Responsibilities: A History of the Recombinant DNA Research Controversy*, 51 S.CAL.L.REV. 1019 (1978), which provides an excellent account of the early debates.

DNA by observing its expression in the bacteria generated an immediate, concerned response from cancer researcher Robert Pollack, who shared his doubts with Berg. The SV-40 virus had shown the ability to transform normal human cells into tumor cells *in vitro*, and the bacteria into which it was to be delivered can readily take root in a human population. The possibility could not be ignored that a novel organism produced in this way might be highly pathogenic. If such possible pathogens were inadvertently released from the laboratory, the potential for catastrophe was clear.

The proposal's circulation among molecular biologists raised sufficient questions that Berg agreed to defer his experiment and, in fact, helped lead the call for a meeting on the hazards of such research. That meeting, which took place in January 1973 at Asilomar, California, was the first scientific meeting to consider the hazards of recombinant DNA research.

In February 1975, a second meeting held at Asilomar, California, became famous as the beginning of the public oversight over recombinant DNA experimentation, rather than simply a debate among scientists. While this Asilomar meeting was an unofficial, interdisciplinary gathering of scientists, philosophers, ethicists, and a few lawyers, a majority of the conferees agreed that the National Academy of Science and the National Institutes of Medicine should establish a study committee to consider the hazards of recombinant DNA research and recommend appropriate actions or guidelines. The result was the formation of the National Institutes of Health Recombinant-DNA Advisory Committee (NIH-RAC) and the promulgation of the NIH-RAC Guidelines for Research Involving Recombinant DNA Molecules (the NIH-RAC Guidelines).

§7.2 Regulation Under the NIH Recombinant DNA Advisory Committee Guidelines (NIH-RAC)

The mission of the National Institutes of Health is the promotion of human health through basic research. The NIH accomplishes its mission through research done at the various Institutes themselves, as well as by funding research done at private and not-for-profit laboratories throughout the United States. Much of the medical and biological research done in not-for-profit university laboratories in the United States is supported by funding from the NIH. As an adjunct to its primary mission of supporting basic research, the NIH has taken on the responsibility for ensuring the safety of that research.

The basic rules by which the risks of research with recombinant DNA are kept within appropriate bounds are set forth in the NIH-RAC Guidelines.[2] The Guidelines, first promulgated in 1975 and frequently revised and amended in the ensuing years, were developed in the context of Berg's proposal to use recombinant techniques with a potentially pathogenic virus. Accordingly, the Guidelines set a variety of limits on recombinant DNA research in an attempt to ensure the safety of those in the laboratory and the public. The Guidelines remain as one of the basic pillars of biotechnology regulation today.

Compliance with the Guidelines is mandatory for research involving recombinant DNA conducted at any institution receiving funds from the NIH. As a practical matter, this covers virtually all research involving genetic engineering done at any not-for-profit institution in the United States. Primary responsibility for monitoring compliance with the Guidelines rests with the Institutional Biosafety Committee (IBC), which must be set up at any institution performing covered research. At the same time, while not mandatory for research done in commercial laboratories not receiving NIH funding, virtually all companies maintain a policy of strictly adhering to the NIH-RAC Guidelines,[3] motivated at least in part by the fear that failure to do so would lead to certain liability in the event of any mishap.

The basic structure of the Guidelines is a classification of possible experiments into risk categories, with appropriately rigorous safety and regulatory review procedures prescribed for each level of risk. The level of risk is determined first by the nature of the genetic material being manipulated and then by the nature of the carrier (or "vector") used to transfer the genetic material and the nature of the "host" organism to which the genetic material is transferred.

The Guidelines rely on three mechanisms for containing or controlling the risks of recombinant DNA experimentation—regulatory control, physical containment and laboratory practices, and biological containment (for example, by transferring potentially dangerous DNA only into an organism which is incapable of reproducing or surviving outside of the laboratory). Regulatory control can entail prior approval by the NIH-RAC, the NIH, and the local Institutional Biosafety Committee (IBC), or merely notification to the IBC when the experiment begins.

---

2. The most recent version of the Guidelines is *available at:* http://www4.od.nih.gov/oba/rac/guidelines_02/NIH_Gdlnes_lnk_2002z.pdf.

3. According to the Biotechnology Industry Organization (BIO) website:
"BIO member companies have voluntarily adhered to these guidelines since their inception." http://www.bio.org/speeches/pubs/er/ethics.asp.

The NIH divides experimentation with recombinant DNA into six categories (listed in decreasing order of risk):

(i) those that require prior approval by the Institutional Biosafety Committee (IBC), and review by the RAC and NIH Director;
(ii) those that require prior approval by NIH Office of Biotechnology Activities (OBA) and Institutional Biosafety Committee;
(iii) those that require prior approval by the Institutional Biosafety Committee and Institutional Review Board, in addition to RAC review before research participant enrollment (essentially those involving human subjects);
(iv) those that require prior approval by the Institutional Biosafety Committee;
(v) those that require only notification to the Institutional Biosafety Committee at the time of beginning the experiment; and,
(vi) those that are exempt from the *NIH Guidelines.*

Category i, which is reserved for experiments involving the "deliberate transfer of a drug resistance trait to microorganisms that are not known to acquire the trait naturally ... if such acquisition could compromise the use of the drug to control disease agents in humans, veterinary medicine, or agriculture," requires prior review by the IBC, the NIH-RAC the NIH Director, publication in the Federal Register and a public comment period.

Category ii research requires prior approval by the NIH Office of Biotechnology Activities (OBA) and Institutional Biosafety Committee and as of this writing applies to proposed experiments involving the "Deliberate formation of recombinant DNA containing genes for the biosynthesis of toxin molecules lethal for vertebrates at an LD50 of less than 100 nanograms per kilogram body weight."[4]

Category iii research, requiring prior approval by the Institutional Biosafety Committee and Institutional Review Board and RAC review before research participant enrollment, applies primarily to human gene therapy trials.

---

4. LD50 represents Lethal Dose for 50 percent, and is a measure of a substance's toxicity defined as the concentration required to kill 50 percent of the population exposed at that dose. Thus the LD50 toxins regulated under category ii would be lethal to at least half of the vertebrates that were exposed to 100 nanograms of the toxin per kilogram of body weight. Examples of such toxins include botulinum toxins, tetanus toxin, diphtheria toxin, and *Shigella dysenteriae* neurotoxin.

The current Guidelines leave most recombinant DNA research involving potential pathogens or genetic material with pathogenic or toxic potential in review category iv, which requires only the prior approval of the Institutional Biosafety Committee. For this category of experiments, the key variable is the degree of physical laboratory or biological containment required. There are four levels of laboratory containment, or biosafety, required for different experiments, based on their levels of risk as determined by the NIH-RAC Guidelines. The risk levels are based on both the pathogenicity of the organism from which the gene product is derived and the nature of the host/vector system into which the genetic material is being transferred. For example, the lowest level of risk, level 1, applies to genetic material from an organism which does not ordinarily cause any disease in healthy human adults, while risk level 4 applies to organisms which generally cause serious disease for which there is no treatment, such as Ebola virus, are in.

Thus IBC approval for category iv experiments should ensure that the appropriate Biosafety Level precautions will be used by the researcher. Biosafety Level 1 (BL1)[5] can be achieved in laboratories equipped with standard facilities by careful laboratory practices. BL2 requires additional safety practices and the use of some specialized equipment, such as Biological Safety Cabinets. BL3 requires very strict laboratory practices, special clothing worn only inside the lab, and additional specialized equipment, such as a HEPA (high efficiency particulate air) filter for venting the laboratory experimental areas. BL4, which is used only for the most dangerous of experiments, requires a separate building or isolated zone within a building, with access only through a double-doored changing and shower room. Thus, both the amount of vigilance and care taken, as well as the expense of laboratory facilities, increases substantially when BL3 or BL4 containment is required. BL4 laboratories are now rarely required under the Guidelines, and the significant expense involved in their construction would discourage many scientists from designing experiments that necessitate BL4 procedures.

Category v includes experiments that only require IBC notification includes some research on transgenic plants and animals, but that pose little or no potential for risk to health or the environment. Category vi experiments that are exempt from the Guidelines and do not even require notification of the Institutional Biosafety Committee include research that involves only the transfer of DNA from one non-pathogenic prokaryote into another prokaryote of the same or a closely related species.

---

5. The Guidelines also contain parallel biosafety levels for research involving some risk involving animals (BL1(2,3,4)-N, and plants BL1(2,3,4)-P.

§7.3 THE FDA AND HUMAN HEALTH CARE APPLICATIONS OF BIOTECHNOLOGY

The principle mission of the FDA is to ensure the public's health, through its power to accept or reject new health care products and through its responsibility for ensuring the purity and safety of the nation's food supply. Most of the FDA's statutory authority is derived from the Food, Drug and Cosmetic Act, Chapter 21 of the United States Code. Chapter 8 examines the FDA's jurisdiction over human health care products in considerable detail. This section provides a brief overview of the FDA's major regulatory responsibilities in approving human therapeutics, vaccines, diagnostics, medical devices, food additives, and new animal drugs, as well as in guarding against the adulteration of food products, all with an emphasis on the problems of biotechnology.

§7.3(A) The FDA Approval of Human Therapeutics and Diagnostics

There are two words that sum up the FDA's statutory mission for all of these kinds of products, whether produced by biotechnology or by conventional manufacturing technologies—"safe and effective." The significance of that statement is two-fold. First, it recognizes that the FDA has quite clearly expressed its intention to regulate biotechnology products by the same mechanisms and with the same scrutiny as other products. As Dr. Henry Miller, former Director of the FDA's Office of Biotechnology, repeatedly stated, the FDA regulates products, not processes. Thus the FDA principal concern about a drug produced by genetic engineering rather than traditional methods is whether the two methods result in products that vary in risk or benefit. Second, it underscores the difference between the traditional mission of the FDA and the mission of the other key agencies regulating biotechnology, the EPA and the USDA. The FDA's regulatory mission is straightforward and relatively unambiguous—the promotion of human health. This is unlike the EPA and the USDA, which are statutorily commanded to balance the interests of industry against interests in the environment and health, respectively.

The problem then, for biotechnology products as for all drugs, is to determine what will be necessary to prove to the FDA that a human health care product is safe and effective. The first lesson to be learned by any newcomer to the field is that NOTHING is absolutely safe and that EVERYTHING is a question of risk versus benefit. Aspirin poses a risk of stomach bleeding, while Acetaminophen (Tylenol) poses a risk of liver toxicity. Each has a risk that is deemed acceptable, when used in accordance with the labeled instructions, in light of the benefit the drug provides. So safety is a relative term, as is "effective." There are relatively few drugs that quickly "cure" a disease, but all approved drugs have been shown to be more likely than not to have a positive effect on the severity or duration of a disease. To use the same examples, we all know that Aspirin and Tylenol are useful for headaches, but not all headaches; and there are few if any headache sufferers for whom either drug provides fast and total relief. Additionally, effectiveness even for headaches might be measured in different ways: one drug might act by reducing the intensity of pain while another might act by shortening the overall duration of pain.

In any case, the basic concept is clear: safety and efficacy are to be determined by data collected from well-controlled clinical studies of the product. In the case of products for which clinical studies involve experimentation on human beings (as opposed to *in vitro* diagnostics, for example), clinical studies may not begin until, after sufficient preclinical data, collected in the laboratory and by toxicological studies of animals, the FDA has been provided with sufficient evidence to justify the risk of human experimentation and begin human clinical trials.[6]

For many companies, their first meeting with the FDA comes at the time that they are seeking approval to begin human clinical trials. As a practical matter, however, consideration of problems involved in obtaining final FDA approval should begin much earlier. The paramount question is always, "How can we demonstrate that our product is safe and effective?" Each of the two parts of the standard has its own range of nuances, which are derived from the concept of risk-benefit. For a disease that is incurable and usually fatal, a drug may be "safe" if it affords the possibility of curing fifteen percent of afflicted individuals while hastening or even causing the death of five percent. For example, for a liver cancer drug, the concept of "safe" means less toxic than the currently used highly cytotoxic (chemotherapeutic) drugs, while for a hayfever drug, "safe" means the complete absence of significant side effects in even the most sensitive individuals. Some slight drowsiness or temporary stinging of the nasal mucosa is probably the outer limit for acceptable side effects for a drug intended to treat an indication that inflicts only minor discomfort or inconvenience.

---

6. Technically, the FDA does not authorize clinical trials, but rather, after receiving the lengthy filing which accompanies a proposed human clinical trial, the FDA either: responds by requesting that the trial be briefly postponed pending clarification of particular issues; or, by suspending the clinical trial pending substantial changes or new data ("clinical hold"); or, by doing nothing, permits the trial to go forward.

Similarly, the concept of efficacy is a highly variable one. For a liver cancer drug, any measurable increase in the percentage of patients living one year after diagnosis would likely be considered proof of efficacy, while for an antihay fever drug efficacy might be determined from physician examination and scoring of the condition of the nasal mucosa during pollen season, or from patients' diaries of sneezing episodes, or by blood sera measurements of histamines or related substances. Just as the standard for safety is higher for less serious diseases, the standard of efficacy may be both more rigorous and less straightforward. Furthermore, the priority that the FDA will assign to the review of a product and the speed with which the agency will reach a final decision are also based on the seriousness of the disease and the extent to which there are available treatments. For that reason, what safety and efficacy will mean for a particular product and how it will be measured in a clinical setting are questions that ought to be seriously examined from the outset. The problem of designing optimal clinical trials is an extraordinarily difficult exercise in turning science into medicine, and is dealt with as a business strategy problem in Chapter 5 and in detail as a regulatory strategy problem in Chapter 8.

Safety and efficacy are also the standards for the approval of devices, including *in vitro* diagnostics; the largest category of FDA regulated human health care applications of biotechnology products currently in the marketplace. Biotechnology products other than diagnostics will rarely be classified as devices, which are defined so as to exclude any product which achieves any of its "intended purposes through chemical action within or on the body of man" or which is "dependent upon being metabolized to achieve any of its intended purposes" (21 U.S.C. 321(h)). For most *in vitro* diagnostics the concept of safety is almost identical to that of efficacy; safety problems arise out of false readings, whether positive or negative, which is the same criterion by which efficacy is measured. A diagnostic that is highly accurate is both effective and safe, while greater margins of error are problematic for both safety and efficacy. The more serious the consequences of an error, the more accuracy that should be required for approval. Thus a false positive in a test for a bacterial ear infection will lead to less serious consequences than a false positive in a diagnostic for determining whether or not a sample of breast tissue is malignant. As with drugs, the FDA regulatory process can be the most time-consuming and expensive aspect of product development, and an early focus on the desired endpoints is extremely important. Strategic planning for a company developing human health care products must include a focus on precisely how safety and efficacy will be measured and proven to gain approval to market the company's products.

§7.3(B) The FDA and New Animal Drugs

The FDA also has control over the introduction and sale of new animal drugs. For biotechnology-derived new animal therapeutics, one of the difficult issues is whether the new product will be classified as a biologic, in which case the USDA will have initial jurisdiction, or as a new animal drug, in which case the FDA will regulate it under 21 U.S.C. §360b. Animal biologics are defined as:

> viruses, serums, toxins, and analogous products of natural or synthetic origin, such as diagnostics, antitoxins, vaccines, live microorganisms, killed microorganisms and the antigenic or immunizing components of microorganisms intended for use in the diagnosis, treatment, or prevention of diseases of animals (7 C.F.R. 205.2 (2006)).

The Memorandum of Understanding between the USDA and the FDA (47 Fed. Reg. 26458-03, June 18, 1982) sets forth the agreement between the two agencies about their procedure in the case of new animal drugs and biologics. If a veterinary product has as its primary mode of action a specific immune process (as is the case for all animal vaccines), it will likely be classified as a biologic and the USDA will be responsible for approving it for interstate commerce, under the Virus-Serum-Toxin Act (VSTA), 21 U.S.C. 151–158. If the primary mechanism of action is other than through an immunological process, the product will be classified as a new animal drug and the FDA will be responsible for approving it. At the same time, the USDA and the FDA agreed to appoint liaison officers for both agencies to coordinate actions in this area and to set up a standing committee with three members from each agency to address regulatory responsibilities for new products and other issues.

If a biotechnology product is classed as a new animal drug (as is the case with bovine growth hormone, the first recombinant product of this type for which FDA approval was sought)[7] then the regulatory process is divided into two phases. First the FDA reviews the product from the perspective of veterinary medicine. In much the same way as for human therapeutics, Section 512(3)(b)(1) of the FDCA (21 U.S.C. 360b(3)(b)(1)) requires applicants for the approval of a new animal drug to submit "full reports of investigations which have been made to show whether or not such drug is safe and effective for use." In addition, the manufacturer must also submit a "proposed toler-

---

7. 58 Fed. Reg. 59946-02 (1993).

ance or withdrawal period or other use restrictions ... to assure that the proposed use of the drug will be safe" from the perspective of human consumption of any food derived from a treated animal (id at (H)). The key concept here is that of tolerance level, for it requires that the manufacturer demonstrate not only that the drug is safe and effective for the animal, but also that the amount that will be found in any derived food will be at a level safe for human consumption. For purposes of this overview it is sufficient to note that the testing required to prove that the residue (tolerance amount) is safe for human consumption can be vastly more expensive and burdensome than that necessary to show that the product is safe and effective from the perspective of veterinary medicine. As a practical matter, then, it may well be easier to plan for the introduction of new animal drug products that will be sufficiently metabolized with a short enough period of time (the "withdrawal period") so that there is no detectable residue in any food product and establishing a tolerance level is unnecessary.

§7.4 The EPA and Agricultural Biotechnology

The EPA and the USDA both have jurisdiction over various agricultural applications of biotechnology, with the EPA's jurisdiction based on its authority, under the Federal Insecticide, Fungicide, and Rodenticide Act (FIFRA) to regulate the testing and use of pesticides. To date, the two biggest applications of biotechnology to agriculture have been to the introduction of the genes encoding the insecticidal protein bacillus thuringiensis toxin ("Bt") into a great many varieties and the introduction of the gene encoding an enzyme that confers resistance to the herbicide "Roundup" (glyphosphate) into a number of others. These newly-engineered strains of corn, soy, cotton, and other commercially important crops have been enormously successful in the U.S., but created enormous controversy in Europe, where "GM" or genetically modified crops are viewed with great suspicion. A full understanding of the popular distrust of agricultural biotechnology in Europe is beyond the scope of this book,[8] but its breadth and strength are certainly significant for anyone contemplating developing new agricultural applications of biotechnology.

---

8. *See* Edward Alden & Jeremy Grant, *WTO rules against Europe in GM Food Case*, Fin. Times, Feb. 8, 2006, at 6.

§7.4(A) EPA's FIFRA Procedures for Field Testing Genetically Engineered Organisms

The Federal Insecticide, Fungicide and Rodenticide Act (FIFRA) (7 U.S.C. §§136–136y) and the Food Quality Protection Act of 1996 (FQPA, Public Law 104-170, 1996) are the statutes governing pesticides, which are defined to include "any substance ... intended for preventing, destroying, repelling, or mitigating any pest and ... any substance ... for use as a plant regulator, defoliant or desiccant." The EPA's supervision of the development of new bio-pesticides and plant regulators, like the FDA's supervision of human drug development, begins with the testing of new products, not with the sale or use of such products. By regulation, the EPA has long exempted "small-scale experimental uses of new pesticides" (generally tests involving less than ten acres in area) from the requirement that the manufacturer obtain EPA approval for an experimental use permit (EUP). That small-scale exemption from the EUP requirement is still in effect and applies to "inert" biotechnology products intended as pesticides, such as insecticidal proteins produced by genetic engineering. For example, the biotechnology company Mycogen, which is now part of Dow Agrosciences, first brought to market more persistent forms of the naturally occurring bacterial protein bacillus thuringiensis endotoxin, commonly referred to as Bt toxin. However, the EPA requires prior notification for small-scale field test experiments involving live genetically engineered plants that produce a "plant incorporated protectant" or PIP. Thus new crops that express Bt toxin or another PIP derived from a non-sexually compatible plant could not be tested without prior EPA notification. It is sufficient for this overview to note that the notice requirements for EPA's reviews of tests are extremely detailed and time-consuming.[9]

Further complicating matters is the fact that the EPA and USDA share jurisdiction over many of the agricultural applications of biotechnology. The overlapping jurisdiction of the EPA and the USDA over genetically engineered plants producing PIPs, or that otherwise are produced by biotechnology, creates substantial uncertainty and additional expense for companies planning the development of products of this type. Although the agencies have pledged to share information and otherwise attempt to coordinate their reviews, in those cases where that require approval by both agencies, such dual review is inevitably time consuming and costly. The jurisdictional relationship is complex enough to have required lengthy explanations in the EPA's and USDA's Coordinated Framework.

---

9. http://www.epa.gov/pesticides/biopesticides/regtools/biotech-reg-prod.htm.

Thus, a company developing a lawn grass genetically engineered to produce a substance that is toxic to crabgrass requires approval from the EPA under FIFRA and from the USDA under the Plant Pest Act in order to test and market its product. No prior approval is required to do small scale field-tests of a genetically engineered, more persistent form of the protein as a separate product or in a chemical solution, and marketing the product requires only EPA approval under FIFRA. Thus the regulatory pathway is much easier for isolated substances than for whole plants and for varieties produced by sexual reproduction or selection pressure rather than genetic engineering. As a result, the regulatory difficulties and international market resistance might well dictate the basic product development strategy of biotechnology companies interested in this general area.

§7.4(B) The FIFRA Standard for the Approval of Pesticides[10]

Although the EPA has been extremely cautious in granting permission to field test genetically engineered plants that combine genetic material from relatively dissimilar or non-sexually compatible plants, the FIFRA standard for the approval or registration of new pesticides is a relatively lenient one. In contrast to the FDA standard for new human or even animal drugs, which requires the applicant to submit data proving the safety and efficacy of the product, FIFRA requires the EPA to register a pesticide when, under the restrictions proposed for its use, "its composition is such as to warrant the proposed claims for it"(7 U.S.C. Sec. 136a(c)(5)(A)) and "it will perform its intended function without unreasonable adverse effects on the environment" ((7 U.S.C. Sec. 136a(c)(5)(C)). Furthermore, once approved, the registration may be canceled only if the product does cause such unreasonable adverse effects on the environment and the EPA has "taken into account the impact of the action ... on production and prices of agricultural commodities, retail food prices, and otherwise on the agricultural economy" (7 U.S.C. Sec. 136d(b)). Thus the intent of Congress in enacting FIFRA was to balance health and environmental concerns with the needs of agriculture. Although pesticide use and its impact on health and the environment have become more controversial of late, at this point changes in FIFRA's basic standard for pesticide approval, whether from biotechnology or conventional chemicals, are not imminent even though the EPA's approach to interpreting the statute may become stricter.

---

10. The EPA's position is officially set out in the Final Rule: Regulations Under the Federal Insecticide, Fungicide, and Rodenticide Act for Plant-Incorporated Protectants (Formerly Plant-Pesticides), 66 Fed. Reg. 37772 (July 19, 2001).

It should also be noted here that even after EPA registration of a pesticide (or PIP), food products containing detectable residues of the pesticide cannot be sold until the EPA establishes a tolerance level or exemption for such residues. The tolerance level, below which the presence of the pesticide substance does not constitute adulteration, is to be set at a level

> necessary to protect the public health.... [G]iving appropriate consideration, among other relevant factors, *(1) to the necessity for the production of an adequate, wholesome, and economical food supply;* [emphasis added]. (21 U.S.C. 346a).

The EPA requires significant toxicity testing to determine the potential health risks of a proposed new pesticidal substance, whether a PIP or traditional chemical. The resulting data is assessed in light of the FIFRA standard, which clearly sets forth a balancing requirement. The EPA must weigh health risks against the economics and requirements of food production. This balancing standard provides a greater tolerance for pesticide residues in food than for other types of food additives or contaminants (*see* §7.7). Here, unlike the basic pesticide approval standard, it is likely that the statute governing pesticide residues will be changed in the near future. The public outcry over Alar, which was used in the growing of some varieties of red apples, indicates the enormous public sensitivity to the problems of potential toxins and carcinogens in food. This sensitivity creates a dual edged sword for producers of biopesticides and genetically engineered pesticide resistant plants. On the one hand, if such products are clearly environmentally superior to conventional pesticides and less toxic than their chemical counterparts, as they generally are, the result may be swift public acceptance and market penetration. On the other hand, if a particular biopesticide or transgenic plant raises safety or environmental concerns, or fails to make a strong case for a health or environmental benefit to allay the general public distrust of "frankenfood," it will inevitably face stiff public resistance. The environmental and human health effects of these products should be a much greater concern for emerging biotech companies than for the huge chemical companies against whose products the new biotech products will compete.

§7.5 EPA's Regulation of Non-Agricultural Biotechnology—TSCA

The EPA has general regulatory authority over the manufacture of new "chemical substances" and "significant new uses" of existing chemical substances, other than pesticides, foods, food additives, drugs, cosmetics, and medical devices. The purpose of TSCA (15 U.S.C. §2601 *et seq.*) is to provide some regulatory control over the introduction of potentially hazardous chemicals into the workplace and the environment. By definition, TSCA is a "catch-all" statute, which fills in the gaps left by other agencies' authority over new chemical substances, such as pesticides or food additives that are regulated under specific statutes. The intent of Congress in enacting TSCA was to provide at least a minimum level of regulatory consideration of the environmental and health effects of new chemical substances prior to large-scale production for industrial and other purposes. The need for such a back-up statute has most recently been clearly demonstrated by the looming global environmental disaster involving the depletion of the earth's ozone layer by chlorofluorocarbons, which have been used as refrigerant and aerosol propellant gases. Another example of the staggering costs that can arise from large-scale production of chemicals that slip through the cracks of other regulatory statutes is the continuing cost of cleaning up PCBs, which were primarily used as insulating fluids in electrical applications. Thus, TSCA represents Congress's recognition that industrial chemicals can have significant health and environmental consequences even without intended uses that involve direct human or animal consumption.

In its statement in the Coordinated Framework, the EPA announced that it would also use TSCA as a back-up authority over the manufacture and release into the environment of genetically engineered microorganisms. To support its assertion of TSCA jurisdiction over this area of biotechnology, the EPA asserted that it would consider genetically engineered DNA sequences and the whole microorganisms containing genetically engineered DNA or RNA sequences to be new chemical substances. Its principal impact will be on the development of genetically engineered microorganisms for industrial processes. One such potential use is for bioremediation of hazardous waste *in situ*, or at the site of contamination. Because that application requires the environmental release of genetically engineered microorganisms, it will encounter particularly rigorous scrutiny from the EPA under TSCA. The basic TSCA procedures and standards are surveyed in this section.

§7.5(A) TSCA's Scope, Procedures, and Standards

TSCA applies, as previously indicated, to the manufacture or importation of new chemical substances or to significant new uses of chemical substances. Section 2602 defines chemical substance as "any organic or inorganic substance of a particular molecular identity...." As previously indicated, the EPA is interpreting that definition to include whole, living, genetically engineered microorganisms or, at least, the genetically engineered nucleotide sequences within such organisms. A "new chemical" subject to the requirements of TSCA is any chemical not already listed under 15 U.S.C. 2607(b), the EPA's list of existing chemicals, while 15 U.S.C. §2604 (a)(2) requires determination of whether or not there is a "significant new use" by reference to such factors as changes in the amount of the substance manufactured, changes in the amount of the substance entering into the environment, and changes in the amount of human exposure.

§7.5(B) Requirement of a Premanufacture Notice (PMN)

Once a company intends to produce a new chemical substance or commence production of a chemical substance for a significant new use, it must give the EPA at least 90 days prior notice in the form of the Premanufacture Notice, or PMN (Sec. 2604(1)). By regulation, persons who manufacture or import small quantities solely for research and development, or are classed as small manufacturers (less than 500 kilos annually at a specific plant), are among those exempted from the requirement of filing a PMN (40 C.F.R. 712.25). However, in its statement in the Coordinated Framework, the EPA announced that the R&D exemption would not apply to any research involving release into the environment of genetically engineered microorganisms and the small manufacturer exception would not apply to any manufacture of such organisms for commercial purposes (although the agency announced that it was considering exempting some contained uses of such "new chemical" microorganisms).

The information required in the PMN is set forth in Section 2604(d)(1) and the EPA's form entitled "Premanufacture Notice for New Chemical Substances" is included in the Code of Federal Regulations as 40 C.F.R. 720 Appendix A. The heart of the PMN is the requirement of the manufacturer to provide "any test data in the possession or control of the person giving such notice ... and a description of any other data concerning the environmental and health effects of such substance, insofar as known ... or ... reasonably ascertainable" (15 U.S.C. 2604(d)(1)(B–C)). If the data submitted is not adequate for the EPA to predict its health and environmental effects and further testing is needed to develop adequate data, Section 2603 requires the EPA to respond to the PMN with a "test rule" which requires that "testing be conducted ... to develop data with respect to the health and environmental effects for which there is an insufficiency of data" (15 U.S.C. 2603(a)(B)(2)). Until there is sufficient information "to permit a reasoned evaluation of the health and environmental effects of a chemical substance with respect to which notice is required ... the Administrator may issue a proposed order to ... prohibit or limit the manufacture, processing, distribution in commerce, use, or disposal of such substance ..." (15 U.S.C. 2604(e)(1)(A)(i–ii)).

One of the more controversial and difficult areas of biotechnology regulation has been the EPA's attempt to formulate a general policy for the application of TSCA to genetically engineered microorganisms. The assertion of jurisdiction has placed the Agency in the position of needing to formulate appropriate "test rules" for generating the data necessary to predict health and environmental effects of the environmental release of genetically engineered microorganisms. Although most such releases involve organisms without any pathogenic characteristics and thus present minimal cause for concern over direct health effects, there is enormous uncertainty over the effect of introducing a new microorganism into any ecological system. The uncertainty extends to a fierce debate over the proper criteria to use in evaluating the risks and even whether or not any data would be sufficient to allow a TSCA finding of "no unreasonable risk" to the environment. For purposes of this overview the implications are relatively clear. As in other areas, the regulatory obstacles to the use of genetically engineered microorganisms in the environment might well necessitate the consideration of alternative methods of achieving the same ends. In the case of bioremediation of hazardous waste, for example, commercial success may be more easily obtained by screening microorganisms produced under the "natural" selective pressure of laboratory exposure to environments, which contain increasing levels of hazardous waste contamination. Microorganisms that had evolved to metabolize toxins under those laboratory conditions could be commercialized with less scrutiny from the EPA, than by organisms engineered to produce similar characteristics. The EPA did, after years of debate, release its "Microbial Products of Biotechnology: Final Regulations Under the Toxic Substances Control Act" in 1997.[11]

§7.6 The USDA and Agricultural Biotechnology

Biotechnology applications for agriculture include veterinary biological products, defined in §7.3(A), such as animal vaccines produced by a variety

---

11. 62 Fed. Reg. 17910 (April 11, 1997). *See also* 40 C.F.R. §172.43 *et seq.*

of biotechnology techniques, as well as the genetic engineering of plant and animal species themselves. Such techniques promise substantial gains in agricultural productivity while raising significant environmental, economic, and human health concerns. The testing and marketing of animal biologics is regulated by the USDA under the Virus-Serum-Toxin Act (VSTA) (21 U.S.C. §§151–158), while the testing of genetically engineered plants and some microorganisms is regulated by the USDA under the Plant Protection Act (7 U.S.C. §7701 *et seq.*), which supercedes the earlier Plant Quarantine and Plant Pest Acts.

Experimentation with genetically engineered livestock, while raising significant issues of public policy and food safety, are not regulated at the experimental stage, but are regulated as food additives or adulterants before approval for market, which requires participation by the USDA under the Food Safety Inspection Service (FSIS). The food safety issues are discussed in §7.7 Food Products Affected by Biotechnology. This section briefly examines the USDA's regulatory authority over the testing and marketing of veterinary biological products and genetically engineered plants and microorganisms.

§7.6(A) USDA's Regulation of Veterinary Biological Products—VSTA

The USDA's regulatory authority under VSTA extends to all veterinary biologics, defined in 7 C.F.R. §205.2 as "all viruses, serums, toxins, and analogous products of natural or synthetic origin, such as diagnostics, antitoxins, vaccines, live microorganisms, killed microorganisms, and the antigenic or immunizing components of microorganisms intended for use in the diagnosis, treatment, or prevention of diseases of animals." As previously noted, the USDA and the FDA have agreed to give the USDA primary authority over those biotechnology-derived products within the definition of biologic that have an immunological mechanism of action, while giving the FDA jurisdiction over recombinant products that do not involve an immunological mechanism of action. In practice, vaccines have been the principle concern of the USDA under VSTA while other protein products have been the concern of the FDA. In its policy statement (51 FR 23339) the USDA divides potential biological products into three categories, based on the underlying technology.

Category one includes vaccines containing no "live" virus, produced by cloning the gene for one or more viral coat proteins and using those proteins to stimulate the immunological response in lieu of the whole virus, or by producing anti-idiotypic antibodies (see Chapter 2—The Basic Science of

Biotechnology) that mimic the shape of viral antigens. Such products raise no new safety or environmental concerns.

Category two consists of vaccines produced by the genetic engineering of live viruses to produce attenuated forms of the virus which retain its antigenic or immunological potential without its pathogenic properties.

Category three products are vaccines produced by using a single, non-pathogenic virus, such as vaccinia, as a "vector" to carry the genes for one or more antigenic coat proteins from another pathogenic virus or viruses. Only those vaccines based on genetically engineered live viruses, categories two and three, raise any real concerns about the spread of disease or potential environmental impact. Nevertheless, the USDA approved a category two vaccine for pseudorabies, a swine disease, as early as 1991.[12] Category three products present the most novel questions of safety and environmental impact, are naturally more difficult to test; however, field tests of a category three vaccine have been conducted for a vaccinia-based rabies glycoprotein vaccine (56 FR 19635-01 April 29, 1991) and ultimately its wider use.[13]

Under VSTA, the USDA's regulatory authority is invoked by the transport, or shipping, of any veterinary biological products. Because VSTA is expressly limited to the movement, whether interstate or intrastate, of veterinary biologicals, the development of a veterinary biological is essentially unregulated until it is necessary to conduct testing of the product at a site outside the research laboratory at which it is produced. For obvious reasons, animals can be brought into the laboratory and kept there for vaccine testing, with only the approval of the institution's animal welfare committee, unlike human vaccines, which cannot be tested on humans without prior FDA approval. After initial laboratory testing, permission to ship an experimental veterinary biological may be granted only upon a showing that the experiment will be conducted with appropriate safeguards against the spread of disease (9 C.F.R. §103.3 (a)–(g)). A similar regulatory structure, which includes testing of a veterinary biological by the USDA itself, governs the importation of veterinary biologicals into the United States (9 C.F.R. §104.2).

As a practical matter, the principal "spread of disease" issue confronting the transport of an unlicensed experimental veterinary biological product arises in the context of vaccines using whole, live viruses and therefore raising the

---

12. *See* BIOTECHNOLOGY BUS. NEWS, Oct. 25, 1991, *Syntro given go-ahead for pseudorabies vaccine*.

13. *See, e.g.,* BF Blackwell et al., *Exposure time of oral rabies vaccine baits relative to baiting density and raccoon population density*, 40 J. Wildlife Diseases 222–229 (April 2004) (abstract available on PubMed).

possibility of the transmission of the vaccine being tested to non-test animals. The production of live, attenuated virus vaccines antedates the introduction of modern biotechnology techniques and, to a large degree, genetically engineered or recombinant viral vaccines do not raise qualitatively different questions from those produced by traditional techniques. Nonetheless, the USDA has proceeded with some caution in determining that the outdoor testing of recombinant viral vaccines poses no health or environmental risks. The approval by the USDA of a veterinary biological product for commercial, non-experimental transport and marketing takes the form of a Product License, which is granted on the basis of purity, safety, potency, efficacy, and labeling of the product. In essence, the issues are the safety and efficacy of the product and the quality control procedures of the facility licensed to produce it. In addition to data concerning the results of tests of the vaccine in animals, the USDA requires applicants for biological licenses to submit an "outline of production" describing the procedures used to manufacture the vaccine and to demonstrate their quality control by preparing three separate, consecutive batches of the vaccine, produced in the facility to be licensed, for testing at the National Veterinary Services Laboratory. Once the field test data and production methods have been reviewed and the samples have proven consistent and pure (free from contamination), a U.S. Veterinary Biological License may be issued to authorize further shipments for commercial purposes. The track record of the USDA in granting licenses is now sufficiently well established that biotechnology entrepreneurs should understand that the principal concern of companies considering the manufacture of veterinary biologicals is more likely to be market size than regulatory problems.

§7.6(B) USDA's Regulation of Genetically Engineered Microorganisms and Plants—PPA

The USDA also has regulatory authority over the development of genetically engineered microorganisms, invertebrate animals, and plants under the Plant Protection Act (PPA) (7 U.S.C.§7701 *et seq.*). Under these statutes, which replaced the prior Federal Plant Pest Act and Plant Quarantine Act, the USDA is given the authority to control the movement into or within the United States of any plant or article that may be or may contain a plant pest. The purpose of the PPA is to provide the Secretary of Agriculture the authority to facilitate commerce in plants and potential plant pests such as noxious weeds while reducing, "to the extent practicable ... the risk of dissemination of plant pests or noxious weeds." Just as the experience with PCBs provides justification for EPA's authority to regulate new chemicals under TSCA, the experience of the invasion of the southeastern U.S. by kudzu, among many other examples, provides ample justification for the USDA's authority under the PPA.

§7.6(C) The Scope, Standards, and Procedures of the PPA

The PPA applies to any plant pest, defined as "any living stage of ... [protozoan, nonhuman animal, parasitic plant, bacterium, fungus, virus or viroid, infectious agent or other pathogen] that can directly or indirectly injure, cause damage to, or cause disease in any plant or plant product." The issue for genetically engineered organisms is whether the organism may be a plant pest, that is, in some way harm or injure plants or plant products. While the USDA reserves the right to make a determination that any new or previously unclassified organism is a plant pest, the rigorous scrutiny is automatic "when an organism or product is altered or produced by genetic engineering and one or more of its constituents (donor, vector/vector agent, or recipient) comes from a family or genus of organisms known to contain plant pests." The USDA has revised its regulations (7 C.F.R. §340.0 *et seq.*) to make it clear that any genetically engineered organism which either contains genetic material from a genera or taxa designated in 7 C.F.R. §340.2 as a plant pest or is an unclassified organism or product which the USDA has reason to believe is a plant pest is a "regulated article" under the FPPA. Any such "regulated article" cannot be shipped within the United States or released into the environment without a USDA permit.

Under 7 C.F.R. part 340, anyone seeking to move within the United States a genetically engineered organism that may be a plant pest must submit a permit application to the USDA. In the case of a proposal to release into the environment any genetically engineered organism that may be a plant pest, the application for a permit must be filed at least 120 days prior to the intended release. Within 30 days of the initial application, the USDA must determine whether or not its review requires additional information. The USDA then must either grant or deny a permit within 120 days of the receipt of a completed application or of any further information required within the 30-day initial review period (unless an environmental impact assessment is required).

The USDA has absolute authority under the FPPA to withhold a permit or to condition its granting of a permit on compliance with such conditions as the USDA believes necessary to protect against environmental harm from the introduction or spread of a new plant pest. There has been no litigation over the scope of the USDA's power under the FPPA, and it seems clear that in the event a permit is denied or an applicant believes the permit conditions are unduly restrictive, the principle recourse is to appeal the USDA's decision within ten days. Alternatively, an applicant for an FPPA permit may attempt to provide further evidence sufficient to support a petition under 7 C.F.R. §340.2 to remove an organism from the class of plant pests. Here again, a company that seeks to produce a product that might be considered by the USDA as a plant pest may wish to consider consulting with the USDA very early in the product planning process so as to minimize the problems that might be encountered in obtaining regulatory approval for the testing and marketing of the product.

§7.7 The Regulation of Food Products from Genetically Engineered Plants

All food is subject to some form of safety regulation. Even common, long-standing foods such as canned peaches are subject to regulation for good food manufacturing practice, contamination, and accurate labeling. If a food is of "natural biological origin," was commonly consumed in the United States prior to 1958, and has not been modified by any process that was introduced after 1958, then it is essentially only regulated for manufacturing practices and labeling. However, marketing a food that has been modified by a process introduced after 1958 (obviously genetically engineered foods come under this category) may entail more elaborate regulatory review of some kind. Understanding the basic regulation of food safety by the FDA for such "new foods" requires a brief review of three fundamental food safety concepts: adulterated food, food additive, and generally recognized as safe.[14]

§7.7(A) Adulterated: The First Concept of Food Safety

Adulterated food may not be sold or transported in commerce. Under 21 U.S.C. §342(a)(1) a food is adulterated if it bears any added harmful substance other than an approved food additive, an approved residue of a new animal drug, or on raw foods, a permitted level of pesticide residue. In turn, under 21 U.S.C. §346, all poisonous or deleterious substances added to food are statutorily defined as unsafe except to the extent that their use is required or their presence cannot be avoided despite the use of good manufacturing practices. In the event that an added poisonous substance cannot be avoided, then the Agency (the FDA in the case of substances other than pesticides, the EPA

---

14. The discussion here assumes that the plant has previously been approved, as outlined above, by the USDA under the Plant Pest Act and by the EPA under FIFRA.

in the case of pesticides) must set a limit on the amount of the substance that can be contained in any food product, at a level necessary to protect human health. Therefore any substance that is not a natural component of a food renders the food adulterated if it may be harmful to human health, unless the added substance is either a permitted additive or residue.

Thus the producer of a food product containing an ingredient, new chemical component (as for example a plant oil that is modified in structure by genetic engineering), or added substance faces a fairly straightforward decision tree, although particular decisions may be difficult to make. If the product is a biopesticide or new animal drug, it can only be used in conformity with the tolerance set for it. If it is any other added substance that "may be harmful to health," it can only be used if it is a permitted food additive or a type of added substances clearly not potentially harmful to human health and, therefore, generally recognized as safe (GRAS), see §7.7(C) below. The most difficult decision is whether or not an added substance "may be harmful to health," or creates a reasonable probability of harm. For food plants affected by genetic engineering, that decision is considered in some detail in the FDA Statement of Policy (§7.7(D) below).

§7.7(B) Food Additive: Added Substances That Do Not Adulterate

A food additive is any substance that will become part of or affect the characteristics of any food (other than a permitted residue of new animal drugs or pesticides[15]) that is not generally recognized as safe by experts in the field (21 U.S.C. §321(s) (2006)). The purpose of genetically engineering a food plant or food animal generally would be to achieve an effect which is included within the statutory definition of food additive, that is to become a component of the food or to affect its characteristics or composition in any other way.

In the case of genetically engineered food products, the added substance is the genetic material and the resulting protein (or in some cases, a changed level of protein, or changed product of a protein-regulated metabolic pathway). Thus, the result of genetically engineering a food plant or food animal would appear to be one of three statutory alternatives: either the resulting food is adulterated; the gene or its resulting product is the subject of a food additive regulation permitting and prescribing its use; or, the product is neither adulterated nor the subject of a food additive regulation because the gene and its resulting product are generally recognized as safe.[16]

---

15. There are a few more exceptions, but not relevant to this overview.
16. It is possible to argue that a genetically engineered plant contains no added substances at all, but is merely a new variety of an existing species. For example, it could be

§7.7(C) Generally Recognized as Safe (Between Additive and Adulteration)

The concept of generally recognized as safe (GRAS), by experts qualified by scientific training and experience to evaluate the safety of food additives (*see* 21 C.F.R. §170.30 (2003)) is the final basic concept in food safety regulation. A food is not adulterated, despite containing an added substance not subject to a food additive regulation, if the added substance is GRAS. Under the regulations, GRAS status can be achieved in one of two ways, both of which are arguably relevant to food biotechnology.

The first route to GRAS status can be based on evidence that a substance was commonly (and safely) consumed in the United States prior to January 1, 1958. For example, if one were to engineer a food plant, for example corn, to contain a protein usually found in another plant food such as soybeans that was commonly consumed prior to 1958 the resulting food product may fall within this first category of GRAS. While corn containing soy protein did not exist prior to 1958, the soy protein gene and its related protein were commonly consumed in food prior to 1958. The common consumption experience must still be of the sort that would be accepted by experts as establishing the safety of the food additive, but it may obviate the need for independent, prospective, controlled studies of the food additive. Thus, if experts would agree that the soy protein must be safe because of its widespread consumption prior to 1958, the engineered corn containing soy protein should be GRAS.

The other route to GRAS status is through complete and thorough scientific studies of the proposed food ingredient and the resulting food product (21 C.F.R. §170.30(b) (2006)). For obvious reasons, this route to GRAS is considerably more expensive, and therefore less desirable, than basing GRAS status on widespread consumption prior to 1958. The scientific studies required for food additives, incorporated by reference into the GRAS process, "should include detailed data derived from appropriate animal and other biological experiments ..." 21 C.F.R. §171.1(c) (2006). Nevertheless, for some biotechnological food products, this alternative may be, when feasible, a better route to market approval than a full-scale petition for a new food additive regulation. If one were to engineer a novel enzyme into corn to produce a low-calorie corn oil, it is likely that the resultant food product would need to gain GRAS sta-

---

argued that Bt producing corn contains no added substance but is just an unusual strain of corn. However, that position, which would foreclose any FDA role in the oversight of the introduction of these new varieties, has been rejected and is not a viable interpretation at this time.

tus through scientific studies of the toxicology of the oil in mammals (or the enzyme and resulting oil would need to be approved by the food additive petition process). These studies are, in part, the sort of high-dose bioassays commonly done in rodents and occasionally on other species.

If the studies required for GRAS status (other than for a substance commonly consumed before 1958) are in many ways the equivalent of those required for a food additive petition, what are the advantages of seeking GRAS status rather than submitting a food additive petition? There are three reasons why GRAS status is preferable. First, GRAS is self-executing. Although the FDA has adopted the principle that GRAS determinations require rigorous testing and publication of studies, the sponsor of an ingredient is the one who will decide whether the studies justify the conclusion that the substance is GRAS. Second, GRAS approvals are generally broader than the very narrow uses permitted for food additives (and subsequent expanded uses of GRAS substances are more easily obtained). Third, when the sponsor of an ingredient seeks FDA affirmation of its GRAS status, the FDA review process itself, is likely to be quicker and less cumbersome. According to James T. O'Reilly, the author of the treatise FOOD AND DRUG ADMINISTRATION (§11.05), the agency-created affirmation procedure is most often used for pre-1958 food substances altered by a process developed since 1958, a category which may well include a great many genetically engineered foods.

Under the FDA's Statement of Policy for New Varieties of Food Plants (57 Fed. Reg. 22984 (1992), the self-determination of GRAS, or in some cases GRAS affirmation, will be the preferred path for any biotechnological product that can qualify for such treatment. This self-determination that an added substance is clearly not one that "may be harmful to human health" requires no regulatory pre-market review whatsoever, although the producer who introduces such a product obviously does so at its own peril. To reduce that risk somewhat, the FDA announced a process by which GRAS self-determinations could be submitted for a speedy regulatory response although the FDA's response would not provide formal regulatory acceptance of the submitter's determination.[17] This GRAS notification process, which applies to all foods, plays a particularly significant role in the FDA's policy on foods from plants created by biotechnology, discussed in the next section.

---

17. The FDA made this clear in the 1997 Federal Register notice of the proposed rule: "This response, however, would not be equivalent to an agency affirmation of GRAS status because FDA would neither receive nor review the detailed data and information that support the GRAS determination." 62 Fed. Reg. 18939 at 18951.

§7.7(D) Foods Derived from New Plant Varieties Created by Genetic Engineering—the FDA Statement of Policy

Many applications of genetic engineering to food plants will be intended to affect the taste, quality, or growing characteristics of the plant (other than for pest-resistance, discussed in §7.7(E)) and may be covered either as a food additive or conversely as an adulterate under the Food, Drug and Cosmetic Act (FDCA), 21 U.S.C. §§348 *et seq.* (1988). The FDA Statement of Policy (57 Fed. Reg. 22984 (1992)) and the Proposed Rule for Premarket Notification, 66 Fed. Reg. 4706 (2001)) are an important effort by the agency to clarify the rules under which such products will be regulated. The Statement provides an analytic framework for genetically engineered food plants that contains three possible outcomes: "no concern"; "consult FDA" (which includes both the possibility of informal regulatory approval as well as the requirement for formal food additive treatment); and, "new variety not acceptable."

The basic positions taken in the Statement are simple. First, genetically engineered food plants will not be subjected to a *per se* requirement of special labeling, rather any labeling requirements will depend on the effects of the genetic engineering.[18] Second, the FDA declared that nucleic acid sequences, or genes themselves, when introduced into plants to produce an effect, are presumed to be safe (57 Fed. Reg. 22984, 22990 (1992)). Thus the FDA will be looking at the nucleic acid sequences' effects on the composition of the food product rather than at the DNA itself. With respect to the effects on composition, the FDA outlined its position on four major categories of possible effects: (1) alteration of protein levels of proteins native to the plant; (2) introduction of a protein not native to the plant; (3) changes in carbohydrates or introduction of new carbohydrates; and, (4) changes in or introduction of new fat or oil constituents of the plant. Each of these is described below.

(1) Alteration of the levels of a protein native to the plant. Some efforts at genetically engineering food plants may be thought of as intra-generic, where the objective is to increase or decrease the production of a particular protein native to the plant, rather than the introduction of a new gene sequence to

---

18. Although the no "*per se*" labeling requirement was the position taken in the FDA Statement originally published in May of 1992, the FDA elaborated on the labeling issue raised by the converse question, which is whether labeling is permitted to indicate that food is not produced from genetically engineered food varieties. *See* Guidance for Industry: Voluntary Labeling Indicating Whether Foods Have or Have Not Been Developed Using Bioengineering, Draft released for comment January 2001 (http://www.cfsan.fda.gov/~dms/biolabgu.html). Such voluntary labeling is permissible and increasingly common.

produce a new constituent not previously found in that species. An example of that kind of intra-generic genetic engineering was Calgene's use of an "antisense" nucleic acid sequence to decrease the production of the tomato enzyme that causes the tomato to rot. Thus Calgene used antisense genetic engineering to control the levels of substances ordinarily found in that edible plant species. In general, when the effect of genetic engineering is to alter the percentage composition of proteins native to that plant, the FDA's concerns are whether or not such changes may have an effect on toxicants present in that species and whether or not the result is a significant change in the nutritive value of that food. In the case of intra-generic genetically engineered plants, when there is no resulting increase in toxicant levels or the overall nutritive value of the food, the FDA's position is one of "no concern."

(2) Introduction of a protein not native to the plant (inter-generic genetic engineering). A variety of desirable characteristics may be transferred to a plant from another plant species, or even from a bacterial or animal species. For example, gene sequences encoding proteins conferring drought-tolerance, ordinarily found in a desert plant species, might be engineered into a variety of domestic wheat. For such "inter-generic" genetically engineered food plants, the Statement again quite properly looks to the characteristics of the introduced protein to guide the regulatory process. If the protein will be found in foods produced from the plant and is not one with a history of safe use in food (as may well be the case with the hypothetical desert plant "drought proteins"), then the principal questions raised by the FDA focus on the effect of the new protein on levels of any toxicants native to the engineered plant and the allergenicity and toxicity of the introduced protein. As the FDA Statement notes, proteins are, with few exceptions, broken down quickly in the digestive tract into small polypeptides or amino acid chains, regardless of their amino acid sequence or three-dimensional folded structure.

In many cases, the issue of any resulting increase in a plant's allergenicity or toxicity will be determined from the similarity of the introduced protein to other proteins, its biological function, the extent to which it resists being broken down by digestive enzymes, and any reports of its toxicity. What is most significant is the FDA's initial position that, with the exception of a limited list of known allergenic proteins and an even more limited list of known toxic proteins with enzymatic function, introduction of new proteins into food crops does not automatically require Agency review of a manufacturer's food safety data. In regulatory terms, although the introduced protein is clearly an "added substance," the producer of the genetically engineered food may make its own determination that the food is not one that "may be harmful to human health" and therefore is "generally recognized as safe." Thus, although the FDA does request premarket notification, the FDA does not require a full pre-market review of all inter-generic genetically engineered food plants, even those containing proteins not previously found in foods (in statutory terms, without a "history of safe use in food"). Instead, the Statement allows the producers of such foods to determine the safety of the resulting food.

Since the original policy was announced, three important changes have been made. The first, mentioned in §7.83 above, was to allow manufacturers to submit to the FDA a notice of their GRAS self-determinations and receive an agency response, albeit one that is short of an agency ratification of that determination. The second change was announced in late 2004 in the form of the Draft Guidance: Recommendations for the Early Food Safety Evaluation of New Non-Pesticidal Proteins Produced by New Plant Varieties Intended for Food Use (http://www.cfsan.fda.gov/~dms/bioprgui.html). According to the Draft Guidance, the FDA is asking for prior notification and an opportunity to evaluate the field-testing of any new non-pesticidal protein that has been engineered into any food plant variety. The FDA's reasoning for seeking this early role was the risk that field-testing of a corn plant with a novel protein, even though not intended for human consumption, could produce pollen that would spread the trait to food corn crops elsewhere in a large area. That possibility was of sufficient concern that the FDA wants an opportunity to preview that any open-air testing of food plants incorporating novel proteins.

The third change in the FDA's biotech food policies, announced in the 2001 Proposed Rule and recommended in the Guidance for Industry: Recommendations for the Early Food Safety Evaluation of New Non-Pesticidal Proteins Produced by New Plant Varieties Intended for Food Use, June 2006 (http://www.cfsan.fda.gov/~dms/bioprgu2.html), would require manufacturers to submit a premarket notification for "any bioengineered food, including a bioengineered food derived from a new plant variety modified to contain a pesticidal substance" unless it had previously been approved or accepted by the FDA. This notice, required 120 days prior to marketing, also would not result in a formal FDA authorization, but rather give the FDA time to express any concerns. Regardless of whether or not the FDA expresses no concern, if a producer of such a food is wrong and the protein turns out to be allergenic or toxic, the FDA can seize the food as adulterated and there is the potential for enormous civil liability (see §11.6 for a discussion of product liability for biotechnology food products).

Since it is unlikely that any producer of genetically engineered food plants would choose to introduce a protein known to be toxic or allergenic, or even one similar to known toxic or allergenic proteins, in most cases intergeneric genetically engineered plants will not undergo formal pre-market review by the FDA. However, for a protein without a history of safe use in food that is likely to be a "macroconstituent" in the human (or animal) diet, the FDA will require premarket consultation in addition to the premarket notification, even if the protein is not from a donor species commonly allergenic, is not reported to be toxic, and is of the type of protein ordinarily well-digested in humans. The basis for this requirement is unclear, since the Statement itself concludes "from a nutritional standpoint, the amount and quality of total protein in the diet, rather than of any particular protein, is of greatest significance." Nevertheless, the Statement clearly distinguishes between proteins likely to be "consumed at a substantial level" and states that "[d]ietary exposure to such proteins should be considered." This is a substantial departure from the traditional approach to food additive safety, which does not exempt additives from premarket review simply because they will be used in only a few foods or in trace amounts. While it is certainly true that the less exposure the less the resulting harm, the Statement's use of the concept of dietary "macroconstituent" has no apparent basis in the statute and may well be a focus for criticism of the FDA's policy or even a challenge by a consumer food safety group.

(3) Changes in carbohydrates or introduction of new carbohydrates. While DNA itself only encodes proteins, many proteins have enzymatic or metabolic functions, operating on other substances to produce carbohydrates or lipids (fats and oils), which are the other major categories of molecules found in plants and animals. Thus one of the results of genetically engineering a food plant can be to affect a metabolic pathway, resulting in a change in a carbohydrate or lipid produced by the engineered organism. Plant varieties might be engineered to contain new or modified carbohydrates because of the characteristics those carbohydrates contribute to the food or, in some cases, the new or modified carbohydrate might be extracted to be used as an ingredient in other foods.

In the case of such carbohydrates, the Statement raises only two questions: does the resulting carbohydrate contain any structural features not ordinarily found on food carbohydrates and, if the carbohydrate is likely to be a macroconstituent of the diet, are there any changes that are likely to affect digestibility or nutritional qualities? Although the two questions present a relatively simple decision tree, the net result may well be that new carbohydrates are treated more stringently than new proteins. There may be several reasons for that greater scrutiny. First, in contrast with the general digestibility of most proteins, some complex carbohydrates with unusual functional groups are not so quickly broken down and, in fact, some complex carbohydrate polysaccharides are toxins. The properties of a complex carbohydrate containing an unusual structural feature or functional group would be difficult to predict.

Small changes in simpler, easily digested carbohydrates (for instance to enhance sweetness) are likely to fall into the "no concerns" category, and allowing the producer to escape pre-market review and make its own determination that the substance is safe. Therefore, a slightly altered, novel, genetically engineered carbohydrate that has no unusual structural groups and is of the same digestibility and nutritional value as common food carbohydrates might avoid FDA pre-market review, based on the manufacturer's self-determination of GRAS.

(4) New or modified fats or oils (lipids). The Statement provides a slightly different regulatory scheme for fats or oils from the one described above for new or modified carbohydrates. The only two questions raised in determining whether or not FDA consultation or pre-market review may be required are slightly different from the questions raised for carbohydrates. For lipids, the first question raised is whether or not the resulting fat or oil will be a macroconstituent of the diet. Consultation is required for such macroconstituents regardless of whether or not there are changes in digestibility or nutritional value. Second, rather than raising the issue of "structural features or functional groups," the Statement asks simply whether the fatty acids produced in the new variety are unusual or toxic. If the new or modified lipid is not unusual or toxic and will not be a macroconstituent of the diet, then the producer can again escape pre-market review and make its own determination of safety.

Since the FDA Statement was first published in 1992, foods from plant varieties produced by genetic engineering have entered the marketplace in enormous quantities. Farmers have found such crops to be economically valuable and, as a result, the American markets for soy and to a lesser extent corn, are now dominated by genetically engineered varieties.[19] Several factors may be credited with the success of agricultural biotech in the U.S. First, as in other areas these products of biotechnology have been regulated within the statutory and regulatory frameworks that were established prior to the development of recombinant DNA technology. Second, the FDA Statement emphasizes the principle that genetic engineering is a method and that regulation of food is aimed at the product, not the method by which the crop strain was originally derived. Third, genetic engineering generally results in more precisely understood and controlled changes in the derived crops, as compared with older methods of plant hybridization and crossing. Nevertheless, despite

---

19. In 2002, 75 percent of U.S. soy and 32 percent of U.S. corn was from biotech varieties, Hannah Wolfson, *Corn-Belt Farmers Find Modified Crops Tough Sell*, B.GLOBE, Feb. 20, 2003 at A3.

the fact that millions of tons of foods containing genetically engineered soy and corn products have been consumed with absolutely no adverse consequences, this record of safety apparently still fails to satisfy consumer advocates whose concerns stem largely from a lay, qualitative risk perception of the risks of genetic engineering.

§7.7(E) Food Plants Engineered to Produce a Biopesticide or Treated with a Biotechnology Derived Biopesticide

One of the most successful applications of biotechnology to food plants has been the introduction of pest-resistance characteristics. Monsanto and others have successfully introduced transgenic plants using the gene encoding the biopesticide bacillus thuringiensis endotoxin (Bt toxin, effective against lepidopteran insects). Food plants that have been genetically altered to resist pests or disease, or which have been treated with genetically engineered biopesticides or biological control agents, are regulated by the EPA under FIFRA (7 U.S.C. §136). The EPA also has jurisdiction over the food safety issues of those plants under the FDCA. The FIFRA definition of pesticide is broad enough to cover all pest-control agricultural uses of biotechnology. Bacillus thuringiensis toxin, the dominant "plant incorporated protectant" or any other PIP is a biological control agent under FIFRA (see 40 C.F.R. §152.3 (2004)) and would be examined from an environmental safety perspective for market approval. 7 U.S.C. §§136–136y (2006).

If a genetically engineered virus-resistant food plant was determined not to threaten unreasonably adverse environmental effects, the next issue would be the food safety question. For pesticides that will be detectable in food products (in our example, the viral-resistance gene and its protein are arguably detectable residues) the food safety regulatory problem is one of setting a tolerance level by determining the residual level that is acceptable. For pesticide tolerance levels for raw foods, the EPA establishes a maximum tolerance level that considers public health and whether or not "[u]se of the pesticide chemical that produces the residue is necessary to avoid a significant disruption in domestic production of an adequate, wholesome, and economical food supply" (21 U.S.C. §346a(b)(2)(B)(iii) (2006)). Pesticide residues that are below the tolerance level do not constitute adulteration. Further, under the so-called "pass-through" provisions of the FDCA (21 U.S.C. §342(a)(1)(c) (2006)), the same tolerance level applies if the raw food containing the pesticide residue becomes an ingredient in a processed food, for example a tomato in tomato sauce or on a frozen pizza. The raw food tolerance standard, which expressly mandates consideration of economic benefits as well as health costs, is clearly less stringent than the standard for general food additives. This somewhat more lenient approach to pesticide residue tolerances than to direct food additives may be due to an implicit assumption that, on the whole, pesticides provide greater benefit to the food supply than do food additives.

§7.7(F) Conclusion: Risk Perception and Food Biotechnology

If glyphosphate resistant soy and Bt-producing corn are so unlikely to pose any risk, why, besides twentieth century Luddism, do genetically engineered foods (commonly "GMO") provoke such controversy? Understanding the controversy fully requires not merely an understanding of the scientific and regulatory issues, but also an understanding of the dynamics of public risk perception and the influence of that perception on the legal and regulatory processes. The FDA's response to the voluntary labeling of products as *not* containing food derived from biotechnology and the desire of companies to promote their foods in that way indicates the problem. In fact, a label that states that food does or does not contain genetically engineered products might be more misleading than no labeling at all. As the FDA's Policy points out, the question is not genetic modification *per se*, but the real difference between the original variety and the new variety. The average consumer would be mystified by labeling that actually disclosed the gene transformation event, the resulting protein, the tests used to assure the protein's safety, and so forth. A label that says "GMO free" on conventionally grown foods containing an undisclosed chemical pesticide residue may mislead the uninformed about relative safety more than the absence of any labeling at all. Nevertheless, the continuing controversy makes it clear that the prudent biotechnology company will must consider the probable public response to its product.

For better or for worse, our system is one in which both the Congress and the executive branch agencies, which design and implement the regulatory framework for biotechnology, are sensitive to public concerns. While it is all very well for a biotechnology company, rooted in a deep faith in science in general and its own science in particular, to take comfort in the sort of objective risk assessment framework exemplified by the FDA Statement of Policy for New Plant Varieties, public risk perception is highly qualitative rather than objectively quantitative.

For example, consider this table, which is an excerpt of a longer table that appeared in an article by Richard Wilson in the 1979 TECHNOLOGY LAW REVIEW.[20] As the table indicates, the risk of dying in an accident by driving 1,000

---

20. The table is excerpted from an article accessible online at http://www.law.nyu.edu/journals/envtllaw/issues/vol3/2/3nyuelj431.html.

miles was more than six times greater than the equivalent risk of dying in a commercial airliner. Yet there are many people who are afraid to fly. Few, if any of those people, are nearly as afraid of driving to the airport. Risk perception and objective risk assessments are frequently divergent and the case of biotechnology and food is another example of such divergence.

Risks That Increase the Annual Death Risk by One in One Million

Activity	Cause of Death
Living 2 days in New York or Boston	Air Pollution
Traveling 6 minutes by canoe	Accident
Traveling 10 minutes bicycle	Accident
Traveling 150 miles by car	Accident
Flying 1000 miles by jet	Accident
Flying 6000 miles by jet	Cancer from cosmic radiation
Living 2 months in Denver	Cancer from cosmic radiation

There is an excellent body of literature that explores the factors influencing this "subjective" risk perception (see e.g. Slovic, Fischhoff & Lichtenstein, *Facts and Fears: Understanding Perceived Risk* in Schwing & Albers (eds.) Societal Risk Assessment (1980), and Committee on Risk and Decisionmaking, National Academy of Sciences, Risk and Decisionmaking: Perspectives and Research (1982)). To summarize the literature on risk perception, it is fairly clear that high up on the list of the factors which "magnify" the public perception of a risk are its voluntariness, controllability, complexity, and the newness of the technology producing the risk. In addition, the perceived benefit of the risk also has a significant effect on public acceptance. Examining the applications of these factors to food safety in general, it is clear that the "hidden" risks of biotechnology-derived food additives[21] or ingredients would heighten the public's perception of risk because of all of those important variables. The risk is involuntary (at least to the extent that there is no special labeling of such foods), uncontrollable, complex, and stems from an extremely new technology. This does not mean that no application of biotechnology to food will find public acceptance. Rather, it means that public acceptance may hinge on the perceived benefit of the new technology to offset

---

21. In recent years, many risk-perception scholars have turned their attention to why biotechnology-produced foods engender the same kinds of risk-perception responses as nuclear power plants and potential terrorist attacks. *See, e.g.,* Jeffrey D. Wolt and Robert K.D. Peterson, *Agricultural Biotechnology And Societal Decision-Making: The Role Of Risk Analysis*, 3 AgBioForum Art. 6 (2000) (http://www.agbioforum.org/v3n1/v3n1a06-wolt.htm).

the perceived risk. Agricultural biotechnology companies were designing products to provide a benefit to their customers, the farmers, rather than for direct public benefit in terms of nutrition or health. Thus applications of biotechnology which fill a strongly perceived public need are likely to win acceptance more easily than those which merely increase producer or farmer profits. For that reason it is most unfortunate that biotechnology's introduction to the public was not with nutritionally-improved rice for developing nations[22] rather than an economically superior soybean (although it has environmental benefits as well). The lesson for the biotechnology industry is clear. Product development that takes into account the subjective nature of public risk perception is likely to pay off both in increased sales and in much faster regulatory approval. The current trade battles with the EU are an example of the cost of ignoring such concerns.

---

22. Andrew Pollack, *Universities to Share Patented Work on Crops*, N.Y. TIMES, July 11, 2003, at 14.

CHAPTER 8

Special Regulatory Issues — FDA Regulation of Drugs, Biologics, and Devices

§8.0 Introduction

Most biotech companies seek to build their businesses on advances in molecular biology and biochemistry, but their business plans can generally be divided into two very different categories. In the first category are the most glamorous: the high profile biotech companies attempting to develop significant new products for human health. The second category is less glamorous: companies that are simply developing tools and reagents for use by other biotech companies and university researchers.[1] For most of the companies developing human health care products, the stakes and risks are extremely high and the biggest hurdle is obtaining the approval of the FDA to market their products. This chapter is intended to present a basic guide to the process of obtaining FDA approval for human therapeutics and diagnostics, providing an overview of the FDA and its basic requirements, rules, and procedures, while focusing on the key issues and s a company will likely face in braving the long and perilous road to developing a new drug.

---

1. There are so few true biotech companies in agricultural biotechnology at this point that they may simply be regarded as anomalies in any general classification of "DBC" biotech companies.

§8.1 introduces Allcure, a hypothetical company used to illustrate the decision-making process for developing a new drug. §8.2 briefly reviews the relevant structure of the FDA and the principal statutory authority of CDER and CBER, the FDA centers that regulate drugs and biologics, respectively, as well as a brief discussion of medical devices. §8.3 summarizes the stages of drug development and includes a practical guide to FDA strategy. §8.4 provides a brief guide to clinical trials, filing for FDA approval, and to a few post-approval issues. Accelerated approval of drugs for serious or life threatening diseases and early, pre-approval access to unapproved drugs are discussed in §8.5. Three important product-exclusivity provisions, the Orphan Drug Act, the Patent Term Restoration Act, and pediatric term-extension are the subject of §8.6. §8.7 reviews the controversy and developments concerning the regulatory framework for "generic" biopharmaceuticals. Marketing and promotion of prescription drugs is discussed in §8.8. §8.9 provides an introduction to international approval for health care products in Europe and Japan and the recent efforts at harmonizing international drug approval standards and processes.

The process of obtaining the FDA's approval to market a new drug in the U.S. is enormously complex and expensive. From discovery to regulatory approval takes an average of about twelve years and over $800 million (including the time-value of the money invested) just to bring one new drug to market. Although this is just an average, there is no indication that for the average product derived from biotechnology significantly less time or money is required. Indeed, it may often take longer for products developed by new biotechnology companies because the products they are developing are often so different from existing therapies that the clinical pathway is substantially less clear than for a new hypertension drug or another antidepressant.

For several reasons, regulatory issues and regulatory strategy must be understood to be the driving force in biotechnology product development. First, regulatory approval is the *sine qua non* of product sales and profits. Without FDA approval, or its equivalent from European or Asian health authorities, biotechnology companies are not businesses, but merely complicated furnaces that burn enormous quantities of money. Second, expenditures aimed at FDA approval are by far the largest single category of expenses a biotechnology company incurs. Thus any regulatory strategy that reduces the cost and time needed for regulatory approval will have a substantial impact on the overall financial health of the company. Third, as a corollary to the preceding, any mistakes made in regulatory affairs can be extremely costly, even terminal for a biotech company. Finally, regulatory approval is not only essential to product revenues, the label indication for which approval is ultimately given is the largest single determinant of the market size for the product and, in many cases, even market share.[2]

§8.1 REGULATORY STRATEGY: AN INTRODUCTORY EXAMPLE

An example, the hypothetical company Allcure Biotech, serves to illustrate the various principles of FDA strategy. Allcure Biotech has a new technology that it believes to be useful in the field of therapeutics for diseases of the central nervous system (CNS), such as stroke, multiple sclerosis, Parkinson's, and Alzheimer's. Allcure's core technology includes a number of different small polypeptides (short chains of amino acids much smaller than a full protein) and these different peptides have neuroprotective activity in different cell types. Allcure has named this group of peptides "Neuroguards." CNS diseases generally involve continuing damage to or death of different groups of nerve cells, or neurons. Allcure must decide which disease and accordingly, which peptide to develop to target for the group of neurons affected by that disease. Preliminary assessment of the various cell-based assays that can be used to test Allcure's Neurogards leads to the conclusion that Neurogard AC12 has the most significant activity on the neurons most common in the cerebral cortex. Therefore stroke and Alzheimer's would be promising targets. Which of those diseases should be the focus of Allcure's development efforts for its lead product? The answer to this question may in fact determine Allcure's fate, for the failure of a company's lead product development effort may bring about the end of the company. Looked at in this light, the guiding principal of regulatory strategy in choosing Allcure's lead product may be minimizing the risk of failure more than maximizing the value of, as opposed to the probability of, success. Picking the optimum target requires balancing a number of factors that include the predictive power of different animal models, the predicted dosage, and method of ad-

---

2. It is important to note here that the term "label" is a formal term that is statutorily defined. It is a lengthy statement of relevant prescribing information about the drug and was originally intended to serve as the prescribing guide for physicians. It is found as a "package insert" in the manufacturer's original packaging as distributed to wholesalers, pharmacies, and so on, as well as reprinted in the standard reference text ,THE PHYSICIAN'S DESK REFERENCE or "PDR."

ministration. After providing an overview of the FDA in §8.2 and a brief summary of the various stages of drug development in §8.3, an analysis of the issues facing Allcure in its strategic planning efforts is provided in §§8.3(A)-(C).

§8.2 THE FDA STRUCTURE AND STATUTORY AUTHORITY FOR DRUGS AND BIOLOGICS (WITH A BRIEF WORD ABOUT DEVICES)

The FDA is literally a bureaucracy within a bureaucracy within a bureaucracy. Statutory authority for the regulation of human health care products is vested in the Secretary of Health and Human Services (HHS),[3] the Cabinet-level officer within whose huge domain the FDA is nestled. In turn, the Secretary has delegated authority under these statutes to the Assistant Secretary for Health, whose domain includes the Public Health Service (PHS). Within the PHS rests the National Institutes of Health (NIH), the Centers for Disease Control (CDC), and the FDA. Thus the Commissioner of the FDA reports to the Assistant Secretary of HHS who reports to the Secretary of HHS, who reports to the President. The products regulated by the FDA (when food, food additives and new animal drugs are included) constitute something on the order of 25 percent of the gross domestic product! Although the FDA is extremely visible and powerful, and its Commissioner rarely overruled, proposals to elevate its formal bureaucratic status have been give scant consideration.

Within the FDA, the Commissioner has a substantial central staff to deal with legal issues, general administration, and overall policy. However, the day-to-day operational authority and principal decision-making power of the FDA is exercised by the directors of the FDA's various Centers, such as the Center for Drug Evaluation and Research (CDER), the Center for Biologics Evaluation and Research (CBER), and the Center for Devices and Radiological Health (CDRH). The Center Directors, in turn, each supervise a substantial number of offices, each of which has direct responsibility for a different subject area within that Center.[4] For example, within CBER there are the Offices

---

3. This footnote serves to acknowledge this first of an inordinate number of acronyms that unfortunately are indispensable in a discussion of FDA regulation. The total possible number of three and four-letter acronyms in the English alphabet is 474,552. The total number actually used by FDA aficionados is very close to that maximum.

4. A link to the organizational charts for all parts of the FDA may be found at http://www.fda.gov/oc/orgcharts/orgchart.html.

of Vaccine Research and Review and of Cellular, Tissue, and Gene Therapies, each headed by its own Office Director. Continuing on down, each Office contains numerous Divisions, such as the Division of Viral Products and the Division of Bacterial, Parasitic and Allergenic Products within CBER's Office of Vaccine Research and Review. Similarly, the Cardio-Renal Drug Products Division and the Oncology and Pulmonary Drug Products Division are within CDER's Office of Drug Evaluation I. Finally, each of these Divisions and their Division Heads exercise more or less direct line authority over the people on the front-line, the medical officers who actually are directly responsible for the various Investigational New Drug (IND), New Drug Application (NDA), and other crucial filings submitted by pharmaceutical and biotechnology companies (heading teams that include biostatisticians, toxicologists and other specialists). Of course, a Division Director, Office Director, and even Center Director, may on occasion take direct responsibility for, as well as review authority over, a particular submission.[5]

§8.2(A) A Brief Excursion into the Distinction between Drugs and Biologics

Until June of 2003, virtually all protein and antibody drugs made by biotechnology companies were classified as biologics and regulated by the CBER. Now, gene therapy and vaccines, including "therapeutic" vaccines for cancer, remain under the supervision of CBER, as will the regulation of stem cell therapies,[6] while most other proteins and antibody-based therapeutics will be reviewed by CDER. The historical and continuing differences between CDER and CBER are not merely a separation into workable units of administrative expertise, as for example, the intra-center divisions into Offices and Divisions. Instead, the separation into Centers reflects the fact that the statutory authority of each Center is different, although overlapping, which adds to the complexity of the regulatory scheme and also permits reorganizations

---

5. Very recently, there have been accusations that the very top officials of the FDA or even HHS have intervened in politically sensitive drug approval issues, particularly whether or not to allow the Plan B, the "morning after" pill to be sold over-the-counter without a prescription. *See* Gardiner Harris, *Abortion*, N.Y. Times, March 5, 2006, at Section 4 p.3.

6. The most recent statement from the FDA indicates that stem cell therapies will be regulated as other cell-based and tissue derived products, but the formulation of guidelines for such products is still in the planning stages. *See* Donald W. Fink, Jr. Ph.D., CBER Presentation: FDA Review of Investigational Cell-Based Therapies: Resource Tools, Framework and Issues, March 14, 2006, (http://www.fda.gov/cber/summaries/stemcell031406df.htm).

such as that just briefly described. The regulatory authority of CDER is derived from 21 U.S.C. §321(g),[7] among other key provisions of the Federal Food, Drug, and Cosmetic Act (21 U.S.C §301 et.seq). The definition of "drug"— "any article (other than foods) intended to affect the structure or any function of the body of man [... **and which is not a device** ...]" is so broad that it clearly includes biological products, which are specifically regulated under the authority of 42 U.S.C §262. In fact, the breadth of the definition of drug has permitted the recent, sweeping change in the FDA's organization. As of June 26, 2003, the FDA began the transfer of direct regulatory authority over many biotechnology products previously regulated in CBER to CDER, as summarized in the following statement found on the FDA's CBER website (http://www.fda.gov/cber/transfer/transfer.htm):

> **The regulatory responsibility, review and continuing oversight for many biologic therapeutic products will be transferred from the Center for Biologics Evaluation and Research (CBER) to the Center for Drug Evaluation and Research (CDER)** [emphasis added]. This

---

7. Under Section 201(g)(1) of the Food Drug and Cosmetic Act (the "Act"), 21 U.S. C. §321(g)(1) :

"(t)he term "drug" means (A) articles recognized in the United States Pharmacopeia, official Homeopathic Pharmacopeia of the United States, or official National Formulary, or any supplement to any of them; and (B) articles intended for use in the diagnosis, cure, mitigation, treatment, or prevention of disease in man or other animals; and (C) articles (other than food) intended to affect the structure or any function of the body of man or other animals; and (D) articles intended for use as a component of any of the articles specified in clause (A), (B), or (C); but does not include devices or their component parts, or accessories."
The term "new drug" is also defined under Section 201(p) of the Act (21 U.S.C. §321(p)), as :

(1) (a)ny drug ... the composition of which is such that such drug is not generally recognized among experts qualified by scientific training and experience to evaluate the safety and effectiveness of drugs, as safe and effective for use under the conditions prescribed, recommended or suggested in the labeling thereof, except that such a drug not so recognized shall not be deemed to be a 'new drug' if at any time prior to the enactment of this Act it was subject to the ... Act of ... 1906..., and if at such time its labeling contained the same representations concerning the conditions of its use; or (2) (a)ny drug ... the composition of which is such that such drug, as a result of investigations to determine its safety and effectiveness for use under such conditions, has been so recognized, but which has not, otherwise than in such investigations, been used to a material extent or for a material time under such conditions.

change in regulatory responsibility will result in the transfer of applications for products belonging to the following product classes:

- Monoclonal antibodies for *in vivo* use;
- Cytokines, growth factors, enzymes, immunomodulators; and thrombolytics;
- Proteins intended for therapeutic use that are extracted from animals or microorganisms, including recombinant versions of these products (except clotting factors)
- Other non-vaccine therapeutic immunotherapies

The following product classes will remain at CBER:

- Viral-vectored gene insertions (i.e., "gene therapy")
- Products composed of human or animal cells or from physical parts of those cells
- Allergen patch tests
- Allergenics
- Antitoxins, antivenins, and venoms
- *In vitro* diagnostics
- Vaccines, including therapeutic vaccines
- Toxoids and toxins intended for immunization
- Blood, blood components and related products

Biological products or biologics are statutorily defined as "any virus, therapeutic serum, toxin, antitoxin, vaccine, blood, blood component or derivative allergenic product, or analogous product ... applicable to the prevention, treatment, or cure of diseases or injuries of man" 42 U.S.C. §262(a). 42 U.S.C. §262(g) specifically provides that nothing contained in the Public Health Service Act providing for the regulation of biological products "shall be construed as in any way affecting, modifying, repealing or superseding the provisions of ... [21 U.S.C. §301 *et seq.*]." So almost all biologics are also drugs and can be regulated as drugs or new drugs (as many now are), but most drugs manufactured by chemical synthesis are not biologics.

What is the practical significance of this distinction beyond determining to which bureaucratic cubbyhole a regulatory submission is sent? The differences stem from the essential characteristics of the two differing kinds of products and are reflected in the core regulatory standards governing each. Drugs are subject to the well-known approval standards of "safe and effective for use under the conditions prescribed, recommended or suggested in the labeling thereof" contained in the definition of new drug, 21 U.S.C. §321(p).[8] Biological products, or biologics, while generally required to meet the standard for new drugs, are also specifically governed by the less well-known standards of "purity, safety and potency" contained in 21 C.F.R. §601.2(a).

Although it would appear that the requirements of "safety and effectiveness" and "purity, safety and potency" contain only one common element that is not the way the FDA interprets the two phrases. The FDA interprets the definition of potency in 21 C.F.R. §600.3(s) " ... the specific ability ... to effect a given result," to be exactly the same as the "effectiveness" requirement for new drugs.[9] Thus the real difference between the regulatory treatment of the two kinds of products is principally derived from the additional emphasis on "purity" in biologics. Biologics are generally derived from or produced by living organisms (and viruses, which are not capable of independent "life"). In dealing with a therapeutic product derived from human blood, or one extracted from a bacterial or mammalian cell culture, the issue of purity, which includes both freedom from contamination and product identity, is a much more complex concern to address than is the case with a traditional pharmaceutical compound synthesized from well-characterized chemical ingredients. The safety and effectiveness of many biologics is more a function of purity and more subject to variation over time or from "batch to batch" than that of a synthetic small molecule. Conversely, the biological activity and safety of a well-characterized protein drug, such as a growth factor or cytokine, is more predictable than that of a novel chemical entity because of the known activity of the natural endogenous protein. Thus the toxicological testing requirements for new drugs, such as newly synthesized calcium-channel blockers, are more extensive than the toxicology requirements for new biologicals that are biotechnology-derived copies of natural human proteins such as colony-stimulating factors.

In short, both drugs and biologics are subject to the requirements of safety and efficacy, but the safety concerns are somewhat different for biologics and

---

8. *See* the previous footnote 144 for the complete text of §321(p).

9. Effective is determined with respect to "use under the conditions prescribed, recommended, or suggested in the labeling thereof" 21 U.S.C. §321(p) which can only be proved by substantial evidence, "consisting of adequate and well-controlled investigations, including clinical investigations, by experts qualified by scientific training and experience to evaluate the effectiveness of the drug involved" 21 U.S.C. §355(d). Furthermore, as interpreted by the FDA, the conditions for which effectiveness can be claimed on the label (indications) must relate to clinically significant improvement with respect to the "diagnosis, cure, mitigation, treatment or prevention of the disease, injury, or other condition for which the drug is promoted" O'REILLY, FOOD AND DRUG ADMINISTRATION, 2nd ed. §15.04.

biologics are subject to the additional requirement of purity, which reflects the different process by which biologics are produced. This somewhat different approach to the issue of safety and toxicology is likely to continue for sometime, regardless of whether the oversight for a biologic such as a monoclonal antibody is done by CDER rather than CBER.

§8.2(B) A Brief Word about Devices

CDRH (the Center for Devices and Radiological Health) is the other FDA Center that has approval authority for health care products, with responsibility for monitoring the clinical development and approval of new devices. Devices are regulated under the Medical Device Amendments of 1976 and the Safe Medical Device Act of 1990. Devices are defined as any:

> instrument, apparatus, implement, machine, contrivance, implant, *in vitro* reagent, or other similar or related article, including any component, part, or accessory...
> (2) intended for use in the diagnosis of disease or other conditions, or in the cure, mitigation, treatment, or prevention of disease, in man or other animals, or
> (3) intended to affect the structure or any function of the body of man or other animals, *and which does not achieve its primary intended purposes through chemical action within or on the body of man or other animals and which is not dependent upon being metabolized for the achievement of its primary intended purposes.* 21 U.S.C. §321(h). (emphasis added).

The italicized clauses of the definition are the key to the distinction between devices on the one hand and drugs and biologics on the other. Like drugs and biologics, a device is regulated under the FDCA if it is used to diagnose or treat disease or otherwise affect human health, structure, or function. If however, it does so by some means other than chemical interaction with the body, and other than by being metabolized, it is a device. If it accomplishes its task by chemical interaction or by being metabolized, it is a drug or a biologic.

The regulation of medical devices begins with the classification of the device into one of three categories. Class I devices include low-risk devices such as crutches, bandages and tongue depressors. For class I devices the regulatory requirement is simply conformity to "general controls" which cover good manufacturing processes, testing processes, quality control, and so on. Class II devices present an intermediate level of risk. For example, contact lenses are Class II devices. Class II devices are subject to both general controls and special controls that are device-type specific. Class III devices are those devices that would pose a serious risk to health if improperly manufactured or designed, such as cardiac pacemakers. Class III devices are subject to individual approval by either premarket notification (PMN) through a 510(k) or through premarket approval (PMA) following clinical studies done under an IDE (Investigational Device Exemption).

Which of these two pathways, PMA or 510(k), a Class III device falls into depends upon whether or not the FDA agrees that the proposed device performs the same function as a device which was marketed prior to 1976 (a "predicate" device) and for which no specific PMA requirements have been implemented by the FDA. If so, then the simpler, faster, and less expensive 510(k) approval pathway is available. For example, an antibody diagnostic for strep can be approved under a 510(k) simply by comparing its performance to a pre-1976 strep test based on older cell culture assays. However, a Class III device that has no pre-1976 predicate, for example an mechanical, implantable heart, would need to go through a process or preclinical development, clinical testing, and regulatory approval that is substantially the same as that for a new drug.

§8.3 Stages of Drug Development and Drug Development Strategy

To fully appreciate what needs to be done for the actual regulatory filings and clinical tests for the approval of a new biotechnology-derived therapeutic, it is important and useful to understand what needs to be done prior to filing an Investigational New Drug application ("IND") or, in the case of a Class III medical device an Investigational Device Exemption ("IDE"), because what occurs in the development phases preceding regulatory filings can be critical in determining the progress and outcome of testing required by the FDA. The filing of an IND or IDE, once accepted by the FDA, begins the period of human testing, or clinical trials, for a new drug, biologic, or device. This preclinical period of research and development, prior to the IND, is the time when a number of the most significant decisions about the desired endpoint of product development must be made: the indication, the formulation, the route of administration, the possible range of doses to be tested in human trials, and the likely duration of use by patients, in both the clinical trials and the real world (such as daily, short-term, one-time, chronic, etc.). The most obvious significance of the preclinical period is to gather data to support the filing of an IND and ultimately NDA (New Drug Application), PLA (Product License Application for those products that will continue to be approved by CBER), or PMA (PreMarket Approval for a medical device). Thus the preclinical data must demonstrate that there is a reasonable scientific basis for believing that the product is sufficiently promising to warrant experimentation using human subjects and to answer some basic concerns about the safety of the product (for example, whether or not it is hepatotoxic in laboratory animals and therefore possibly hepatotoxic in humans). At the same time, the preclinical data helps refine the plans for clinical trials, both by providing some animal pharmacokinetics to aid in initial human experimental dose ranges and by helping to refine the initial patient selection criteria that determine the ultimate indication for which approval is sought (e.g. stroke or Alzheimer's for Allcure's neuroprotective agent).

There are generally ten phases of drug development, briefly described below:

(1) **Discovery**: The class of compounds or individual compounds are discovered or invented.

(2) *In vitro* Testing/Compound Screening: Potential candidates for development are tested in the laboratory in various cell-based assay systems.

(3) *In vivo* Testing in Small Animals: Screening for preliminary toxicity profile and efficacy is done in a variety of animal models, generally in rodents.

(4) **Formulation and Manufacturing Scale-up**: Decisions are made about how the product will be formulated and manufactured, e.g. as a powder to be reconstituted and injected, or in an i.v. solution, or as active ingredient/salt crystals in a standard tablet formulation, etc.

(5) **Preclinical Toxicology**: More formal and much more extensive preclinical toxicology than earlier "proof of concept" studies are required by the FDA before any "first-in-human" trials begin. These require a significant number of animals in two species (often rats and dogs) at a range of doses and for a duration of testing that reflect the anticipated clinical trials. For example, a clinical trial that is expected to be single dose (as would be expected for planned emergency room drugs for heart attack or stroke) requires much shorter preclinical toxicology than a study to support an IND for a drug that will be given for a long period of time (as would be expected for an antidepressant or Alzheimer's drug).

(6) **Phase I Clinical Trials**: After an IND/IDE is filed the first phase of human testing begins. In many cases, these are purely safety trials in which the subjects are healthy volunteers and therefore no efficacy data would be generated. In some cases, particularly for potential breakthrough drugs for serious diseases for which there is no effective therapy, initial testing is done in patients with the disease, in which case the initial trial should be referred to as a Phase I/II clini-

cal trial, as some efficacy data may be generated. In addition to looking for adverse effects, a principal goal of the first phase of human testing is to generate significant information about the drug's *in vivo* metabolism, including absorption, half-life, distribution, etc.

(7) **Phase II Clinical Trials:** These trials are dose-ranging and involve further safety and efficacy testing on a target population. Large pharmaceutical companies have almost always done multiple Phase II trials, with varying dose ranges and patient eligibility criteria. Increasingly biotech companies are learning the value of doing multiple Phase II trials. The purpose of multiple Phase II trials is to establish not just a reasonable basis for designing a Phase III "pivotal" trial, but to learn enough about dosing and patient eligibility criteria to make a successful Phase III trial highly probable. For example, if a Phase II trial in stroke shows efficacy when patients were treated within four hours of the onset of symptoms, but a subgroup analysis shows even greater power among patients who received the drug within two hours of onset of symptoms, it might be worthwhile to do another Phase II trial in which one experimental group ("arm") is restricted to the shorter duration, to determine the optimal conditions for the Phase III trial.

(8) **Phase III Clinical Trials:** These are multi-center studies conducted to prove safety and efficacy and to study adverse reactions, particularly those which may be sufficiently rare that they were unlikely to have occurred in the smaller, Phase II patient populations. It is a matter of straightforward probability that adverse reactions that are likely to occur in 1 percent of the population have only about a 63 percent chance of occurring even once in any particular 100-person clinical trial. Although serious adverse reactions that occur only once in a thousand patients are unlikely to be discovered during preclinical testing, Phase III tests should be sensitive enough to provide at least preliminary evidence of most significant safety issues.

(9) **New Drug Application** (Product License Approval for Biologics): After completion of trials an NDA (or PLA) is submitted, an Advisory Panel reviews it, and the FDA approves it (or the sky falls, investors flee, and management tries to regroup). The volume of material submitted is staggering. The NDA filing not only includes the summaries of data as to safety and efficacy supporting the application, and significant information about manufacturing and quality assurance, it includes the complete records of all data collected from every patient in all clinical trials, from the initial documentation of the patient's eligibility for the trial to every vital sign measured at

every point along the way. The computer form of the application (CANDA, or Computer-Assisted New Drug Application) may require only a small box of compact disks, but the actual records in paper form, shipped simultaneously to the FDA, can fill a large truck.

(10) **Post-Marketing:** Phase IV studies may be conducted to support new indications and may be required in cases where the FDA believes there is a need for additional safety data from a broader population in actual clinical practice.

Preclinical development is the foundation on which a biotech company's drug development program is built. The over-riding purpose of preclinical development is to maximize the probability of success at the end of the development process, with the approval of an NDA. The choices to be made are numerous and of critical importance in the long run. Which of the possible compounds merit selection for more extensive preclinical testing? Which compounds performance *in vitro* screen results demonstrate sufficient promise to merit animal testing? Which animal models should be chosen and should the animal models drive target selection or does target selection drive the choice of animal models? For example, should Allcure first decide on Alzheimer's as a target and pick the corresponding animal model, or should Allcure consider whether animal models for stroke offer advantages over animal models for Alzheimer's and use that as a basis for deciding between stroke and Alzheimer's as a clinical target? Should different requirements for full preclinical toxicology be considered in picking a clinical target and therefore the animal models that precede them? These questions are all-important in the foundational preclinical phase of drug development and are discussed in the next section on strategic planning for drug approval.

§8.3(A) From Preclinical Decision Making to NDA

FDA strategy is more than just complying with all the FDA's requirements. To be sure, success requires a thorough knowledge of the FDA's statutory authority, regulations, and regulatory guidances. But that is like saying that becoming an international chess master requires a thorough knowledge of the rules of chess: absolutely necessary but clearly not sufficient. FDA strategy entails a careful plan to reach the goal, which is 1) FDA approval, for 2) the optimal indication, in 3) the shortest period of time possible, 4) using the least amount of scarce resources (money and people), and with 5) a reasonable likelihood of inclusion in the formularies of, and reimbursement from, the major health care providers for the target market. This §8.3(A) briefly defines and explains each of the five key terms used above. §8.3(B) examines in greater depth how those terms relate to the strategic issues that need to be made at each of the key steps along the pathway from the laboratory to the formulary. §8.3(C) concludes this strategic overview by analyzing some of the mistakes that biotechnology companies have made and which presumably other biotechnology companies should avoid if possible.

§8.3(A)(1) The Five Key Parameters of Strategic Success: FDA Approval

First and foremost, success is measured by successfully obtaining FDA approval. FDA approval, generally approval of an NDA is required before any new drug can be sold in the United States. It is more than simply a yes/no decision. More precisely, it is approval to sell the drug with a very specific label stating the "indication" or precise patient population for whom the drug is "indicated" or approved for marketing. Any other use of the drug is "off-label."[10]

§8.3(A)(2) Optimal Indication

A successful strategy is one that leads to approval for the optimal indication. The optimal indication is one which best combines the desired market, or patient population, with other factors, particularly competitive factors in various potential markets and, perhaps most importantly, the ease and speed of the clinical pathway to the approval process.

§8.3(A)(3) The Shortest Possible Time

Success includes achieving approval in the shortest period of time possible. The importance of speed may seem obvious, but different clinical pathways require different periods of time and different preclinical pathways also can make a difference. For example, while CNS drug trials generally take longer than HIV trials, some CNS trials, i.e. for stroke or head injury, would almost certainly be shorter than an HIV trial aimed at demonstrating that a particular regimen is less prone to resistance or adverse affects than another regimen.

§8.3(A)(4) Conserving Scarce Resources: Time and Money

Time is money, so generally shorter pathways to clinical approval mean less money spent and required to keep the enterprise going during the clini-

---

10. Doctors can prescribe any drug for any patient for any reason, but drug company marketing directly to consumers can address only the labeled indication, and patients' reimbursement from their insurers may be limited to the labeled indication. Marketing to health care professionals may concern unapproved uses, but only within a limited framework, as discussed in §8.8.

cal program; but different indications also entail different costs independent of time. For example, a clinical trial that requires multiple MRIs to assess the relevant endpoint will be more expensive than one that requires a patient to fill out a pain scale self-assessment, to take two examples that diverge significantly in cost.

§8.3(A)(5) *Planning for Reimbursement*

Fifth, success should include being listed in formularies and winning reimbursement. These objectives depend, to an extent, on understanding the target population's likely health insurance status. For example, if the target population has a substantial percentage of patients over age 65, as might be the case for an Alzheimer's or stroke drug, Medicare part A plans do not pay for most prescription drugs or self-injected drugs, although such drugs may now come within the Byzantine world of the new Medicare Part D insurance coverage, but it does pay for 100 percent of the cost of in-hospital or in-clinic administered intravenous drugs. Accordingly, developing an i.v. formulation for a stroke drug, particularly one that is likely to be administered in an emergency room, may well make more economic sense than an oral formulation, which is preferred for most other indications.

§8.3(B) Strategic Issues in the Development Process and the Terms of Success

For biopharmaceutical companies and medical device companies, a clear understanding of the core concept underlying FDA approval, "safe and effective" and the how those terms translate into clinical trial design and data are of primary importance. First, "safe" does not mean risk-free. Nothing is absolutely safe: acetaminophen is liver toxic; aspirin causes gastrointestinal bleeding; and, Celebrex may reduce the risk of gastrointestinal bleeding but may increase the risk of heart attack compared to traditional NSAIDs. Safe is always a relative term, a comparison of risk and benefit. A great deal of risk is tolerated in a drug for malignant brain tumors; very little risk is tolerated in an allergy drug.

Second, "effective" is also a term of art and one that can change over time for any particular indication. In general, "effective" must mean a significant positive change in a clinically meaningful endpoint. In turn, "clinically meaningful" means a parameter that doctors and patients would agree is of significance to the patient. For example, in cancer trials, tumor size reduction might be considered as a possible primary endpoint in a clinical trial; it is relatively easy to measure and significant changes might well be expected in a reasonably short period of time. However, for some cancers, experience has shown that tumor size reduction does not impact survival and is not necessarily correlated with a patient's general quality of life. Survival and quality of life are what cancer patients care about; they are not necessarily interested in tumor size reduction for its own sake. Accordingly, the FDA might not be willing (and perhaps ought not to be willing) to accept tumor size reduction as a primary endpoint in a clinical trial for some cancers. Understanding how different possible clinical endpoints, such as the number and size of lesions in an MS trial, or infarct size in a stroke trial, relate to patient benefit is crucial and having the FDA agree to the endpoints selected is essential.

§8.3(B)(1) The Preclinical Development Process and the Optimal Indication

Many companies may well look at the choice of indication, or target, as one of relative market size. That certainly is one factor, but it is a great mistake to consider relative market size alone as determining the optimal indication for a particular drug development program. If a company has a significant new product for an unmet medical need, then the revenues are almost certain to justify the investment. The risk of failing to get to the goal is much greater than the risk of getting to the goal and being disappointed in the results. It is generally better to choose a target that has a higher probability of success, a shorter time period to approval, and a lower cost to get there than one which is riskier, or longer, or more expensive, or all three. This may be obvious advice, but how does one accomplish these three objectives: less risk, shorter time period, and lower cost?

First, reducing risk has three components: knowing the biology, having a good animal model and, if possible, having a reasonably clear clinical pathway. Knowing the biology: big drug companies can afford to be second entrants into a new market, or even the fifth or sixth statin, or SSRI, or ACE inhibitor. Biotech companies frequently must aim for diseases for which there is no satisfactory treatment, the biological target or pathway is unproven in the clinic, and therefore the biology is less than certain, or the animal models less convincing, or both. For these path-breaking trials, the risk is correspondingly higher. Scientists often mistake the quality or elegance of their science for the soundness of the "medical" biology. Examples of biotech drugs that failed when the biology was perhaps less clear than the initial scientific analyses are numerous, but sepsis and rheumatoid arthritis provide interesting case studies.

When the human inflammatory protein TNF was first explored for its role in disease, the initial target was sepsis because the science plausibly led to the view that the inflammatory cascade and TNF are clinically important targets. However, anti-TNF antibodies failed to provide evidence of clinical benefit in human trials. Years later, however, an anti-TNF antibody was approved for the treatment of rheumatoid arthritis, another disease in which the inflammatory process and TNF are scientifically plausible targets. Interestingly, the biology of RA may still not be well enough understood to remove the risk from other approaches to treatment. Genentech failed in a clinical trial of an anti-T-Cell antibody that targeted CD11a to neutralize T-cells. Then Biogen Idec risked clinical trials of an antibody that is B-cell depleting and did demonstrate efficacy in RA, gaining FDA approval. Is RA a T-cell mediated disease, a B-cell mediated disease, or both? Right now the B-cell targets have the edge, but the biology is still unclear.

While all drug development programs must rely on animal models in the preclinical process, not all animal models are equally reliable or convincing. In general, the better the understanding of the biology of the disease and the more closely the animal model follows the same biological pathways as the human disease, the better the animal model. In the CNS field, the biology of Alzheimer's disease is not completely understood and there is no animal model that is truly convincing. Not surprisingly, such adventurous approaches as a beta-amyloid "vaccine" have failed in the clinic, at least so far. In comparison to Alzheimer's, the biology of stroke is very well understood and, at least for occlusive stroke, there are reasonably convincing animal models. The mechanism of occlusive stroke is simply the blockage of an artery to the brain and it is an easy process to mechanically replicate in a laboratory animal. So if Allcure has an agent that could be used in either stroke or Alzheimer's, there may well be less risk in pursuing Alzheimer's.

The third factor in choosing the optimal indication, a reasonably clear clinical pathway, is somewhat more complicated. As has already been stated, biotech companies generally are looking for unmet medical needs. Since this often means that there is no successful therapeutic for that indication, it may also mean that the best design for a clinical trial, that is what to measure and when and how to measure it, may not be clear. Breakthroughs do not always require the complete lack of any approved therapeutic. Although drugs have been approved for stroke and Alzheimer's disease, none of those drugs has provided a major breakthrough for those medical problems. Therefore, there is plenty of room for a more effective agent and there are also some trial designs to study and improve upon. Whatever the indication, the endpoints chosen ought to be agreed to by the FDA as proof of clinic benefit and measurable with relative precision (measuring the rate of wound healing in diabetic ulcers, for example, is not necessarily easy).

In the end, the optimal indication will be the one with the shortest, least risky pathway to ANY relatively significant market, not the one with the biggest patient population. Even a market of 20,000 patients in the U.S. can be sufficient, if the disease is serious or life threatening and there is no effective alternative therapy. At a "mere" $10,000 per treatment, such a drug would provide $200 million in U.S. sales alone and the price could easily be more with worldwide sales likely to be two to three times greater. That may only be a small fraction of Lipitor's sales, but it is a substantial reward for most biotech investors nonetheless.

§8.3(B)(2) Conserving Time and Money

This measure of success can also involve significant choices. First, some indications will clearly take longer to produce results than others. In general, the slower a disease progresses, the longer it will take to measure whether or not a drug is effective. For example, returning to Allcure's strategic planning, it is almost certain that a stroke trial will take less time, per patient and overall, than an Alzheimer's trial. Stroke damage or recovery could be measured at two weeks to thirty days and provide a significant clinical endpoint for approval, while it is difficult to imagine that even a quickly regenerative super drug could produce a meaningful result in the condition of Alzheimer's patients in that time period.

While it would seem obvious that a shorter trial is a cheaper one, there are other considerations besides time. For example, a drug administered only acutely, for example immediately post-stroke or for a short time period thereafter, would require less drug to be produced for clinical trials and significantly less preclinical toxicology (at a savings of approximately a million dollars in animal toxicology alone) than a drug which requires long-term administration. Alzheimer's, Parkinson's, and Multiple Sclerosis are all indications likely to require chronic administration rather than single or short-term administration. Both the cost of clinical materials and the cost of preclinical toxicology can be substantially higher in trials for chronic, rather than acute, conditions.

Furthermore, among targets, whether short-term or long-term, costs of clinical trials can vary significantly depending upon the clinical endpoints and other significant attributes of the trial protocol. For example, a trial of an MS drug is likely to require multiple MRIs over a substantial period of time, as that clinical trial design was proven acceptable in the trial of the various Interferon-Betas. However, a trial of a Parkinson's drug, while possibly of the same duration, will involve clinical endpoints that are largely functional and more cheaply measured. So the clinical endpoints are a factor to be thrown into the complex decision making that already includes the degree to which the biology is well understood and whether or not there is a reasonably good animal model.

The number of patients required for the Phase III trials is one of the most important variables in the cost of development; and, while not always predictable at the very beginning of the trial, some educated guesses are possible at the early, strategic decision-making stage. Virtually without exception, the phases of clinical trials increase in the number of patients enrolled: Phase I trials enroll 10–20, Phase II trials 40–200, and Phase III trials 100 to several thousand. The increase in the size of the trials reflects the additional goals of showing efficacy with a statistically significant difference between the test group and the control group and with a sufficient sample size to find adverse events of sufficient frequency to raise safety questions given the purpose of the drug. Rare events are of more concern in a headache remedy than an end-stage cancer drug. One basic consideration in planning the number of patients for any clinical trial depends on what is to be measured and, based on pre-clinical data, how easily the effects of the drug will be detected. The term for this is "signal strength." Other variables are the practical constraint of the actual size of the patient population pool (it is impossible to do a very large trial in a rare disease) and the seriousness of the disease (one can be less concerned about one in a thousand adverse events in a trial of a drug for a late stage cancer than in a trial of a hay-fever drug).

Continuing to use Allcure's stroke vs. Alzheimer's therapy choice as an example, both are arguably of roughly equal "seriousness" and both have large patient potential population pools. As to other factors, the primary clinical endpoint for Phase II might be a performance assessment scale that measures motor coordination and speech in stroke patients and a performance assessment that measures cognitive impairment in Alzheimer's patients. If many stroke patients have significant impairments at two weeks and the Phase II data shows a significant improvement in a group of only fifty patients given the drug, a Phase III trial of only a few hundred patients might provide sufficient data, particularly if the nature of the drug and the toxicology would seem to minimize any real safety issues. Alternatively, when testing a new drug for Alzheimer's, the very complex biology of the disease may mean that the signal strength will be less and the time needed to measure significant differences longer. There is substantial variability in the rate of the progression of Alzheimer's and there can be significant variations for an individual patient from day to day, thus making it even more difficult to distinguish the drug's effect from the general experimental "noise." If what is hoped for is a ten percent increase in performance on a cognitive function scale for the test group, it would likely require a much larger trial to find that difference with statistical significance.[11]

---

11. For those readers unfamiliar with basic statistical concepts, the difference between two groups can either be due to chance or to the experimental effect. To use the familiar

§8.3(B)(3) *Planning for Reimbursement*

All other things being equal, it is certainly advisable to make choices that result in a drug paid for someone other than the patient. Reimbursement, whether from managed care, Medicaid, Medicare, or one of the new Medicare Part D plans, is almost certainly going to translate into increased sales. By and large, managed care plans will pay a reasonable price for any drug indicated for a condition for which there is no effective alternative treatment. So if there truly is a breakthrough therapy for Parkinson's or Alzheimer's, then patients with a managed care pharmaceutical benefit will be reimbursed for much of the cost of the drug. On the other hand, Medicare generally does not pay directly, as opposed to the Part D insurance subsidy, for prescription drugs. While there often is little choice of formulation, as previously mentioned, Medicare does pay for in-hospital or in-clinic IV drugs, are also covered by private insurance. So formulation, a choice which is made at the very beginning of the preclinical process, can be a profoundly significant decision at the very end of the road, when a product enters the market and patients have to find a way to pay a sufficient price to justify the significant risk and investment.

In addition, inclusion in the "formulary" of various managed care providers preferred drugs for an indication is another important commercialization objective. Keeping in mind that inclusion in the formulary is easiest when the economic benefit of a drug is clear can also influence clinical trial strategy. For example, while the number of hospital days required for a given patient population is unlikely to be your primary endpoint for a clinical trial, keeping track of hospital days is relatively easy additional data to collect and a reduction in hospital days can be persuasive "pharmacoeconomic" evidence that helps to persuade a managed care organization to include a product in its preferred formulary.

§8.3(C) Drug Development Strategy: Learning from the Mistakes of Previous Companies

Biotechnology companies have been pushing back the frontiers of medicine long enough to have made numerous mistakes and, even if it is of no con-

---

example of coin-flipping, a result of seven heads in ten trials could either be just chance or could indicate that the coin has been altered in such a way as to produce a greater likelihood of producing heads just as a difference in seven out of ten patients might be chance or could be due to Allcure's drug, for example. Seven out of ten is more likely to be chance than seventy out of one hundred.

solation to the founders and investors in those companies, we can nevertheless learn from them. The first mistake is choosing a "moving" target. A "moving" target is one the endpoint varies with the quality of care and the quality of care is either generally increasing or likely to be substantially better at clinical trial sites than in the general population. One example of a failure that may have been due to the selection of a moving target is that of Gensia's drug "Protara" which was tested for its ability to reduce the incidence of heart attacks and other adverse cardiovascular events in conjunction with coronary artery bypass surgery. It is clear that the rate of adverse events in such surgeries varies widely from surgeon to surgeon and hospital to hospital and that the rates of such complications are extremely low at the best hospitals using the best surgeons. Of course, a clinical trial of such a drug will only be done at leading hospitals in the field of cardiovascular surgery, making a significant improvement much more difficult to show. The lesson: choose an endpoint that is not likely to be dramatically affected by the standard treatment at a leading medical center.

Another, related type of mistake is aiming at what might be a "soft" target, one that is significantly affected by real adherence to the standard of care. Telios Pharmaceuticals' failure in demonstrating benefit in healing diabetic foot ulcers might be attributable to just such a failing. Here the problem was not that the standard of care at leading centers is necessarily much better than at other sites, but apparently that many patients do not receive proper care. Debridement and general proper wound care is often not given to patients with diabetic foot ulcers. When it is, otherwise stubborn wounds frequently do heal. Naturally, when enrolled in a clinical trial, patients in the placebo arm of such a trial do get good standard treatment, making the effect of a test drug more difficult to measure. The lesson: pick an indication for which the endpoint remains a big target with substantial room for improvement even though the patients are likely to have received the standard of care.

Rushing to Phase III is another mistake often made by small, cash poor companies. Major pharmaceutical companies generally do multiple Phase II trials to determine the optimum dosage and patient eligibility criterion to order to fine-tune the indication before moving on to Phase III. The key is to make sure that a Phase II trial has pinpointed the precise patient population that will be the focus of Phase III. For example, if ultimately an MS drug is likely to be used in combination with a beta-Interferon, then the Phase II study needs to test the combination against interferon alone and, if successful, Phase III should seek to replicate that finding in much the same patient population. Corixa learned the hard way that designing a Phase III trial around a real-world clinical situation may be essential and that rushing to

Phase III can be very costly. Corixa did ultimately generate additional positive data and receive approval. Compare the Phase III data originally submitted by Corixa for Bexxar, which was found wanting by the FDA, with the Phase III data submitted by IDEC Pharmaceuticals in support of their successful application for Zevalin:

> **Bexxar**: Single arm, open-label, "historical" control:
>
>> N=60 patients who had been treated with at least two protocol-specified qualifying chemotherapy regimens (median 4) and who had not responded or progressed within six months after their LQC were treated with a single course of iodine I 131 tositumab.
>
>> **Patients responding completely or partially to treatment (Overall Response Rate)**
>
>> 39 (65 percent) patients showed a partial or complete response
>
>> 17 (28 percent) patients had responded to their last qualifying chemotherapy.
>
>> $p<.001$ (meaning that the difference between the number of patients responding to Bexxar and the number who had responded to their previous chemotherapy was highly unlikely to be due to chance, if all other variables were random and did not skew the outcome).
>
>> **Median duration of response:**
>
>> 6.5 mo. after treatment with Bexxar
>
>> 3.4 mo. after LQC
>
>> $p<.001$ (same statistical significance as for Overall Response Rate).

Zevalin: Two armed, open label but assessment-blinded study of Zevalin following Rituxan versus Rituxan alone.

N=143 patients with relapsed or refractory low-grade, follicular, or transformed CD20(+) transformed NHL.

73 received one dose of Zevalin following two doses of Rituxan

70 received four doses of Rituxan

Overall Response Rate (ORR)

80 percent for the Zevalin group (90)Y ibritumomab tiuxetan group

56 percent for the rituximab group

p =.002 (highly statistically significant)

Complete Response (CR) rates

30 percent for the Zevalin group

16 percent for the rituximab group

p =.04 (statistically significant)

Rituxan had become the standard of care for chemotherapy refractory patients and had been approved for some time. What doctors and patients needed to know was if a new drug would provide an advantage over Rituxan in that patient population. That is what the IDEC study clearly showed. Corixa enrolled fewer patients and tested their drug against the population's response to a previous standard therapy, rather than the actual standard of care, which had become Rituxan. While Rituxan is IDEC's product, by testing against the last standard chemotherapy, Corixa failed to demonstrate to the FDA that Bexxar provided a real clinical benefit in a patient population that would ordinarily get Rituxan. While in theory a drug need not be better than its competition, the FDA is not keen on single arm trials that use a previous history as a control group and deny any of the patients the current standard therapy.[12] To paraphrase an old cliché, the FDA always believes that two "arms" are better than one. The lesson: don't proceed to Phase III without successful Phase II data in that patient population and discuss the Phase III design with the FDA.

---

12. Consult with the FDA!! If you have a pre-Phase III meeting, the FDA can generally be held to any agreed-upon trial design.

§8.3(D) Conclusion: Planning the Clinical Development of a New Therapeutic Begins at the Early Preclinical Stage

Most problems stem from a failure to understand the true end-game and failure to consult the FDA about trial design at an early stage. Success may not be easy, but failure is not an option.

§8.4 A Practical Guide to Clinical Trials and the NDA

There is a simple way to think about the clinical development of a new drug or biologic. "What are you going to prove and how are you going to prove it?" Every drug development program is a long and expensive scientific experiment designed to prove that a drug is safe and effective. But "safe" is a relative term and "effective" is a term of art. So with a target indication in clear focus, and the preclinical data issues resolved, drug development can focus on the clinical issues that need to be addressed in the Phase I IND. Although the first phase of clinical testing, aptly known as "Phase I," is generally limited to the collection of initial safety and human pharmacokinetic data on a small number of patients, most of the basic clinical trial issues are involved and need to be addressed, including the patient selection criteria, clinical end-points, the kinds of measurements that will be taken, and the intervals at which each measurement will be made. Decisions on each of these points is necessary to prepare the required investigator brochures, which are materials for the investigators, detailing the exact set-by-step "drill" or protocol to be followed by the investigators performing the studies, including the case report forms for reporting the data.

The IND will ordinarily identify the centers at which the Phase I clinical trials will take place (Phase I is often single site, but subsequent phases, particularly Phase III, are generally multi-site), along with the number of patients and the identity of the principal investigators at each site. Although there is no FDA rule that determines the number of patients that should be enrolled in each center, the applicant should test as many patients as it anticipates will result in a large enough statistical sample to establish the preliminary safety of the drug so it can move on to the next phase of testing.

Each clinical trial site must obtain Investigational Review Board ("IRB") approval before the FDA will accept the IND. The IRB is a medical institution's internal review board for evaluating and approving the institution's participation in the clinical study, with a mandate to protect human subjects from any unnecessary risk and to insure the ethical propriety of the study, including the informed consent procedures. Without IRB approval, an institution cannot commit to performing clinical studies. Furthermore, an investigator's failure to obtain such approval may not only be a violation of that investigator's terms of engagement or employment contract with the institution, but more importantly may result in disqualifying data obtained from those studies from being included in the applicant's NDA. The IRB approval is generally requested and obtained by the investigator who will be performing the studies. The application for IRB approval must contain the exact patient entry criteria and protocols that will be submitted to the FDA in the applicant's IND application as well as the informed consent form and materials to be given to patients prior to their final enrollment.

Patient entry criteria and the investigators' compliance with protocols are very important in all clinical trial phases. The data derived from studies performed on patients who were not eligible under the enrollment criteria will likely have to be excluded from the final report of test results and the NDA. For example, if a criterion for patient eligibility in a study is that the patient must not have had more than one prior chemotherapy regimen, inclusion of a patient who had previously undergone two chemotherapy regimens will disqualify the results from the study, even if the criterion turns out not to be directly relevant to the performance of the drug and the results obtained during the study.

Similarly, if an investigator fails to comply with a protocol in some material way, the results of the study will not be valid, even if failure to comply did not actually affect the performance of the drug or test results. For example, if a protocol requires the patient to lie down during the administration of the drug, but the study is performed while the patient is seated, this could disqualify the results of that study.

Generally, the company's clinical research associates ("CRAs") are charged with assuring that patient inclusion criteria are adhered to and that the investigators comply with the study protocols. Many biotech companies cannot justify building the internal capacity to manage the clinical trials directly so they contract with Contract Research Organizations (CROs) to provide additional expertise and staff for much of this work (including the drafting of materials, such as case report forms, which are included in the IND). Even large pharmaceutical companies will turn to CROs when they are blessed with a bulge in their pipeline, for example with four products entering Phase III at more or less the same time but in-house capacity to manage only three.

The FDA has thirty days to review an IND application. The IND is assigned to a specific office of CDER or CBER (IDEs go to CDRH). The particular office to which the IND is assigned can make a great deal of difference in the nature, extent, and length of the review. If the FDA responds with a rejection of the application (technically an IND to which the FDA objects is on "clinical hold") and some further questions, the thirty-day clock starts running again from the date that the applicant's response is received by the FDA. It is not unusual for several cycles of application/question/response to occur before an IND is accepted. After the IND is accepted, the applicant can begin Phase I clinical studies. In all three phases, the FDA's objective is to assure the safety of all subjects. In Phases II and III an additional objective is to assure that the scientific evidence being compiled in the studies is adequate to support an evaluation of safety and efficacy and to assess the likelihood that the studies will yield data capable of supporting market approval. However, without a formal consultation in a Pre-Phase III meeting, the FDA is not bound to accept data from a protocol that it allowed to be performed.

There is no hard and fast rule as to how much information and data must be collected and analyzed in to meet statutory standards of safety and effectiveness. The focus of the initial IND submission should be on the development and articulation of a general plan and protocols for human investigation. Subsequent amendments to the IND (which, by the way, stays open through all three phases of investigation until an NDA is filed), should build logically on the scientific evidence previously derived. Thus, changes in protocols from one phase to another, or even within phases, should be supported by additional information garnered from the human studies under the IND, from additional animal studies, particularly related to toxicology, and from studies, if any, that are being or have been conducted in other countries.

§8.4(A) The New Drug Application or Product License Application

At the end of the clinical trials, if all has gone well, the sponsor needs to file a New Drug Application ("NDA"),[13] the formal request to the FDA for approval to market a new drug. Once an NDA is submitted, the sponsor is formally referred to under 21 C.F.R. §314.3(b) as an "applicant." The NDA must be sub-

---

13. For the remainder of this section, only the term New Drug Application (NDA) will be used. It should be remembered that the submission for the approval of a biologic is actually referred to as a Product License Application (PLA). The treatment of the two is sufficiently similar that the description in this section can be applied to both.

mitted in triplicate[14] and must contain all the information required under 21 C.F.R. §314.50. If the drug is a new chemical entity, the NDA will generally contain an application form, index, summary, "five or six" technical sections, case report tabulations, case report forms, drug samples, and proposed labeling. If the drug is not a new chemical entity, the NDA can include less information, but must include enough to support the particular submission.

The heart of an NDA is the *summary*, in which the applicant describes the NDA in enough detail to enable the reviewers to obtain a general understanding of the data and information contained in the NDA. The summary should be a well-structured, cogent synthesis of the NDA and should be written at the level of detail required by relevant scientific and medical journals. Data in the summary should be presented in tables or graphs. Guidelines for NDA preparation are found in 21 C.F.R. §314.50 "Content and Format of an Application." The summary may be used later by the FDA or the applicant to prepare the *Summary Basis of Approval*, which is made public after approval. Under 21 C.F.R. §314.100, the "review clock" for an NDA is established. The FDA has 180 days from date of "receipt" of the NDA to review the NDA and send the applicant an *approval* letter, an *approvable* letter, or a *not approvable* letter. However, the review clock may be extended to permit review of any major amendments or additional data submitted.[15]

The actual minimum duration of the review period is almost always longer than 180 days from the FDA's receipt of the NDA due to the requirement of 21 C.F.R. §314.101 that an NDA be "filed." After receipt of an NDA, the FDA has sixty days to determine if the NDA is administratively acceptable and thus

---

14. Although paper NDAs are still filed, the FDA accepts (and encourages) "Computer Assisted New Drug Applications" or CANDAs, which allow the reviewer to access the enormous amount of data in an NDA much more easily and to quickly move from points in summary tables of data to actual case report forms.

15. 21 C.F.R. §314.60 (2006) provides that during the course of a review the applicant can submit amendments to an NDA that has been filed, but not yet approved. It is important to note that the submission of a major amendment, such as an amendment which contains significant new study data or analyses, whether prompted by the FDA or not, will constitute an agreement by the applicant to extend the statutory date by which the agency is required to reach a decision on the NDA. This extension may be for as long a period of time as it will take for the FDA to adequately review the amendment, up to a maximum of 180 days. In practice, the FDA sometimes deliberately uses requests for major amendments to lengthen its statutory review times. Furthermore, since the applicant has very little recourse if amendment review times are exceeded, the FDA may exceed them, though the FDA is committed to acting within its statutory review time on most applications in order to meet its PDUFA goals. PDUFA is described in the next paragraph in this subsection.

ready for a substantive review.[16] If the FDA finds a technical deficiency in an NDA, it informs the applicant and the review period does not begin until the deficiency is corrected and the FDA accepts the NDA as "filed." Under §505(c) of the Act, the FDA then has 180 days from the date of filing (plus any extensions because of major amendments) either to (i) approve the NDA or (ii) issue a notice of opportunity for a hearing, if the applicant asked for it in response to an "approvable" or "not approvable" letter.[17] As a result, the FDA really has at least 240 days from receipt of the NDA to respond to the applicant (60 days to "file" the NDA plus 180 days to review it substantively). If the FDA sends the applicant an approvable or not approvable letter, the review period may be longer because of the response and review cycles involved in those cases where the applicant requests a hearing.

The FDA has made significant strides in shortening its average review times in recent years due to the Prescription Drug User Fees Act (PDUFA).[18] Under PDUFA, sponsors of NDAs pay very substantial "user" fees and that money is allocated to increasing the number of reviewers at the FDA and is intended to provide shorter review periods. The FDA has rather detailed goals to meet for various categories of approvals and they are set out on the FDA's website at http://www.fda.gov/oc/pdufa3/2003plan/default.htm. However, the FDA is also being criticized for the recent "premature" approval of several drugs such as Rezulin and Tysabri, which were recalled for safety reasons (Tysabri has since been reintroduced), and Vioxx, which was withdrawn from the market. As a result, although timeframes for FDA decisions are much shorter than they were ten to twelve years ago, the continuing decreases in review times seem to have come to a halt, and there is even evidence of some increases in review times.

When the FDA does issue its determination, it can take one of three forms: approved, approvable, and not approvable letters.[19] An approvable letter, under 21 C.F.R. 314.110, is issued when the FDA feels that an NDA can be approved, provided that certain issues are resolved, additional specific information is submitted, or certain specified conditions are met. As a practical matter, approvable letters have traditionally been used to swiftly address "loose

---

16. Thus it is incorrect for a company to state that it has "filed" an NDA for a product when the documents have been trucked off to the FDA and received in Maryland. The NDA has merely been "submitted" until the FDA has indicated that it is in acceptable form for review, at which point it then receives the status of "filed."

17. These terms are discussed in the text accompanying note 156.

18. Prescription Drug User Fee Act of 1992, Pub. Law 102-571.

19. In biologics, the analogous term for a "not approvable" letter is a "deficiencies" letter.

end" issues, such as labeling changes, in response to applications presenting data that appear to be sufficient to support approval.

The applicant then has ten days after receipt of the approvable letter within which to (a) communicate its intent to make the changes or meet the conditions prescribed in the approvable letter, (b) withdraw the NDA by affirmatively indicating this intent or by not responding within ten days, (c) ask for a hearing on the issues presented in the approvable letter, or (d) request an extension of the review period in order to determine how to proceed.

If the applicant indicates within ten days that it is prepared to amend its NDA to resolve the issues in the approvable letter, then the FDA has 45 days *from receipt of the amendment* to review it and, presuming the response is satisfactory, send an approval letter to the applicant.

Under 21 C.F.R. §314.120, if the FDA sends the applicant a not approvable letter, then the applicant has ten days within which to notify the FDA of its intent to address the issues raised in the not approvable letter by filing a major amendment under 21 C.F.R. §314.60. The review period would then be extended from the date of receipt of the amendment for a period of time adequate to review the additional information, up to a maximum of 180 days.

The FDA can also refuse to approve an NDA. Under 21 C.F.R. §314.125 the FDA must provide the applicant notice of an opportunity for a hearing on such a denial. The grounds for refusing approval are specified under 21 U.S.C. §355(d) and include insufficient evidence of safety or efficacy, manufacturing deficiencies, labeling problems, or any material misstatement of fact.

Under 21 C.F.R. §314.65, an applicant may withdraw an unapproved application without prejudice to its right to refile it. It is not uncommon for an applicant to withdraw and refile a revised NDA with the tacit encouragement and agreement of the reviewers. When the new NDA is resubmitted, it generally includes additional material, similar to that which an applicant would submit in an amendment. In this way, the reviewers can get credit for "dispositioning," which is reviewing and approving the resubmitted NDA in a much shorter period of time.

§8.4(B) Beyond Approval: Post-Marketing Issues

After an NDA has been approved, an applicant can market a new drug. Under 21 C.F.R. §312.85, the FDA may condition approval upon the sponsor's agreement to conduct post-marketing, "Phase IV" studies to obtain additional information about the drug's risks, benefits, and optimal use. Such studies could be designed to explore different doses, study different population groups or patient subsets, study different stages of a disease, or examine the effects of the drug when used for longer periods of time.[20]

Whether or not the FDA requires a Phase IV study, the producer of a drug or biologic faces a number of continuing obligations after receiving market approval. The applicant must notify the FDA of any changes made to an approved NDA. An applicant may do so either by submitting a supplement or by reporting the change in an annual report. In general, under 21 C.F.R. §314.70, any major changes, such as proposed changes in labeling, must be approved through such a submission. Prior notification, but not prior approval, is required if the labeling changes are more restrictive, such as the addition of a warning, precaution or adverse reaction, the strengthening of an instruction regarding dosage or administration designed to increase the safety of the drug, or the deletion of false, misleading or unsupported indications or claims (which begs the question of how they got there in the first place). Any changes of the type submitted in an annual report do not need to be approved prior to implementation. Generally, minor or non-material changes can be made and implemented without prior approval or submission to the FDA in the form of a supplement, as long as they are included in progress reports, which must be filed at least annually.

Under 21 C.F.R. §314.81, the applicant is required to submit other postmarketing reports including data on additional animal toxicology, new manufacturing control or analysis data, later published clinical trial data, any unpublished clinical trial data the sponsor may have, data on pediatric use, and any post-marketing study progress. Submitting and collecting reports is an effort that companies should undertake very seriously and thoroughly. Under 21 C.F.R. §314.81(d), if an applicant fails to make a required report, the FDA may withdraw the approval of an NDA and prohibit continued marketing of a drug.

§8.4(C) Reporting Adverse Drug Experiences (ADEs)

After a new drug is on the market, a sponsor must report adverse drug experiences ("ADEs"). An adverse drug experience is "Any adverse event associated with the use of a drug in humans, whether or not considered drug related" 21 C.F.R. 310.305(b). If an ADE is serious or unexpected, then 21 C.F.R. §314.80(c)(1) states it must be reported as soon as possible, but no later than

---

20. The FDA's enforcement of the commitments made to do postmarketing studies was the subject of recent criticism. *See, e.g., FDA Should Impose Financial Penalties for Postmarket Study Delays, Critics Say*, 5 DRUG INDUS. DAILY 45, March 6, 2006.

fifteen working days after the initial report of the ADE. These are called "fifteen-day alert reports." A serious ADE is one results in: "[d]eath, a life-threatening adverse drug experience, inpatient hospitalization or prolongation of existing hospitalization, a persistent or significant disability/ incapacity, or a congenital anomaly/birth defect" 21 C.F.R. §314.80(a). An unexpected ADE is "[a]ny adverse drug experience that is not listed in the current labeling for the drug product." The sponsor must investigate any serious or unexpected ADE and must submit any new information regarding that ADE within fifteen working days of the receipt of such information. An applicant is obligated to track the frequency of serious and unexpected ADES, and must report any increase in frequency.

§8.5 Accelerated Approval and Early Access to Unapproved Drugs: Drugs for Severely Debilitating and Life Threatening Illnesses

Though the same standards of safety and effectiveness apply to all drugs, the FDA formally provides for greater flexibility in applying these standards to drugs designed to treat life-threatening and severely debilitating diseases, where patients and physicians may be more willing to accept greater risks or side effects than with less serious illnesses. The FDA has also created regulatory procedures to allow terminally ill patients to obtain access to promising drugs that are still in clinical trials, albeit in very limited circumstances. Both of these programs are discussed in this subsection.

§8.5(A) Fast Track, Accelerated Approval, and Priority Review

The FDA Modernization Act of 1997 (FDAMA) codified and expanded upon the FDA's regulatory practices with regard to drugs being developed to significantly improve the treatment of a serious or life-threatening disease, which the FDA terms an "unmet medical need." There are three key differences in the FDA's handling of such severe diseases compared to less serious ones: early and frequent sponsor consultation with the FDA on clinical trial design; early and expanded patient access to an experimental drug to allow treatment use of the new drug to patients who are not enrolled in the controlled clinical trials; and, expedited review with greater latitude for "surrogate" clinical endpoints (and in some cases after only two distinct phases of clinical trials rather than three).

Surrogate endpoints are used to provide clinically-measurable results when measuring the ultimate health outcome would take a long time, thus delaying the drug's approval and availability to desperately ill patients. In the case of accelerated review for serious or life-threatening diseases, "the FDA may grant approval based on an effect on a surrogate endpoint that is *reasonably likely* to predict clinical benefit ('based on epidemiologic, therapeutic, pathophysiologic, or other evidence')."[21] These surrogates are less well established than surrogates in regular use, such as blood pressure or cholesterol for cardiovascular disease. For example, in HIV trials, the real outcome desired is prolongation of survival with a reasonable quality of life. However, because it was quickly and easily measurable and was believed to be predictive, viral load was used as an endpoint even before it was definitively shown to be strongly predictive of the patient's future health. On that basis, AIDS patients received access to a number of life-saving drugs.

Similarly, blood pressure was an accepted clinical endpoint for the approval of anti-hypertension drugs, long before it was definitively established that the lowering of blood pressure in persons with mild-hypertension would affect morbidity and mortality. The difficulty is that some surrogate endpoints turn out to be non-predictive. While the FDA has provided guidelines on surrogate endpoints in a number of disease areas, close coordination with the FDA is imperative in planning any clinical trial in which anything less than the most clear-cut measures of morbidity or mortality will be submitted as a basis for approval.

To the extent that FDA resources allow, the FDA will set up pre-IND meetings to review and reach agreement on animal studies needed to initiate clinical trials, to discuss the scope and design of Phase I clinical studies, and to devise the best approach for presentation and formatting of the IND.[22] For those drugs which the FDA agrees are appropriate for the accelerated review "fast-track" program, the Phase I trial will certainly involve subjects who have the disease and will be designed to provide preliminary data about efficacy

---

21. FDA, Guidance for Industry, Clinical Trial Endpoints for the Approval of Cancer Drugs and Biologics (www.fda.gov/cder/guidance/6592dft.doc).

22. In some cases, drugs can actually be approved with no human trials at all. Under 21 C.F.R. §314.600 et.seq. drugs may be approved on the basis of animal studies only when the human trials "cannot be conducted because it would be unethical to deliberately expose healthy human volunteers to a lethal or permanently disabling toxic biological, chemical, radiological, or nuclear substance; and field trials to study the product's effectiveness after an accidental or hostile exposure have not been feasible." This so-called "Animal Rule" was developed in part as a response to the threat of bioterrorism, but would have other applications as well.

and dosage as well as initial safety data. Such trials are sometimes more accurately referred to as Phase I/II trials, since they combine the goals of both.

At the end of the Phase I/II trial of a "fast-track" drug, promising initial results would lead to another FDA meeting with the sponsor to design controlled Phase II studies that will produce data sufficient enough to support approvability. The clearly intended result of such intensive involvement of the FDA in the clinical trial design is to permit approval, in at least some cases, after only two distinct phases of clinical trials (the latter phase being a "Phase II/III").

At the end of the fast track trial, FDAMA permits the sponsor to file the NDA in stages, to enable the FDA to expedite its review. The FDA, in turn, will designate a drug that has shown efficacy during clinical trials for priority review and accelerated approval. If at the end of the clinical trial process the FDA concludes that the drug is promising, it will grant the drug priority review and accelerated approval. In general, such drugs are being approved in less than six months after the NDA filing and the approval is conditioned on the sponsor's willingness to do postmarketing studies to provide more complete data or to validate the surrogate endpoints used in the approval process.

§8.5(B) Early Access to Experimental Drugs

There are two basic types of early access to experimental drugs, which are drugs in clinical trials but not yet approved; single patient access protocols (commonly referred to as "compassionate use"), and treatment INDs. The single patient protocol, which was long known as and is still sometimes referred to as "compassionate use," can actually follow one of two paths—the Single Patient IND and the emergency use. In a Single Patient IND, either the sponsor or a physician can propose to use a drug that is in Phase II or Phase III to treat an individual patient not qualified or otherwise unable to enroll in an ongoing clinical trial. When a physician initiates the IND, the sponsor must still be willing to supply the drug and cooperate with the request. Generally, the FDA will quickly consent if the patient is seriously or terminally ill, the drug has shown substantial promise of efficacy in early human testing, there is no satisfactory alternative therapy, and the safety of the drug justifies the risk. For example, a patient dying of brain cancer who might be helped by a drug in testing for pancreatic cancer might try to go the Single Patient IND route.

The problem for many Single Patient IND requests is in getting the sponsor's cooperation. The sponsor's concerns are that any adverse effects in a patient outside the clinical trial can create difficulties for the ultimate approval of the drug and are part of the drug's safety history. Nevertheless, if the toxicity of the drug in clinical testing has been very low, sponsors might well agree. While the efficacy data cannot help them formally, it can provide interesting anecdotal information about other indications if there is a clinically beneficial response. In the related Emergency Use, a single patient whose condition does not allow time for the paperwork and IRB approval necessary for a Single Patient IND can ask a sponsor and the FDA for emergency access to a drug in Phase II or III trials. While the sponsor's concerns are still the same, the FDA will respond almost immediately to such requests and allow dying patients access to promising drugs if the sponsor is also willing to provide the drugs to a qualified physician.

The other pre-approval distribution route for experimental drugs is the Treatment IND.[23] If the preliminary analysis of Phase II results are promising, under 21 C.F.R. §312.83 the FDA may ask a sponsor to submit a "treatment protocol" for "treatment use" which, if granted, would remain in effect while the complete data necessary for a marketing application can be assembled by the sponsor and reviewed by the FDA. Under a treatment protocol, a sponsor is essentially allowed to distribute a new drug being tested under an IND on patients who are not enrolled in the clinical trials. Under 21 C.F.R. §312.34, it is clear that the FDA will allow the use of a new drug to treat desperately ill patients as early as possible in the drug development process before general marketing begins. For serious diseases, treatment use may begin during Phase III, while for immediately life-threatening diseases, "treatment use" (including diagnostic use) may begin earlier than Phase III, but ordinarily not earlier than Phase II. Under 21 C.F.R. §312.34(B)(3)(ii) "immediately life-threatening" means a stage of a disease in which there is a reasonable likelihood of death within months or of premature death without early treatment.

Several points merit some elaboration. First, the treatment IND is intended to allow clinical investigators ("qualified experts") to provide the drug to additional patients on an unblinded, uncontrolled basis, while recording minimal data beyond safety reporting. Second, the rules permit a sponsor to recover its costs for the treatment IND patients by charging them for the drug (but only in an amount sufficient to cover costs, not at a "market" rate). Third, the requirement that the clinical trials be continuing or completed reflects the

---

23. Technically, "Parallel Track" expanded access, which is limited to experimental HIV drugs, is still on the books, but is primarily of historical interest as the number of effective antiviral drugs has grown. For an excellent discussion of the topic of early access and single patient use, see J. Morehead, *Single Patient Access to Investigational Drugs: Friend or Foe?*, 21 BIOTECH. L. REPORT 231 (2002).

major problem with treatment protocols, which is that expanded access outside the clinical trial can make it difficult to recruit patients to the blinded, controlled trial where they run the risk of getting the placebo or control drug rather than the promising experimental drug. Of course, the patients in the treatment IND pay for the drug and their medical care while it is ordinarily free to clinical subjects, but that is far from an adequate incentive in many cases. Thus sponsors should carefully consider the difficulty that a treatment IND would pose for completing their clinical trials before submitting a treatment IND protocol to the FDA.

§8.6 Product Exclusivity and the FDA: Orphan Drugs, Patent Term Extension, and Pediatric Study Extensions

While the primary form of intellectual property protection against competition comes from the United States Patent Office and is discussed in Chapter 4, there are several ways in which interaction with the FDA can result in either the extension of the term of a patent or protection against competition independent of any patent rights (or the lack thereof). The first is the Orphan Drug Act, which provides special incentives and exclusivity for drugs developed for smaller patient populations; the second is the general Patent Term Extension provisions of the Hatch Waxman Act; and, the third is the Pediatric use extensions given for testing already-approved drugs for pediatric indications. Each of these is discussed in this section.

§8.6(A) The Orphan Drug Act

The enormous investments required to bring a new pharmaceutical through the regulatory process to the market make the market exclusivity afforded by the patent system a major factor in a company's decision to pursue the development of a potential product (see Chapter 4, Intellectual Property Protection for Biotechnology). This creates a particularly difficult problem for the development of compounds that cannot be protected by patent, particularly if the drug's chief potential is for the treatment of diseases that affect a relatively small number of people.[24] Congress has attempted to address this

---

24. 21 U.S.C. §360bb provides a statutory presumption that diseases affecting less than 200,000 persons annually in the United States are entitled to orphan drug certification.

problem with the enactment of the Orphan Drug Act (21 U.S.C. §360aa *et seq.*). The Act creates several incentives to develop orphan drugs (so-called because the compounds have been "abandoned" in the public domain, in many cases because of the small potential markets involved). The developer of an orphan drug receives Orphan Drug exclusivity, significant tax incentives, and may also receive direct funding for part of the cost of clinical trials.

The most significant incentive is the creation of seven-year periods of market exclusivity for the developer of an orphan drug, during which the FDA may not approve any other application to treat the same indication with the same compound.[25] This provides significant protection, even in the absence of patent protection, when there is no other indication for the compound, thus precluding physician or pharmacy substitution of another manufacturer's drug. In addition to the seven-year market exclusivity provisions of the Orphan Drug Act, the Act also provides three other incentives: significant tax credits for the costs of development, FDA assistance in the design and planning of clinical trials, and limited grant funds in some instances.

There are at least two significant problems that have arisen in the context of the market-exclusivity feature of the Orphan Drug Act. First, there have been instances where two or more biotechnology companies have undertaken to develop the same compound hoping to receive Orphan Drug status for their products. The first to receive NDA or PLA approval precludes the approval of all others for the seven-year period, making the development process a very high-stakes all-or-nothing race. This risk runs counter to the intention of the Act to serve as an incentive. Second, the number of such instances of multiple companies hot on the trail of orphan drug development indicates that rather than serving as a minimal incentive for drug development for rare diseases, some such diseases become extremely lucrative target markets. This has brought repeated criticism of the Act. There have been instances when the manufacturer of an orphan drug has appeared to take unfair advantage of orphan drug status by charging extraordinarily high prices.[26] All of these prob-

---

25. There are exceptions to this rule if the original sponsor either cannot produce enough of the drug to meet the needs of those with the disease or if the original sponsor consents to the approval of additional manufacturers.

26. One of the most prominent such examples is Genzyme's Ceredase therapeutic for Gaucher's disease.

A few medicines, like Ceredase, a treatment for Gaucher disease (pronounced go-SHAY) from the biotechnology company Genzyme, can cost as much as $500,000 a year for some patients. Gaucher disease is a rare metabolic disorder whose symptoms include anemia. (Alex Berenson, *A Cancer Drug Shows Promise, At a Price That Many Can't Pay?*, N.Y. TIMES, Feb. 15, 2006 at A1.)

lems led to much discussion and Congressional passage of amendments to the Orphan Drug Act (vetoed by President Bush in 1990). This illustrates how difficult it is to create by statute a finely balanced set of incentives intended to affect private sector decision making.

For biotech companies, it can be very good business to use the power of proteomics and genomics to develop orphan drugs. The various incentives and additional exclusivity have made developing drugs for orphan targets a popular pursuit for the biotech industry. However, the exclusivity is awarded only at the end of the race, can be engineered around by determined competitors[27] and subjects a company's pricing decisions to unusual scrutiny. The best advice for a biotech company is to consider the probable breadth of orphan drug exclusivity in much the same way one evaluates the degree to which patent claims protect against competition. A company that has substantial patent protection for a protein used as an orphan drug therapeutic is well protected against competition by both its patent and the orphan drug exclusivity. A company that has developed a clinically effective antibody to an antigen that is in the public domain and is known to be of clinical utility in treating an orphan disease will get orphan drug protection (and patent protection) for that particular antibody. Competitors could well develop competing antibodies for the same orphan disease without violating either the patent or the orphan drug exclusivity. Orphan drug incentives are valuable, but don't bet the farm on the exclusivity.

§8.6(B) Patent Term Restoration

The Hatch Waxman Drug Price Competition and Patent Term Restoration Act (1984, subsequently amended numerous times) was sponsored by Senator Orin Hatch and Congressman Henry Waxman, two legislators who could not be further apart philosophically and politically. Not surprisingly, the Act that bears their names is one of the great political compromises and balancing acts of modern times. The Act was intended to promote the introduction of generic drugs and do something to appease the developers of "branded" drugs in exchange. The problem that the Act was intended to solve was to provide some measure of balance in promoting generic drugs without unduly reducing the incentive to develop pioneer drugs. The solution provided by the

---

27. For a more detailed discussion of biotechnology and orphan drugs, see R. Bohrer, *It's the Antigen Stupid: A Risk/Reward Approach to the Problem of Orphan Drug Act Exclusivity for Monoclonal Antibody Therapeutics*, 5 COL. SCI. & TECH. L. REV.1 (2003) (available at http://www.stlr.org/html/volume5/bohrer.pdf) and R. Bohrer and J. Prince, *A Tale of Two Proteins: The FDA's Uncertain Interpretation of the Orphan Drug Act*, 12 HARV. J. OF L. & TECH. 365 (1999).

Act was to allow generic manufacturers to begin testing their copycat drugs before the pioneer's patent expired and to gain approval by submitting only data which showed that their generic version was "bioequivalent" to the pioneer drug by comparing the two drugs' metabolic properties and rate of clearance in a limited number of patients. In return, to limit the impact on the incentives for innovator companies, the Act gave the developers of new pharmaceuticals an extension of their patent terms under a complicated formula. The formula adds 1) one-half of the time taken in actual clinical trials (from Phase I IND to NDA filing, less any time in which the sponsor delayed the process by lack of diligence, plus 2) the time taken by the FDA in regulatory review from the time of NDA submission to approval. This sum of one-half the clinical trial time plus all of the FDA review time is then added to the patent, up to a maximum extension of five years, with a further limitation that the resulting remaining patent term cannot exceed fourteen years. Sound complicated? Try reading the statute—35 U.S.C. §156. The following hypothetical illustration might help.

- Sponsor Allcure gets the go ahead from the FDA to begin the Phase I clinical trials on January 2, 1996.
- Without ever delaying progress due to lack of diligence, an NDA is filed on January 2, 2004.
- The FDA issues Allcure its approval on July 2, 2004.
- Allcure's original patent application for the active ingredient in the drug was filed on July 2, 1995 and would expire on July 2, 2015.
- Allcure's testing time was 8 years, one-half of which is 4 years.
- The FDA's approval time was 8 months, leading to an extension of 4 years, 8 months.
- The five-year maximum cap does not apply. On the date of approval, there are exactly 11 years left on the patent term.
- Adding in the 4 years 8 months would yield a term of 15 years 8 months.
- The fourteen-year maximum does apply, thus the maximum extension is 3 years 8 months.

All of this also depends on Allcure applying for its patent term extension promptly within the sixty-day period after it receives notice of approval. Still, if sales at the end of the drug's patent term are $50 million a month, the extension provides a tidy $2.2 billion in sales during the extra 44 months.[28]

---

28. It is also possible for Allcure to go back later and do an Alzheimer's trial to get Alzheimer's (or some other additional indication) on the label of the drug that was approved for stroke. In that case, Allcure will receive an additional three years of protection,

Hatch Waxman and generic drugs are generating enormous controversy at this time, primarily about the battles that go on when pioneer drug manufacturers add additional patent protection to their original patent, for example on the formulation or the method of manufacture. These are beyond the scope of this book, but are almost certain to lead to future changes in this complicated and economically critical area of the law.

§8.6(C) Pediatric Study Patent Extensions

Kids are not just small adults. That statement is not just a psychological or philosophical comment on childhood, but an important truth about drug metabolism and action in children. Predicting whether or not a drug is safe and effective in children and at what dose, cannot be determined simply by a formula in which a child's weight or body surface area is the numerator and an adult weight or body surface area is the denominator. Nevertheless, the great majority of drugs, whether cancer drugs, antibiotics, or antidepressants, are given to children with insufficient clinical trial data to provide adequate information about safety, efficacy, or dosage.

To remedy this alarming lack of crucial information, Congress enacted 21 U.S.C. §355a Pediatric Studies of Drugs, which is not to be confused with 21 U.S.C. §355(a) of the FDA Modernization Act of 1997. Under the terms of §355a the Secretary of HHS may request that the sponsor of an approved drug covered by a patent conduct additional studies in pediatric age groups to the extent that the Secretary has determined that health benefits would be achieved by using the drug in those age groups. In other words, if the Secretary determines that a drug may be useful in children, he can request additional studies on a drug that is already marketed or is still in clinical trials before marketing. The sponsor then has 180 days from receipt of the Secretary's request either to agree to do the studies and provide a date by which the studies should be completed or to respond by refusing to do pediatric studies.

The incentive for a drug manufacturer to do the studies is an additional six-months of additional protection against generic competition added on to their patent or orphan drug exclusivity. Manufacturers should promptly con-

---

after their patents on the stroke drug expire, which would prevent the FDA from approving any other company to produce a generic version of Allcure's drug with Alzheimer's on the label (21 U.S.C. §355(j)(5)(F)(iii)). However, that additional protection will not be worth very much, as the generic versions of Allcure's stroke drug would then be sold at a major discount and will quickly be used "off-label" for stroke without much recourse by Allcure.

fer with the FDA in order to ascertain whether or not the FDA formally agrees that their proposed study protocols will provide sufficient data to support appropriate labeling for pediatric use and then conduct the studies according to agreed-upon protocols. The sponsor then submits the results of the studies and the FDA determines whether or not the studies were in fact performed in accordance with the agreed upon protocols. In exchange for performing the agreed upon studies and submitting the data from the protocols (and any necessary labeling changes), the sponsor is rewarded six months added to their patent-based or orphan drug-based exclusivity. If the drug has two or more indications (e.g. strep infections and staph infections) it is possible for the sponsor to be asked to perform two or more such studies to receive a maximum of two six-month extensions.[29]

Are the six-month extensions worth the cost of designing and running pediatric trials? In general, yes. The trials need not be large or long in duration. The FDA is even more concerned about the differences in pharmacokinetics (half-life and drug metabolism) between children and adults than it is about statistical significance in efficacy endpoints. So the studies may well cost a couple of million or more, but almost any drug worth studying because of significant use in children earns more than a couple of million dollars each month in net profits, while some earn many times that. Six months of additional protection from generic competition generally provides the opportunity to recoup the additional investment many times over. Why would some sponsors balk at the Secretary's request to do such studies? It is hard to speculate, but perhaps some drug manufacturers really do not want to know how one of their drugs works in kids or even worse, perhaps learn that the drugs should not be used in kids at all, which might be the result of the study. Still, the Secretary can ask the Foundation for the NIH (an entity statutorily created to perform such studies) to study the drug if the sponsor rejects the Secretary's 355a request. In addition, the Secretary is statutorily authorized to make public the names of the drugs and the manufacturers who have refused to do a pediatric study, which in itself ought to limit the pediatric market. It

---

29. (g) Limitations. A drug to which the six-month period under subsection (b) or (c) has already been applied—
    (1) may receive an additional six-month period under subsection (c)(1)(A)(ii) for a supplemental application if all other requirements under this section are satisfied, except that such a drug may not receive any additional such period under subsection (c)(2); and
    (2) may not receive any additional such period under subsection (c)(1)(B). 21 U.S.C. §355a(g).

is no wonder that most manufacturers are eager to do the small studies required to gain their six months of additional brand life.

§8.7 "Generic Biopharmaceuticals"

One of the most difficult issues surrounding the biotech industry today is what the FDA will require for "generic" versions of the first generation of biotechnology protein drugs when their patents expire. What type of data will be necessary to secure approval from the FDA to sell a "generic" erythropoietin or interferon-beta? Will full-scale human efficacy trials be required or could relatively simple *in vivo* metabolic and immunologic data combined with *in vitro* structural analysis of the "generic" protein suffice? This could be a $64 billion dollar question (and that is just per year)! In other words, the entry barriers to new copies of erythropoietin will either serve to protect Amgen's multi-billion-dollar-per-year drug for some time beyond the expiration of its patent (as well as the branded lives of other protein therapeutics like it) or allow speedy entry of cheaper "generic" versions of the drug soon after the principal patents expire. At the time of this writing, the rules that will govern this enormously high-stakes issue are still uncertain.[30] This section briefly introduces the scientific and legal framework regarding generic biopharmaceuticals and concludes with a brief survey of the possible outcomes.

The problem, from a scientific perspective, is how to determine whether two large complicated glycoproteins are clinically identical. Erythropoietin is a 165 amino acid protein with multiple glycosylation sites. There are several technical issues raised by any would-be "generic" erythropoietin that has the same linear amino acid structure as the approved protein. The first problem is that the number of glycosylation sites and the precise structure of the carbohydrate molecules which are attached at those sites may vary in different recombinant protein expression systems, not only from E. coli to yeast to CHO to human, but even within a particular species between different cell types at different fermentation conditions. Because of this, there can be differences in the three dimensional shape of proteins produced by different ex-

---

30. However, in the EU, the EMEA (European Medicines Evaluation Agency) has announced its first approvals of such drugs, a copycat human growth hormone (*see* http://www.emea.eu.int/pdfs/human/press/pr/3179706en.pdf), and also has begun issuing guidelines for further such approvals (http://www.emea.eu.int/pdfs/human/biosimilar/9452605en.pdf). The FDA followed the EMEA with its own approval, *see F.D.A. Approves A Generic Biodrug*, N.Y. Times, June 1, 2006 at C5.

pression systems or in different conditions. From a regulatory perspective, the problem is that two versions of, for example, erythropoietin, made by two different companies, can vary in glycosylation and folding, and therefore the sponsor of a "generic" protein must be required to provide data that clearly pinpoints any difference between the two proteins in their glycosylation or folding. In addition, if there are any such differences, the current state of science cannot predict whether such differences will or will not result in a difference in biological activity and therapeutic safety and efficacy. The third issue then is to determine the clinical significance, if any, of those differences that may arise between two different versions of the same protein. Of course, this third issue exists only if there are, in fact, differences at the chemical/structural level.

The bottom line is that there are rapid advances in the science necessary for solving the second problem. There are increasingly powerful laboratory techniques being developed to determine whether or not two proteins are structurally identical.[31] If two proteins can be rapidly determined to be structurally identical, then there is no real scientific reason not to allow a second "generic" version of a protein therapeutic to be rapidly introduced, as long as the purity and freedom from contamination of the manufactured product is also scrutinized. If there are any differences, it may not require full-scale clinical trials to determine their significance, but precisely what should be required to determine "bioequivalence" for two similar but not identical proteins is a tough problem.

What then are the legal issues? The answer to this question calls to mind a classic scene from the 1974 film directed by Roman Polanski, CHINATOWN:

> Jake: (He slaps her.) I said, 'I want the truth.' Mrs. Mulwray: She's my sister. (He slaps her again.) She's my daughter. (Slap.) My sister, my daughter. (Slap. Slap.)

---

31. See, e.g., M. Ohta, N. Kawasaki, S. Itoh, T, Hayakawa, [*Study on evaluating methods for the quality control of glycoprotein products. (IV)—erythropoietin products. Part 4*], 120 Kokuritsu Iyakuhin Shokuhin Eisei Kenkyusho Hokoku. 89-97, (2002). (Article in Japanese, the citation and following quotation are taken from the Pubmed Abstract: PMID: 12638188):

> We demonstrated that the mass spectrometric peptide mapping can elucidate the identity of protein part, and differences in site-specific carbohydrates heterogeneity due to acetylation and sulfation among the three rhEPOs. Our method can thus be useful in comparability assessment of therapeutic glycoproteins.

Jake: I said, 'I want the truth!' (He throws her against the sofa.) Mrs. Mulwray: ... **She's my sister *and* my daughter!**.... [32]

Now, how does the regulatory problem call to mind CHINATOWN? It's a drug. It's a biologic. It's a drug. **It's a biologic AND a drug.** The FDA originally decided that all the new recombinant drugs (except for Insulin and Growth Hormone, for reasons that are not germane here) would be regulated as biologics. However, §8.11 of this chapter discusses the transfer of jurisdiction over therapeutic proteins from CBER to CDER and noted that the broad definition of drug included biologics, so that all biologics could be regulated as drugs. The reorganization of CBER and CDER notwithstanding, the FDA's position has long been that the scientific complexity of therapeutic proteins justified regulating them separately as biologics and precluded any "generic" copies of proteins. Of course, the science justifying that position is rapidly changing.

The FDA's statutes may complicate the issue regardless of the changing science. Hatch-Waxman's simplification of the procedures for approving generic drugs in 1984, as was described in §8.62, did two things: first, it allowed generic manufacturers to begin developing and testing their copycat drugs before the patents on those drugs expired; and, second, it gave the pioneer manufacturers an extension of their patent life for part of the clinical and regulatory development time. The problem is that the patent term extension section of the Act (Title II) specifically granted patent term extension to both biologics licensees and the makers of approved new drugs, while Title I, simplifying the generic approval procedures, ONLY mentioned new drugs approved under NDAs (remember, biologics receive PLAs). So while the FDA *could*, if it chooses, approve future therapeutic proteins with NDAs, all of the existing therapeutic proteins have PLAs (or the predecessor BLA), thereby allowing Amgen and other pioneer biotechnology companies to argue that Hatch Waxman's clear statutory intent was to allow generic approval procedures only for NDA-approved drugs and not PLA-approved proteins.

The regulatory framework is still being hammered out and, whatever the result, will likely be subjected to some combination of court challenges and possible legislative reaction to settle the issue. In the meantime, the FDA has taken a baby step[33] towards generic biopharmaceuticals by putting CDER in

---

32. Tim Dirks, Review, *Chinatown*, 1974, http://www.greatestfilms.org/chin4.html (last visited April 4, 2006).

33. In a recent review of the biogenerics issue, GENERIC BIOLOGICS: OUTLINE OF LEGAL AND REGULATORY ISSUES, Biotechnology Law 2003 Biotechnology Patents & Business Strategies (PLI, 2003), Wayne Matelski quoted the following excerpt from BIO-

charge of drugs. Whether or not that turns out to be a giant step may well hinge on how CDER handles the first application for a "generic" therapeutic protein filed under 505(b)(2).[34] While the FDA is continuing to formulate a general policy, the EU announced its approval of its very first biogeneric drug, a "copy" of the human growth hormone protein and the FDA followed with its own approval of that protein.[35]

The 505(b)(2) Guidelines present a potentially major shortcut for would-be generic biopharmaceutical manufacturers. §505(b)(2) was part of the original Hatch Waxman tradeoff, but it has been largely ignored. The Act gave generic drug manufacturers two routes to the market: §505(j) and §505(b)(2).[36] The preferred route for virtually all generic drugs since 1984 has been through §505(j), which permits ANDA approvals of generic drugs that have identical active ingredients and comparable bioavailability. §505(b)(2) provides a route for the approval of not-quite-identical drugs, through:

> An application submitted under paragraph (1) [NDA] for a drug for which the investigations described in clause (A) of such paragraph and relied upon by the applicant for approval of the application were not conducted by or for the applicant and for which the applicant has not obtained a right of reference or use from the person by or for whom the investigations were conducted ... 21 U.S.C. §355(b)(2).

In other words, a §505(b)(2) NDA can be used when some or all of the safety and efficacy data was submitted to the FDA by another party, the pioneer manufacturer. Note that this solution gets around the problem of that ANDAs were authorized only for copies of drugs approved by NDA, because the FDA has full authority to allow biologics manufacturers to market their drugs under *either* an NDA or a BLA. If a would-be copycat manufacturer of erythropoietin submits, in a 505(b)(2) NDA, detailed data establishing that

---

CENTURY EXTRA, on April 2, 2003, at 1:
   Human insulin and growth hormone present opportunities for approving generics under current law.... I have a vision that includes effective and safe biogenerics potentially being available in the very long-term.... We are taking some baby steps now ... toward creation of a biogenerics approval mechanism.

34. 21 U.S.C 355(b)(2), *see also* FDA, Guidance for Industry, Applications Covered by Section 505(b)(2) DRAFT GUIDANCE, October 1999 (*available at* http://www.fda.gov/cder/guidance/2853dft.pdf).

35. The EMEA approval is posted at http://www.emea.eu.int/pdfs/human/press/pr/3179706en.pdf). The FDA's subsequent approval received wide press coverage, *see* F.D.A. *Approves A Generic Biodrug*, N.Y. TIMES, June 1, 2006 at C5.

36. 21 U.S.C. §355(j) and §355(b)(2).

their molecule is chemically and structurally identical to Amgen's, theoretically the FDA has the statutory power to supplement the chemistry, manufacturing and controls data with safety and efficacy data filed years before by Amgen! Now, the theoretical possibility is still not one that the FDA is firmly committed to implement at this time (it is only a "Draft Guidance"), but it does point to a possible solution, albeit one that, like any other solution, is going to be put through a litigation and lobbying minefield before it could become accepted practice.

§8.8 Advertising and Promoting Prescription Drugs (Herein Also a Brief Word about Dietary Supplements-DSHEA)

While a company is awaiting word from the FDA that yes, the Phase III data is wonderful and its breakthrough drug is approved, the marketing campaign should be already mapped out and ready to go. This section provides an introduction to the rules and regulations that govern the advertising, marketing, and promotion of prescription drugs and biologics. §8.81 discusses the legal background of the FDA's marketing and promotion regulation. §8.82 summarizes the FDA rules about marketing to physicians and formulary committees or similar organizations about a drug's approved use and any off-label uses for which there is reasonable scientific evidence, as well as label information about the economic advantages of a drug, commonly referred to as "pharmacoeconomics." §8.83 summarizes the FDA rules about marketing directly to consumers in print and in broadcast media and provides a brief look at a representative enforcement action—a warning letter from CDER's Division of Drug Marketing, Advertising and Communications (DDMAC) to Johnson & Johnson. §8.84 provides a very brief introduction to the very different marketing universe of dietary supplements, a booming industry targeting many of the same markets as biotech and pharmaceutical companies yet with almost none of the marketing and promotional limitations.

§8.8(A) The Legal Background of the FDA's Marketing and Promotion Regulation

The basic statutory authority for the FDA's regulation is the prohibition against misbranding, found in 21 U.S.C. §352(a), defined as any false or misleading labeling and in §352(n) requiring advertising to contain adequate information from the drug label to avoid being false and misleading. Remember, the label of a drug is the full statement of information that is approved (as part of the NDA final approval) about the drug's chemical name, ingredients, approved uses, warnings, contraindications, side effects, and adverse effects. Thus Congress has required that prescription drug advertising contain adequate information about the drug to avoid being misleading. Further statutory authority regulating prescription drug marketing and promotion is found in 21 U.S.C. §§ 360aaa through aaa-5.

The FDA's statutory and regulatory authority over prescription drug advertising and promotion have been the subject of significant litigation in recent years, with a major Supreme Court decision[37] as well as significant decisions by the Court of Appeals for the District of Columbia.[38] The issue in these cases was the extent to which prescription drug advertising is speech protected by the First Amendment and whether or not the FDA's regulation of prescription drug promotion violated the First Amendment's protections. In *Thompson v. Western States Medical Center* the Supreme Court, in a 5-4 ruling, held that pharmacy advertising of their compounding of particular drugs was commercial speech that received substantial

First Amendment protection and that the FDA's regulations went too far in restricting that advertising. The commercial speech doctrine permits government regulation of speech only so long as it meets the so-called "Central Hudson" tests:

> In *Central Hudson* ... we articulated a test for determining whether a particular commercial speech regulation is constitutionally permissible. Under that test we ask as a threshold matter whether the commercial speech concerns unlawful activity or is misleading. If so, then the speech is not protected by the First Amendment. If the speech concerns lawful activity and is not misleading, however, we next ask "whether the asserted governmental interest is substantial." *Id.,* at 566, 100 S.Ct. 2343. If it is, then we "determine whether the regulation di-

---

37. Thompson v. Western States Medical Center, 535 U.S. 357 (2002) (holding that the FDA's regulation of pharmacy advertising of compounding of prescription drugs violated the 1st Amendment).

38. Washington Legal Foundation v. Henney, 202 F.3d 331, (D.C. Cir. 2000) (vacating a District Court injunction barring the FDA from enforcing Guidance documents on off-label use and Continuing Medical Education programs, on grounds that the FDA's interpretation of FDAMA mooted the controversy).

rectly advances the governmental interest asserted," and, finally, "whether it is not more extensive than is necessary to serve that interest." *Ibid.* Each of these latter three inquiries must be answered in the affirmative for the regulation to be found constitutional. Thompson v. Western States Medical Center, 535 U.S. 357 at 367.

In *Thompson v. Western States Medical Center* the Supreme Court makes it clear that the FDA's regulations in question, concerning the advertising of pharmacy compounding services, is subject to the First Amendment and that the appropriate analysis is the *Central Hudson* four-part test for commercial speech set out in the quoted excerpt. However, it does little to clarify what would happen if other FDA advertising and marketing regulations were to be challenged. Even though the current statute and regulations are largely a response to prior lower court decisions applying the commercial speech doctrine to a former ban by the FDA on off-label promotion, future challenges might result in further curtailment of the FDA's regulatory authority. It is interesting to note that the significant First Amendment challenges discussed here have been not been made by pharmaceutical companies, but by a conservative legal think tank (Washington Legal Foundation) and a pharmacy (Western States Medical Center). The current framework does not significantly impair the ability of pharmaceutical companies and biotech companies to promote their products; so the best advice to a biotech company clearly is to follow the rules and assume that they are constitutional.

§8.8(B) Marketing to Physicians and Formulary Committees

Major pharmaceutical companies and major biotechnology companies (that is biotechnology companies with significant therapeutic products) do a great deal of marketing and promotion to physicians. This marketing takes two forms: continuing medical education programs (often in the form of all-expenses paid trips to seminars in desirable locations); and, direct contact by pharmaceutical company representatives known as "detailers" (who often have slick promotional items and free samples to hand out). Industry-supported educational programs, regardless of the location or duration, whether a full week in the Bahamas or an evening at a steakhouse, are considered by the FDA to be advertising. The FDA does generally not scrutinize advertising directed at physicians as long as it promotes the use of an approved product for an approved use. It is only when unapproved uses are involved that the FDA becomes concerned, whether the promotion is done by an after-dinner speaker or a pharmaceutical company representative. If the event is a scientific event that is "controlled" by a pharmaceutical company, in the 1997 Guidance[39] the FDA has expressed its concern that the sponsor's interest be disclosed, that the discussion of products be balanced, that the sponsor ought not to have control over speaker selection, and so forth. An FDA attempt to sanction a pharmaceutical company for organizing a scientific event that does meet the concerns of the Guidance might or might not survive judicial scrutiny if challenged as violating the free speech rights of pharmaceutical companies.

Contact by detailers directly to doctors or by pharmaceutical companies at booths at meetings are much more tightly controlled when it comes to off-label uses and are scrutinized by the FDA for false and misleading statements as well. While the eminent (hypothetical) Dr. Neuron, a faculty member at Prestige U., is entitled to speak at a meeting concerning her recent study of Allcure's stroke drug in a group of Parkinson's patients, employees of pharmaceutical companies are more limited. If Allcure wishes inform physicians of the possible use of its drug in Parkinson's patients, Allcure can only insure that it will avoid a misbranding enforcement action by the FDA by adhering to several criteria:

1) the information provided to physicians is contained in a peer-reviewed scientific or medical journal,
2) the company provides an unabridged copy of the Neuron paper (and sends one to the FDA sixty days before outside distribution),
3) the company discloses that Parkinson's use is not approved by the FDA,
4) provides balance by also furnishing or disclosing any negative studies concerning that use, and,
5) is proceeding with reasonable diligence (generally defined as within 36 months, unless an exemption from this requirement is granted by the FDA) to complete its own studies in preparation for filing an SNDA with the FDA for the Parkinson's indication.[40]

Off-label promotion to physicians can, in appropriate circumstances, be a very important part of a company's business plan, but it requires good studies, peer-reviewed publications, a good-faith effort to follow the FDA Guidelines, and an effort to put the use on the label within a reasonable time.

---

39. Final Guidance on Industry-Supported Scientific and Educational Activities, 62 Fed. Reg. 64074 (Nov. 24, 1997).

40. Washington Legal Found. v. Henney, 202 F.3d 331 (D.C. Cir. 2000); Dissemination of Information on Unapproved/New Uses for Marketed Drugs, Biologics, and Devices, 63 Fed. Reg. 64556 (1998).

Increasingly important in pharmaceutical marketing are efforts to win inclusion on managed care formularies[41] by providing studies that demonstrate the favorable health care economics (pharmacoeconomics) of a product,[42] such as a reduction in the number and cost of hospitalizations by a new asthma drug compared to other treatments or a reduction in the number of fractures and the cost of their treatments due to a new osteoporosis drug. Under §21 U.S.C. 352(a) such communications are safe from FDA enforcement actions so long as they concern an approved use and are also supported by "competent and reliable scientific evidence."

§8.8(C) Direct-to-Consumer (DTC) Advertising in Print and by Electronic Media

Direct to consumer advertising is big business for major pharmaceutical companies and often features drugs licensed from biotechnology companies, such as ICOS's Cialis. Television shows are filled with commercials for drugs for heart disease, acid reflux, high cholesterol, erectile dysfunction, and virtually every other category in which U.S. sales can be a billion dollars a year. While it is clear that such ads are commercially protected speech, the FDA has very particular rules to follow to avoid a product being "misbranded," and such ads have come under increased scrutiny in the wake of the controversy over Vioxx's cardiovascular risks.[43] The most important of these rules is that any ad, whether in print or broadcast, provide the consumer with a "brief summary relating to side effects, contraindications, and effectiveness" as required by the FDA under regulations authorized by 21 U.S.C. §352(n). For print ads, that usually means putting the contents of the relevant portions of the label in small type at the end of the ad.

Frequently, sponsors print in small type, verbatim, the risk-related sections of the approved product labeling (also called the package insert, professional

---

41. A formulary is the list of preferred medications for prescription or reimbursement by a government payor, managed care organization or other health-care provider.

42. 21 U.S.C. §352(a) defines health care economics:

health care economic information" means any analysis that identifies, measures, or compares the economic consequences, including the costs of the represented health outcomes, of the use of a drug to the use of another drug, to another health care intervention, or to no intervention....

43. *See*, e.g. FDA Week, *FDA Happier With Drug Ads; Companies Preclearing Commercials* February 24, 2006; Dean Starkman, *Pfizer to Change Drug-Ad Policy; FDA, Doctors To Get Early Look*, WASH. POST, Aug.12, 2005, D02.

labeling, prescribing information, and direction circular). This labeling is written for health care professionals, using medical terminology. The FDA believes that this is one reasonable way to fulfill the brief summary requirement for print advertisements directed toward health care professionals, but may be difficult for consumers to understand.[44]

In broadcast ads that mention a particular product, the "safe harbor" requirements that insure against an enforcement action are laid out in the FDA's 1999 Guidance for Industry: Consumer-Directed Broadcast Advertisements. In broadcast ads, drug sponsors can either include the brief summary that would be provided in the print ad or, alternatively, since that is generally impractical on TV and radio, may make "adequate provision" for disseminating the full labeling information (21 C.F.R. §202.1(e)(1)). In addition to this "adequate provision" requirement (discussed below), the ad must also:

- ... not [be] false or misleading in any respect. For a prescription drug, this would include communicating that the advertised product is available only by prescription and that only a prescribing healthcare professional can decide whether the product is appropriate for a patient.
- Present a fair balance between information about effectiveness and information about risk.
- Include a thorough *major statement* conveying all of the product's most important risk information in consumer-friendly language.
- Communicate all information relevant to the product's indication (including limitations to use) in consumer-friendly language. http://www.fda.gov/cder/guidance/1804fnl.pdf.

The "adequate provision" requirement is intended to provide the consumer with information concerning ALL the products risks, unlike the "major statement" which covers only the most important risks, meaning the risks of the most serious adverse effects or the most frequent side effects. How can the "adequate provision" requirement be fulfilled? There are several possibilities, including providing a toll-free number to call (at which point the caller has

---

44. (DRAFT) GUIDANCE FOR INDUSTRY: Using FDA-Approved Patient Labeling in Consumer-Directed Print Advertisements (http://www.fda.gov/cder/guidance/4114dft.htm) (2001). Although the last portion of the quote suggests that the FDA does not look favorably upon the practice, it does allow it, and the draft guidance on the subject is intended to raise the possibility that where more patient-friendly labeling that discusses the serious and frequent risks has been approved, it is probably a better choice for DTC print ads.

the choice of having the information read aloud or mailed) or providing a website URL where the information posted or providing handouts at pharmacies and other locations.

A major enforcement issue in DTC advertising has not been the adequate provision requirement, but the problems created by the media—the power of suggestion used in so many subtle ways. For example, in an FDA DDMAC enforcement letter to Johnson & Johnson concerning a 30-second television ad for an acne medication, one of the FDA's concerns was that the ad overstated the product's efficacy by using two actors with clear skin, while the clinical trials showed only that the drug group had a 32–45 percent reduction in total lesion count.[45] The reason that pharmaceutical companies are spending billions on DTC ads is clearer than the average teenager's complexion: the ads work. However, while complying with the FDA's regulations would seem to be relatively straightforward, the temptation to mislead consumers about efficacy in subtle ways is very hard to resist. If a biotech company is going to advertise its rheumatoid arthritis drug on television, it better be prepared to have the FDA review the ad in advance to scrutinize exactly how spry those folks in the commercial appear to be.

§8.9 DSHEA—Anything Goes (Almost)

In 1994 Congress enacted the Dietary Supplement Health Education Awareness Act (DSHEA).[46] In the decade since, the vitamin and supplement market has boomed, growing 28 percent from 1996 to 2001 to reach $17.8 billion in 2001 U.S. sales according to The Wall Street Journal.[47] Some dietary supplements are now widely recommended by mainstream physicians; for example the glucosamine-chondroitin supplements to relieve arthritis reached $600 million in 2001 U.S. sales.[48] This section provides a brief introduction to two major aspects of dietary supplement regulation: the definition of a dietary

---

45. Letter from Rebecca Williams, Pharm.D., review officer at CDER/DDMAC to George Latyszonek, Director, Regulatory Affairs, Johnson & Johnson Consumer Companies, Inc. *available at* http://www.fda.gov/cder/warn/2002/11125.pdf (last visited April 24th, 2006).
46. Dietary Supplement Health and Education Act of 1994, Pub. L No. 103-417 (1994).
47. Andrea Petersen, *States Grant 'Herb Doctors' New Powers*, Wall St. J. Aug. 22, 2002, at D1.
48. Robert J. Davis, *Aches & Claims: Other Ways to Ease Pain From Arthritis*, Wall St. J., June 11, 2002 at D4.

supplement and permissible claims or labeling, and discusses the significance of DSHEA for the biotechnology industry.

DSHEA provides a lengthy definition of a dietary supplement that permits a wide range of dietary supplements:

(ff) The term 'dietary supplement'—
(1) means a product (other than tobacco) intended to supplement the diet that bears or contains one or more of the following dietary ingredients:
(A) a vitamin;
(B) a mineral;
(C) an herb or other botanical;
(D) an amino acid;
(E) a dietary substance for use by man to supplement the diet by increasing the total dietary intake; or
(F) a concentrate, metabolite, constituent, extract, or combination of any ingredient described in clause (A), (B), (C), (D), or (E);....
(2)....
(B) is not represented for use as a conventional food or as a sole item of a meal or the diet; and
(C) is labeled as a dietary supplement; and
(3) does—
(A) include an article that is approved as a new drug under section 505, certified as an antibiotic under section 507, or licensed as a biologic under section 351 of the Public Health Service Act (42 U.S.C. 262) and was, prior to such approval, certification, or license, marketed as a dietary supplement or as a food unless the Secretary has issued a regulation, after notice and comment, finding that the article, when used as or in a dietary supplement under the conditions of use and dosages set forth in the labeling for such dietary supplement, is unlawful under section 402(f); and
(B) not include—
(i) an article that is approved as a new drug under section 505, certified as an antibiotic under section 507, or licensed as a biologic under section 351 of the Public Health Service Act (42 U.S.C. 262), or
(ii) an article authorized for investigation as a new drug, antibiotic, or biological for which substantial clinical investigations have been instituted and for which the existence of such investigations has been made public, which was not before such approval, certification, licensing, or authorization marketed as a dietary supplement or as a

food unless the Secretary, in the Secretary's discretion, has issued a regulation, after notice and comment, finding that the article would be lawful under this Act. 21 U.S.C. §321(ff).

This definition is very broad, including any vitamin, mineral, herb, botanical, dietary substance, or concentrate, extract, constituent, or ingredient of any of those (including extracts of herbs, botanicals and dietary substances). The term dietary supplement is further limited in §413 of the Act (21 U.S.C. §350b):

> A dietary supplement which contains a new dietary ingredient shall be deemed adulterated [*i.e. unlawful*] under section 402(f) unless it meets **one** of the following requirements [emphasis added]:
> (1) The dietary supplement contains only dietary ingredients which have been present in the food supply as an article used for food in a form in which the food has not been chemically altered.
> (2) There is a history of use or other evidence of safety establishing that the dietary ingredient when used under the conditions recommended or suggested in the labeling of the dietary supplement will reasonably be expected to be safe and, at least 75 days before being introduced or delivered for introduction into interstate commerce, the manufacturer or distributor of the dietary ingredient or dietary supplement provides the Secretary with information, including any citation to published articles, which is the basis on which the manufacturer or distributor has concluded that a dietary supplement containing such dietary ingredient will reasonably be expected to be safe....
> For purposes of this section, the term 'new dietary ingredient' means a dietary ingredient that was not marketed in the United States before October 15, 1994 and does not include any dietary ingredient which was marketed in the United States before October 15, 1994. 21 U.S.C. §350b(a).

These two lengthy definitions present at least two complexities worth clarification here: the first is with respect to the pre-1994 cut-off for dietary ingredients that are statutorily not-adulterated; and, the second is with respect to the interaction between DSHEA and the drug approval process.

First, the pre-October 1994 cut-off for the statutorily defined non-adulterated dietary ingredients means that any food extract, concentrate, or ingredient has to be derived from a food or food ingredient that was sold in the United States before October 15, 1994, or, at least 75 days in advance of sale, the manufacturer must provide the Secretary of HHS with a premarket notice that supports the manufacturer's conclusion that the supplement will be safe, with adequate evidence documenting the ingredient's "history of use or other evidence of safety." So regardless of the date of its introduction into the U.S., any foreign botanical or herb can be sold under DSHEA if the manufacturer provides adequate evidence supporting its "history of use or other evidence of safety," for example by documenting its long use as an herbal medicine in China.

There are two problems with this premarket notification requirement for dietary supplements that consist of or contain new dietary ingredients. First, there is no list of the foods or constituents of food consumed in the United States prior to October 1994. This is for the manufacturer to determine in deciding whether or not to file the premarket notification and, as in *Pharmanex v. Shalala*, it can be a point of factual dispute.[49] Second, if the manufacturer does submit a premarket notification, the statute leaves largely to the FDA's judgment whether or not the submitted evidence supports the conclusion that the supplement "when used under the conditions recommended or suggested ... will reasonably be expected to be safe." The FDA can be very hard on "new dietary ingredients" that do not have good scientific studies supporting safety of the precise formulation, dosage, and duration of consumption.[50]

In addition to the confusion and ambiguity surrounding the "new dietary ingredient" standards, and more important for pharmaceutical companies and biotech companies, if a substance is either approved as a drug or biologic or is in clinical trials for approval as a drug or biologic before it is marketed as a dietary supplement, it cannot be marketed as a dietary supplement. Nevertheless, if a marketed dietary supplement contains an active ingredient that is later isolated, tested and approved as a drug, the dietary supplement may continue to be sold, in competition with the drug (which more or less makes the prescription drug subject to "generic" competition from the date of its approval). This particular tangle of NDAs and DSHEA was the subject of a courtroom battle between Merck and Pharmanex[51] over the sale of Pharmanex's "Cholestin," a red rice yeast extract which is high in lovastatin, the

---

49. Pharmanex v. Shalala, 221 F.3d 1151 (10th Cir.2000).

50. *See* Memoranda, http://www.fda.gov/ohrms/dockets/dailys/03/Feb03/020703/8004 dff5.pdf and http://www.fda.gov/ohrms/dockets/dockets/95s0316/95s-0316-rpt0204-01-vol154.pdf (second item), for examples of the many memoranda posted on the FDA docket, which reject a premarket submission.

51. Pharmanex, *supra* note 186.

statin-type active ingredient in Merck's Mevacor anti-hyperlipidemia drug. While the 10th Circuit Court of Appeals did not decide the factual issue as to whether red rice yeast supplements had been sold in the United States before the development of Mevacor, the opinion did clarify that if Mevacor had preceded the sale of red rice yeast, the dietary supplements were adulterated under DSHEA, despite the fact that Cholestin was a mixture of which only one constituent was the active ingredient in Mevacor.[52] To conclude this definitional section, virtually any botanical, herb, or any constituent of any food consumed in the U.S. before 1994 or with a history of use or other evidence of safety outside the U.S. is a dietary supplement, unless it contains an ingredient sold as an approved drug or biologic (or in clinical trials to that end) prior to the introduction of the dietary supplement.

Given the extremely broad definition of a dietary supplement, what claims can their manufacturers make about them on their labels? Here too the rules are extremely broad and define what are referred to as "structure/function" claims, a label stating that the supplement helps to maintain or promote the healthy condition of a particular structure or function of the human body.[53] A claim that a dietary supplement helps promote a healthy condition or maintain a normal condition is okay, but a claim that it treats disease is not. Thus the claims "helps promote healthy gums" or "helps maintain a positive feeling of well-being" are acceptable, but "treats gingivitis" or "relieves depression or anxiety" are not acceptable. The only caveat is that supplement-makers who use a "structure/function" claim must also "prominently" place on the label, in bold face type, the disclaimer: "This statement has not been evaluated by the Food and Drug Administration. This product is not intended to diagnose, treat, cure, or prevent any disease."[54] The manufacturer is essentially on its

---

52. *See* Pharmanex, *supra* note 186 at p. 1159:
   To permit manufacturers to market dietary supplements with components identical to the active ingredients in prescription drugs would, as the FDA points out, contravene the incentive structures in place in the FDA for the development of orphan drugs and pediatric drugs.
53. *See, e.g.*, Claims That Can Be Made for Conventional Foods and Dietary Supplements, http://www.cfsan.fda.gov/~dms/hclaims.html (last visited April 4, 2006):
   Statements that address a role of a specific substance in maintaining
   normal healthy structures or functions of the body are considered to be structure/function claims. Structure/function claims may not explicitly or implicitly link the relationship to a disease or health related condition. Unlike health claims, dietary guidance statements and structure/function claims are not subject to FDA review and authorization.
54. 21 U.S.C. §343(r)(6)(c).

honor that it "has substantiation that such statement is truthful and not misleading..."[55] The FDA has no power to examine that data, although it is conceivable that the FTC could, should it choose to, pursue a deceptively advertised supplement.

The border between the wide-open world of botanicals and the heavily-regulated world of pharmaceuticals is absolutely Swiss cheese. Any enterprising supplement company can craft an acceptable claim that dances around just about any disease imaginable—Allcure's stroke drug, were it a botanical, would "maintain and promote the normal circulatory function of the brain" and the FDA would have no say in reviewing it, as long as the ingredient was a botanical or herb with a history of use outside the United States supporting the conclusion that it is safe. Which brings us to the question of what power does the FDA have over dietary supplements?

Other than policing the generous boundaries of DSHEA for new dietary ingredients that might be adulterated supplements and claims that go too far toward disease, the Secretary of HHS may personally order withdrawn from the market a supplement that:

> (A) presents a significant or unreasonable risk of illness or injury under—
> (i) conditions of use recommended or suggested in labeling, or
> (ii) if no conditions of use are suggested or recommended in the labeling, under ordinary conditions of use;[56]

But promptly after any such order, the Secretary must provide a hearing to the maker of the accused supplement and bear the burden of proof as to the risk, which can then be reviewed de novo (without any deference) in any manufacturer's appeal to a federal court.[57]

What is the bottom line about DSHEA for biotech companies? Simply, it opens the door to competition if the active ingredient can be found in any dietary supplement that has a history of use or other sufficient evidence of safety from outside the United States. It is one more detail to consider when

---

55. 21 U.S.C. §343(r)(6)(B).
56. 21 U.S.C. §342(f)(1).
57. To confuse matters further if, in a rulemaking procedure, the FDA determines that a particular substance is unreasonably dangerous and, therefore, supplements containing it are adulterated, the rulemaking itself might *not* be reviewed de novo. NVE Inc. v. Thompson, 436 F.3d 182 (3rd Cir. 2006).

part of a biotech business includes screening against naturally-produced molecules.

§8.10 Postscript: International Pharmaceutical Regulation and Harmonization

Biotechnology companies are a clear example of the globalization of business and the integration of the world's economies and markets. This section provides a very brief introduction to the problem of developing a biopharmaceutical for the global marketplace, with particular emphasis on the U.S., EU, and Japan. We will also examine the current progress towards the harmonization of the process for the approval of new drugs in the U.S., EU, and Japan.

From the earliest phases of its development, a new biotechnology company is concerned about developing its products on a worldwide basis. The initial impact of this global product development focus is generally on the company's financing and patent strategies. Biotech companies typically finance the development of their first products through a series of corporate alliances in which marketing rights to some or all of a product's applications are granted to large pharmaceutical companies for particular geographic regions, in exchange for financial and technical assistance in developing the product (See Chapter 6). Most often, the biotech industry's corporate alliances divide the globe into three parts: North America, the EU, and Japan. This reflects the disproportionate share of the pharmaceutical market represented by those three regions, as compared to other, even more populous regions, such as China and India. Although this will likely change as developing countries increase their per capita use of modern technology and modern medicine, for the near future it will remain true that the major marketing areas are the U.S., EC, and Japan. Although licenses for other areas of Asia, Europe, and Latin America have been negotiated and will continue to be negotiated, the financing and development of biotech products is driven by the requirements of the three currently dominant markets.

Developing biotech therapeutics for the global marketplace adds complications to an already complex business planning process. The problem of obtaining regulatory approval for a new drug or biological drives the process of product development and is made even more difficult by the need to obtain approval in different countries with varying regulatory requirements. Although the basic standards of "safe" and "effective" are virtually universal, for most of the twentieth century different national drug approval agencies interpreted those standards somewhat differently. For example, a drug company might be forced to do several different versions of its preclinical toxicology studies, using different doses and different numbers of animals, each version costing upwards of $500,000 in current dollars. These preclinical cost differences would precede even greater costs imposed by differing standards for human clinical trials and, eventually, even for the reporting of data and the application process. One Pharmaceutical Research and Manufacturers Association (PhRMA) study estimated that the time required to convert a U.S. NDA into an acceptable dossier for the EMEA required three months of work, prior to more recent progress towards harmonization.[58]

To partially alleviate the burden that multiple national regulatory agencies place on the pharmaceutical and biotech industry, for most of the past decade there has been an increased interest in and effort towards the harmonization of regulatory requirements between the EU,[59] Japan[60] and U.S. If each of those authorities were to require the same data and would accept data from experiments done in one of the other regions, the problem of biotech product development would be greatly simplified. The regulatory agencies of the three regimes, the FDA, European Medicines Evaluation Agency (EMEA), and the Japanese Pharmaceutical and Medical Devices Agency (PMDA) have sent representatives to meetings of working groups of The International Conference on Harmonisation of Technical Requirements for Registration of Pharmaceuticals for Human Use (ICH).[61] These working groups have made considerable progress in arriving at mutually acceptable standards for various parts of the drug development process, in the form of ICH standards documents.[62] By

---

58. Sharon Smith Holston, *An Overview of International Cooperation*, 52 FOOD & DRUG L.J. 197, 199. In note 7 of her article Ms. Holston cites an unpublished 1996 PhRMA survey.

59. The EU central drug regulatory agency is the European Medicines Evaluation Agency (EMEA), *see* http://www.emea.eu.int/.

60. The Japanese regulatory agency is the Pharmaceutical and Medical Safety Bureau within the Ministry of Health, Labour and Welfare, http://www.mhlw.go.jp/english/org/policy/p13-14.html.

61. *See, e.g.,* http://www.fda.gov/cder/audiences/iact/iachome.htm#ICH. In this text, when referring to the actual process, the American standard spelling "harmonization" is used, but when referring to the title of the organization, the British spelling "harmonisation" used by the organization itself is retained.

62. *See, e.g.,* Guideline for Industry: Dose Selection for Carcinogenicity Studies of Pharmaceuticals *available at* http://www.fda.gov/cder/guidance/ichs1c.pdf; Guidance for Industry: M-4: CTD—Efficacy: Questions and Answers, (concerning the Common Technical Document, or new drug application material) *Available at* http://www.fda.gov/cber/gdlns/ichm4efficqa.pdf.

1997, 35 ICH technical documents had been completed and 23 had been incorporated into regulatory procedures by the FDA, EMEA, and relevant Japanese Ministry. Given the complexity of the drug approval process, the potential savings for biotechnology and pharmaceutical companies are enormous. The resulting ICH standards offer enormous potential for simplifying the daunting task of international drug approval and substantially reducing costs.

In addition to the ICH process, two other major developments in the organization of international drug approval have simplified the process to a degree. In the same decade from 1993 to the present, the European Union has made the EMEA single-approval process broadly applicable, obviating the need for individual approval efforts for virtually any biotechnology drug. Similarly, the Japanese Ministry of Health, Welfare, and Labor has centralized and streamlined its drug approval process in the PMDA, resulting in substantially shorter review times.

CHAPTER 9

Ethical Perspectives on New Ethical Dilemmas for Biotechnology

§9.0 Introduction

As the pace of scientific advance accelerates and the era of globalization brings new threats as well as new promises, the world of biotechnology is increasingly forced to confront bioethical issues that once would have been merely theoretical discussions of purely speculative issues in the distant future. While the debate about bioethical topics is often emotion-laden and even shrill, biotech company executives and researchers should be prepared to participate rationally and cogently in bioethical decision making and debate involving issues affecting their research or development. The purpose of this chapter is to introduce three recognized ethical systems that can be applied to bioethical issues: Utilitarian, Kantian, and Rawlsian, and to apply those systems to some of the current controversies confronting biotechnology research and development. The intent is not portray any of these as providing clear answers to these difficult current controversies, but rather to contrast these approaches with "revelation-based" religious approaches, to demonstrate that rational analysis of these issues using the dominant Western philosophical traditions is possible, and to enable biotechnology executives and lawyers to participate more effectively in the continuing debate.

This chapter discusses the general problem of ethical frameworks in social decision making and then uses a Rawlsian ethical framework to discuss four such issues:

1) the post-September 11th, post-Gulf War II world in which bioterrorism is a threat raises questions about the ethical limits on scientific research;

2) increasing prices of pharmaceutical prices makes the question of access to affordable health care a serious political and ethical issue for the biotechnology industry;
3) the use of genetic tests in the workplace raises questions about privacy and discrimination; and,
4) Embryonic Stem Cell research raises questions about the ethical status of embryos and research on embryos.

The major scientific advances that are leading to new ethical problems are discussed in §9.1 of this chapter. §9.2 briefly discusses ethical frameworks, particularly utilitarianism, Kantian autonomy, and the ethical framework of John Rawls's *A Theory of Justice*.[1] The chapter concludes with §9.3, which provides an analysis of each of the four ethical problems using the three ethical frameworks.

§9.1 The Scientific Advances and Ethical Challenges to Biotechnology Research and Development

The biggest recent scientific advance in biotechnology was undoubtedly the completion of the sequencing of the Human Genome, which received so much publicity that it is unnecessary to provide an extensive introduction to the Human Genome Project or its importance. It has been described as the beginning of a technological leap that rivals the advent of the industrial age.[2] The NIH and the private genomic effort led by Craig Venter announced the draft completion[3] of the project, which fully sequenced[4] the 23 pairs of chro-

---

1. Revised edition 1999.
2. In an L.A. Times article, physicist Freeman Dyson is noted as saying "Biotechnology, including breakthroughs in the application of discoveries in human genetics, will pose the biggest impact on human development in the next 100 years." Larry B. Stammer, *Physicist Awarded $948,000 Templeton Prize*, L.A. Times, March 23, 2000 at A23 (available on Lexis).
3. *See* Thomas Hayden, *A Genome Milestone: A first draft is completed as two sides call a truce*, Newsweek, July 3, 2000 at 51.
4. To map a chromosome means to determine its location on a chromosome. To sequence a chromosome, or a gene, which is a constituent part of a chromosome, is to determine the order of the nucleotide bases Adenine, Thymidine, Guanine, and Cytosine (A, T, G, C), which are the building blocks of DNA and which encode the genetic instructions of all living things, although some viruses use RNA only. The genetic instructions of RNA

mosomes that contain the complete genetic contents of a normal human cell (the human genome).[5] Over the next decade or more, the raw information about the chemical make up of the human genome will be transformed into an enormous amount of information about the function of each and every gene as well as the significance of a great many of the variations found in most genes. It has brought the era of genetic testing and ultimately gene therapy significantly closer.

The year 2002 also brought two major developments in the field of virology at a time of rising concerns about the ethics of performing and publishing research with a clear potential for use by bioterrorists: first, the announcement of a completely artificially synthesized virus, a polio virus;[6] and second, an Australian group's announcement of the reengineering of a mousepox virus to express Interleukin-4, significantly increasing the virulence of that virus.[7]

Ethical challenges to the traditional business models of biotechnology and pharmaceuticals were raised by controversies over the pricing of pharmaceuticals and the resulting limitations on access to lifesaving drugs. While the cost of pharmaceuticals has always provoked some controversy, it became the subject of heated debate when a new biotechnology-produced anti-HIV drug that was the first in a new class of antvirals, Fuzeon, was marketed at a record price for anti-HIV drugs of $20,000 per year. [8] Even more recently, Genentech announced plans to raise the price of a full treatment regimen of Avastin, a cancer drug, to $100,000, raising again the question of balancing incentives for new drug discovery against the problem of affordable access for patients.[9]

---

follow the same principles but RNA contains Uracil rather than Thymidine. For an excellent introduction to the general science of genetics and molecular biology, *see* WATSON, GILMAN, WITKOWSKI AND ZOLLER, RECOMBINANT DNA (2nd ed. 1992).

5. *See www.genome.gov*, the website of the National Human Genome Research Institute of the NIH, for additional details concerning the Human Genome Project (visited June 8, 2006).

6. Jeronimo Cello, Aniko V. Paul, and Eckard Wimmer, *Chemical Synthesis of Poliovirus cDNA: Generation of Infectious Virus in the Absence of Natural Template*, SCIENCE 1016–1018, Aug. 9, 2002.

7. See **Testimony By** Ronald M. Atlas, Ph.D., *Openness v. Security in Science Research*, Federal Document Clearing House Congressional Testimony October 10, 2002 Thursday (available on Lexis).

8. Geoff Dyer, *How do you price Aids treatment?*, FIN. TIMES (London), March 26, 2003, at 13.

9. Alex Berenson, *A Cancer Drug's Big Price Rise Disturbs Doctors and Patients*, N.Y. TIMES, March 12, 2006, at 1.

The fourth area of controversy discussed here, embryonic stem cell research, followed a breakthrough by a group of Scottish scientists whose techniques increased the viability of embryonic stem cell lines.[10]

The next section of this chapter discusses the problem of ethical decision making and ethical frameworks, particularly the ethical framework developed in John Rawls's *A Theory of Justice*.[11]

§9.2 Ethical Frameworks and *A Theory of Justice*

Ethical decision making is principled and rational. It derives its power from its ability to generate reasoned responses based on explicit premises.[12] It can thus be distinguished from assertions of good that derive from personal preference or religious revelation. There are two longstanding and widely used ethical frameworks: utilitarianism and Kantian autonomy.[13] Utilitarianism is the maximization of the sum of the individual good of all affected individuals.[14] The problems with utilitarianism are both practical and theoretical. On a practical level, it is impossible to measure individual utility or good in any meaningful way and the commonplace substitute of cost-benefit analysis is a

---

10. Ian Chambers, Douglas Colby, Morag Robertson, Jennifer Nichols, Sonia Lee, Susan Tweedie, and Austin Smith, *Functional Expression Cloning of Nanog, a Pluripotency Sustaining Factor in Embryonic Stem Cells,* 113 Cell 643 (2003).

11. John Rawls, A Theory of Justice (Revised edition 1999).

12. Ethics is defined as: a. A set of principles of right conduct. b. A theory or a system of moral values. The American Heritage(r) Dictionary of the English Language, (4th ed. 2000).

13. Utilitarianism, or the notion of maximizing the amount of total good, is widely used and understood. *See, e.g.,* Michael Gove, *Turning the thumbscrews on our liberal conscience,* The Times (London), March 4, 2003 at Features 18 (available on Lexis). Kant is often, and for good reason, thought of as one of the most profound and inaccessible thinkers. However, the difficulty is largely in following his attempts to ground his conclusions in pure logic. His ultimate and most important conclusion, the categorical imperative, is really quite straightforward: "For **Kant**, the Categorical Imperative was the **Golden Rule** in philosophic form, as well as common sense, accepted by every moral person." Thelma Z. Lavine, *Reasonable Doubt,* WASH. POST, May 20, 2001, at T13 (book review).

14. *See* Richard Fallon, *Essay, Should We All Be Welfare Economists?* 101 Mich. L. Rev. 979 (2003). Fallon's essay lays out the general principals of utilitarianism and the major critiques of utilitarianism as a foundation for social policy.

questionable solution to that problem.[15] On a theoretical level, it would allow for the greater good to override what many would consider to be fundamental human rights, including the right to life itself. For example, if five young persons were in need of different organ donations, a utilitarian framework could conceivably require that one middle-aged healthy person be sacrificed so that his organs could be used to save all of their lives. Despite its limitations as a sole form of ethical guidance, risk-benefit analysis, which is a form of utilitarianism, is frequently used in public policy discussions and embodied in various statutes. For example, when the FDA approves a new drug it does so because the drug's overall benefits outweigh its overall risks when used in the indicated population.

Kant's categorical imperative, which insists on respect for the autonomy of each individual, and forbids using others simply as a means to further one's own ends, is subject to different limitations. It is quite possible for there to be no obvious Kantian resolution to common problems. For example, if poor patients were dying for want of a drug, requiring a pharmaceutical company to give it to them for a price they can afford would violate the Kantian autonomy of the pharmaceutical company employees, not least the scientists who worked to develop the drug. Even taxing the wealthy to buy the drug for the poor may violate Kant's imperative.[16] Thus both utilitarianism and Kant's "golden rule" leave us unable to solve important problems, including some of the problems of bioethics and biotechnology. In such cases, the political theory of John Rawls provides a third major ethical framework that can be helpful.

John Rawls' s *A Theory of Justice*[17] is widely considered to be one of the most important contributions to the literature of law and philosophy written in the twentieth century.[18] Rawls's stated objective in *A Theory of Justice* is to explore

---

15. *See* Alyson C. Flournoy, *In Search of an Environmental Ethic*, 28 COLUM. J. ENVTL. L. 63 (2003). Flournoy discusses a number of the problems of utilitarianism and cost benefit analysis in the context of environmental decisionmaking.

16. Suffice it to say that the question of what Kant has to say about taxation and redistribution is one about which serious scholars disagree. *See* Allen W. Wood, *Kant and Fichte: on Right, Welfare and Economic Redistribution* (unpublished paper on file with the author, abstract available online at www.stanford.edu/~allenw/webpapers/KantFichte.doc).

17. RAWLS, *supra* note 210. *See also* Internet Encyclopedia of Philosophy entry on Robert Nozick, http://www.iep.utm.edu/n/nozick.htm, referring to Rawls and Nozick as the two philosophers whose contributions to twentieth century political philosophy are widely considered to be the most significant.

18. A rough measure of its impact can certainly be seen in the frequency with which it is cited. First published in 1971, a search of Lexis's law review database revealed approximately 473 articles citing *A Theory of Justice* in the two-year period ending April 25, 2000.

the meaning of justice by combining the social contract tradition of Locke and Rousseau with the Kantian emphasis on the primacy of individual rights over utilitarian-derived norms. To anchor his contractarian, or social contract exploration of justice, Rawls uses the device of the "original position," which he likens to the "state of nature" conceptions of his philosophical forebears including Locke and Rousseau. However, Rawls's original position adds the constraints of the "veil of ignorance" to the idea of a social contract at the beginnings of human society. A brief exposition of these related concepts should suffice for purposes of this chapter.

Rawls first assumes that a society and the social institutions that comprise it would be fair or just if the rules that govern the society were freely agreed upon by persons who entered into the society as autonomous, self-interested individuals with equal rights to participate in the rule-making. This pre-societal negotiation is the "original position"[19] and is closely related to the philosophical conception of a social contract as the justification for government. Rawls's extension of the social contract tradition rests on his addition of the "veil of ignorance" to the original position. To insure the true fairness of the rules that would be selected, Rawls requires that the parties in the original position must bargain without knowledge of the strengths, talents, weaknesses, and disabilities they will actually possess once they leave the original position and enter into the society whose rules they have chosen. This absence of individual knowledge is the "veil of ignorance."[20]

The veil of ignorance would produce individuals who are self-interested in a very particular way; without knowledge of their own talents or lack thereof, they would bargain for rules that were likely to be to their advantage (or at least minimize their disadvantage) regardless of their actual lot in life. Thus an individual would consider the possibility that she might be gifted with intelligence, leadership, and health and would bargain for rules that would reward persons so blessed. However, *at the same time,* mindful of the possibility that she might be of below average intelligence, socially inept, and afflicted by significant physical disability, she also would bargain for rules that minimized as much as possible the impact of any such disadvantages on her welfare. The fundamental principle that accommodates both the reward for talents and the minimization of disadvantage is Rawls's "difference principle." The difference principle requires that the more talented be rewarded only to the extent necessary to induce them to exercise their talents fully, creating an

---

19. RAWLS, *supra* note 210, at 11.
20. *Id.*

overall structure that works to minimize the disadvantage or maximize the welfare of the least well-off or most-disadvantaged members of society.[21] The great benefit of Rawls's framework is its reconciliation of the competing ideas of utilitarianism and Kantian autonomy. It relies on the utilitarian desires of the parties behind the veil to insure that overall social good not be adversely affected. At the same time, it protects autonomy, albeit in the form of the social contract that autonomous individuals would presumptively choose behind the veil.

§9.3 Applying the Frameworks to the Problems

The application of a Rawlsian analysis provides interesting answers to the ethical questions examined in this chapter. Each of the four ethical dilemmas presented at the beginning of this chapter is explored in the following subsections.

§9.3(A) Bioterrorism and Restraints on Research and Publication

First, the threat of bioterrorism in the Post 9-11 world has raised the question of whether or not scientific research and publication in, for example, the field of virology, should be subjected to significant restrictions to keep the results and information from being used by potential bioterrorists? Analyzing the question from a utilitarian point of view, any particular restriction would have costs, and potentially significant costs, that are difficult to measure but must be acknowledged. For example, the field of virology is an important one for the advancement of human health, and any restriction on research and the dissemination of knowledge among researchers might impact the rate of development of important health benefits from general antiviral drugs to new antiviral vaccines, including vaccines for HIV, SARS, and Ebola. In turn, that slowing of the development of benefits may mean that lives will be lost to those diseases that might otherwise have been saved—a significant cost, but impossible to quantify. The benefit of the restrictions might be significant and equally difficult to measure: the magnitude of the benefit depends on an estimate of how many people might be injured by a bioterrorist with access to information that would not have been available to him had the safeguards been in place. In short, the framework is clear and the answer is almost certainly going to be completely uncertain.

---

21. *Id.* at 68.

Second, analyzing these kinds of restrictions through a Kantian lens would bring into focus the problem of restricting the autonomy of scientists to allay the fears and vulnerability of others who might be at risk from third parties not under the scientists' control. While a voluntary set of guidelines for scientists would not violate a Kantian norm, would scientists violate the categorical imperative by putting others at risk by disregarding voluntary guidelines? Are such scientists using the potential life and health of others to further their own ends of scientific achievement, or are they simply deciding to use as their ends the desires of those afflicted with deadly disease rather than the fears of potential victims of terror? Once again, a powerful ethical framework seems to have no clear answer.

Finally, does Rawls provide a clearer response to this problem? How would persons behind the veil of ignorance deal with the potential restriction on scientists' freedom of action, should they be scientists or sufferers from viral diseases, in order to protect themselves against the threat of others who might be bioterrorists?[22] It is reasonable to assume that persons would want to do what utility in fact requires, that is restrict scientific research only to the point where there was a net gain in risk reduction from the bioterrorist threat that outweighed the potential harm from slowing the rate of advance to help them if they were disease sufferers. While that assumption alone does not provide an answer, it also necessary to ask how those behind the veil would deal with the uncertainty of what their actual position in the world will be. One suggestion is that they would simply look at their chance of being a victim of a serious viral disease, which is significant, at least worldwide. Persons behind the veil would not know if they will be South African miners at great risk of HIV or American citizens of Washington, DC, at some unquantifiable risk of bioterrorism. Would even the most dedicated terrorist seek to unleash a virus that would kill not just the citizens of a large area of a target country but vast populations on every continent, including their own? It seems very unlikely. Given that one assumption, a person entering into life uncertain as to his location and status would most likely face a greater risk of suffering from a "natural" viral disease than being a potential victim of bioterrorism. On this basis a Rawlsian analysis might well caution against any restrictions on research or publication that might impede the rate of progress of medical research.

---

22. For that matter, they would have to consider the likelihood that they would actually be a bioterrorist, but the odds of that are so vanishingly small I think that it can be disregarded in the bargaining.

§9.3(B) Balancing Incentives for Pharmaceutical Innovation against Affordability and Access for Persons with Medical Needs

This is a very difficult problem because there seem to be two very important goals in direct conflict. On the one side, there can be little doubt that the enormous profits to be made from developing important new pharmaceuticals provide a large and effective incentive for the goal of developing new drugs. On the other side, these same profits present increasing difficulty for reaching the goal of providing access to these pharmaceuticals for many individuals with desperate medical needs. The battle still raging over the new Medicare drug benefit is a current and very public manifestation of this problem. How can it be approached under the three ethical frameworks? The utilitarians would attempt to look at the costs and benefits, particularly the benefits of the rate of development of new drugs, including the long-term benefits of the drugs. The long-term benefits go far beyond the period of their patent life, and therefore extend to a period when they become generic and more affordable. On the other hand principal cost may be the unavailability to some persons of lifesaving drugs because of price during their period of patent exclusivity.[23] It is hard to conclude that these short-term costs can outweigh the much longer-term benefits of these same drugs once they become generic. In addition, the logic of utilitarianism and the assumptions of market economics are usually in harmony because the price in a "free market" is determined by the value that the buyer ascribes to its utility, in dollars.[24] Therefore it is difficult for a utilitarian to justify any interference in the current market-based system.

From a Kantian perspective, accomplishing both goals is also not an easy problem. While it is difficult to justify "using" the needs of the critically ill to fulfill the sellers' monetary objectives, it is also impossible to justify any policy that requires talented researchers and their supporting organizations to devote their labors to serve the needs of the ill and ignore their own ends. There are two Kantian autonomies in conflict: that of the sellers and that of the would-be buyers. It is precisely for such conflicts in Kantian positions that Rawls's systemic notions of fairness are most useful. How would the "just" so-

---

23. See Arti K. Rai, *Pharmacogenetic Interventions, Distributive Justice, and Orphan Drugs: The Role of Cost-Benefit Analysis*, 19 J. OF SOCIAL PHILOSOPHY & POLICY 246 (2001). Professor Rai provides a very thoughtful application of cost-benefit analysis to a particular problem in pricing and social policy for research incentives, that of drugs for smaller "orphan" patient populations.

24. *Id.*

ciety approach this problem? Most Rawlsian health policy theorists assume that a right to health is one of the indispensable rights to which everyone is entitled.[25] In fact, Rawls's central tenet is squarely aimed at just such a dilemma, the problem of providing incentives for the talented with due regard for the least advantaged. In a Rawlsian world, incentives for the most talented must always be sufficient to induce them to use their talents willingly. Beyond the point of adequate incentive, the surplus must go to ensure that the disadvantages to the least fortunate members of society are minimized.[26] In our context, that simply means ensuring that the progressive taxation of stock options, corporate profits, and individual incomes from all of the most talented, including pharmaceutical companies and their scientists and executives, is adequate to provide a prescription drug subsidy to provide access to those who truly need those drugs. The Rawlsian solution to the problem is to charge what the market demands for lifesaving drugs and to require those who are wealthy to pay sufficient taxes to provide desperately needed medicines to those who cannot afford them. In a Rawlsian society, as in any other, there is no such thing as a free lunch, but it is mandatory that some of the marginal gains by the more talented be used to feed the starving. It may not be politically neutral in this era, but Rawls is not politically neutral in that sense.

§9.3(C) Echazabal and Workplace Genetic Testing: The Law and Ethics of Protecting Workers against Risks

Over the next decade or more, the raw information about the chemical make-up of the human genome will be transformed into an enormous amount of information about the function of each and every gene as well as the significance of a great many of the variations found in most genes. Inevitably, some of these variations will be found in genes that render us more or less susceptible to carcinogens, whether in the environment, in food, or in the workplace. It has long been suspected that such variations must exist, if for no other reason than the fact that some very potent carcinogens, such as cigarette smoke, cause cancer in some persons at a much earlier age than others. Now, scientific knowledge about the specific pathways by which carcinogens are metabolized within the body and the significance of the genetically-based variations in those pathways is beginning to emerge and is certain to

---

25. Norman Daniels, *Health-Care Needs and Distributive Justice*, 10 PHIL. & PUB. AFF. 146 (1981). Daniels is a leading health-care theorist who uses a Rawlsian framework to argue that equality of health-care access is a fundamental principle of a just society.

26. *See* text accompanying notes 219–220, *supra*.

grow.²⁷ In genetic testing, an individual's blood or tissue sample is examined to determine whether or not the individual has a particular variation in a gene of interest. The technology for doing genetic testing is developing rapidly and is now widespread.²⁸ It is likely to increasingly be the case that at least some of the most important information about risks to an employee's health will come from new genetic tests. §9.3(C)(1) discusses a recent U.S. Supreme Court decision that employers may legally discriminate against a worker when the job would pose a direct threat to the worker's health, which would seem to include threats to health created by an employee's genetic susceptibility as well as other health conditions of the employee. §9.3(C)(2) then uses a Rawlsian analysis to discuss whether or not such discrimination is ethical as well as legally permissible.

§9.3(C)(1) The Legality of Protecting Workers against Risks to Their Own Health

In *Chevron USA, Inc. v. Echazabal*,²⁹ The Supreme Court unanimously reversed the decision of the 9th Circuit Court of Appeals and upheld an EEOC rule. The EEOC had ruled that the Americans with Disabilities Act (ADA)³⁰ permitted employers to exclude an employee or job applicant from a position that would pose a direct threat to the employee's own health, while the statute itself only expressly referred to the right to exclude persons whose employment would pose a direct threat to the health or safety of others.³¹

Mario Echazabal was employed by Chevron's subcontractor, Irwin Industries ("Irwin"), to work at a Chevron refinery, primarily in the coker unit.³²

---

27. *Id.*
28. *See* Helen Pearson, *At-Home DNA Tests Are Here*, THE WALL ST. J., June 25, 2002, at D6.
29. Chevron U.S.A. Inc. v. Echazabal, 535 U.S. 73 (2002).
30. 42 U.S.C. §12101 *et. seq.* (2002).
31. *Compare* the EEOC regulation:
The term 'qualification standard' may include a requirement that an individual shall not pose a direct threat to the health or safety of *the individual* or others in the workplace." [emphasis added]. 29 C.F.R. §1630.15(b)(2) (1999).
with the statutory language:
(b) Qualification standards. The term "qualification standards" may include a requirement that an individual shall not pose a direct threat to the health or safety of other individuals in the workplace. 42 U.S.C. §12113.
32. The facts as presented in this article are drawn from the more detailed description of facts found in the 9th Circuit's opinion in Echazabal v. Chevron U.S.A., Inc., 226 F.3d. 1063 (9th Cir. 2000).

In 1992 he applied for a position with Chevron itself and was accepted for employment pending the results of a required pre-employment medical examination. The examination revealed that he had elevated levels of liver enzymes, leading Chevron to conclude that his liver might be at risk of damage from the airborne chemicals present in the vicinity of the coker unit. For that reason Chevron withdrew their conditional offer of employment.[33] Chevron did not, however, then require their subcontractor, Irwin, to remove him from duties that required him to remain in the vicinity of the coker unit. When Echazabal learned of his medical test results, he consulted private physicians and learned that the liver test abnormalities were due to an asymptomatic, chronic, active infection with Hepatitis C. Echazabal's physicians were aware of his employment circumstances and the chemicals he was exposed to but did not advise him to find other employment. In 1995 Echazabal applied again to work directly for Chevron and the sequence of conditional acceptance, medical examination, and withdrawal of the job offer was repeated. This time, however, Chevron also contacted Irwin and requested that Irwin "immediately remove Mr. Echazabal from [the] refinery or place him in a position that eliminates his exposure to solvents/chemicals."[34]

Echazabal responded to his termination from employment by filing a complaint with the EEOC against both Chevron and Irwin, alleging unlawful discrimination under the ADA. Chevron sought and was granted removal of the case to federal court. After summary judgment for Chevron in the district court, Echazabal appealed to the Ninth Circuit. At issue was whether or not the EEOC was correct in permitting employers to exclude persons from employment when the job would pose a direct threat to their own health or safety, but not to the health or safety of others in the workplace. The Court of Appeals ruled against Chevron, reversed the District Court, and held the EEOC's regulation was contrary to law and therefore invalid.[35]

The Supreme Court upheld the EEOC's expansive interpretation of the statutory "direct threat" defense and unanimously reversed the Ninth Circuit.[36] Of course, the Supreme Court did not decide whether or not exposure to the airborne chemicals in Chevron's refinery actually did pose a threat to Echazabal because of his chronic, active Hepatitis C infection.[37] The reality of the threat to Echazabal's health remained a factual dispute to be resolved by fur-

---

33. *Id.* at 1065.
34. Echazabal v. Chevron, 226 F.3d 1063 (9th Cir. 2000).
35. *Id.*
36. Chevron v. Echazabal, 536 U.S. 73 (2002).
37. *Id.*

ther litigation. Furthermore, the Supreme Court's decision was primarily an exercise in statutory interpretation and administrative agency review. The Court held only that the EEOC's ruling was a permissible interpretation and that such discrimination was therefore legal, not that such discrimination was the best or most wise public policy choice. The significance for genetic testing is that the Court's decision in Echazabal would seemingly permit employers to use genetic testing when the best evidence concerning "direct threats" to the employees' own health may come from such genetic testing.[38] The next subsection discusses whether or not such use of genetic testing is in fact ethical as well as legal.

§9.3(C)(2) A Theory of Justice[39] and Workplace Genetic Testing

The conceptual starting point of Rawls's *A Theory of Justice*, the original position accompanied by the veil of ignorance, has been criticized as resting on a very artificial and unworkable device.[40] That criticism might be least justified in the context of genetic testing its current stage of its development. With most of the information from the Human Genome Project in the raw form of DNA sequences without function or knowledge of the significance of most sequence variations, most people are very close to being in the original position and behind the veil of ignorance as to their genetic advantages and disadvantages. Researchers are just beginning to discover genetic variations that are associated with increased risks of cancer. Many more such variations undoubtedly will be discovered in the next decade. Furthermore, science will begin to provide a great deal more understanding of the risks posed by par-

---

38. *See e.g.,* M.T. Smith, *Benzene, NQO1 and Genetic Susceptibility to Cancer*, 96 PROC. NAT'L ACAD. SCI. 7624 (1999); R.A. Larson, M.T. Smith, et al., *Prevalence of the Inactivating C→T Polymorphism in the NAD(P)H: Quinone Oxidoreductase (NQO1) Gene in Patients with Primary and Therapy Related Myeloid Leukemia*, 94 BLOOD 803 (1999).

These two studies both support the conclusion that particular variations ("polymorphisms") in the NQ1 gene can significantly increase of leukemia in individuals with those genetic variations who are exposed to airborne benzene. Evidence of significant genetic variation in the risk of benzene exposure is even more interesting in the context of Echazabal and his pursuit of employment in an oil refinery, since one of the well-established carcinogens in that environment is in fact, benzene.

39. *Rawls, supra* note 210.

40. *See, e.g.,* Pierre Schlag, *The Empty Circles of Liberal Justification*, 96 MICH. L. REV. 1, 30 (1997). Schlag's criticism is particularly aimed at the circular nature of Rawl's "mythic" person behind the veil: "The 'real' consumer of liberal justification is already deeply ensconced in the circles of liberal justification, for he is himself already a mythified construction."

ticular variations, both with respect to the absolute magnitude of such risks, and also as to particular environmental events that interact with particular variant genes and gene products to increase the risks of those variant genes.[41] Now is the time to set the rules for the post-genome society while still the great majority of people are behind the veil, largely ignorant of their actual lot in the post-genomic order.

Before using a Rawlsian framework, it is interesting to note that using the Kantian and utilitarian ethical frameworks would provide conflicting answers to the problem of workplace genetic testing. A utilitarian framework would support the use of genetic testing in those circumstances where the cost of making a workplace safe far exceeds the cost of excluding abnormally sensitive individuals. A Kantian framework would reach the opposite conclusion, valuing the individual's choice as to whether or not to be tested and whether or not to take the risk as part of respecting individual autonomy. Most ethicists who have looked at this issue have used Kantian autonomy to argue against genetic testing in the workplace.

It is illuminating to reexamine the issues in *Chevron v. Echazabal* and, for purposes of discussion, to add to the analysis hypothetical facts involving possible genetic tests. Mario Echazabal applied to work in an area of the Chevron plant where he would be exposed to airborne solvents, such as benzene.[42] His liver was actually weakened by hepatitis, but in the context of this analysis it is appropriate to assume that a genetic test was available to accurately determine whether or not he would also be additionally vulnerable because of his inability to metabolize benzene at the normal rate.[43] Rather than simply extending the ruling of the Supreme Court in *Echazabal* to include genetic testing, the ethics of that extension can be examined by retreating behind the Rawlsian veil of ignorance to find the just solution.

What are the possible choices? First, a rule could be imposed to require that all employers spend whatever is necessary to make their workplace environments reasonably safe for the individuals who are most genetically susceptible to that environment or, if unwilling or unable to do so, be shut down.[44] It

---

41. *Supra* note 237.

42. Benzene is a solvent of particular interest both because it is one of the airborne solvents regulated in the refinery environment in which Echazabal sought employment and because it was the subject of a long and historic legal controversy, *see Indus. Union Dep't., AFL-CIO v. Am. Petroleum Inst.*, 448 U.S. 607 (1980).

43. *Id. See also* note 237 for scientific studies that suggest that such a test is more science than science fiction.

44. This is substantially the position taken by Elaine Draper in *The Screening of America: The Social and Legal Framework of Employers' Use of Genetic Information*, 20 BERKELEY

is unlikely that Rawlsian bargainers would require such an absolutist approach because of its obvious potential to affect the overall welfare of the group adversely by imposing major economic costs to protect what may be a very small number of individuals.[45] Similarly, it is unlikely that the Rawlsian choice would be to impose rules allowing workplaces to be dangerous to some subgroup of abnormally sensitive workers while prohibiting any testing by either employers or employees to determine who was in fact at risk. Finally, the Rawlsian bargainers would likely reject a rule that allowed employers to test and then to exclude from the workplace abnormally sensitive persons but without any compensation for the disadvantaged and excluded workers. This violates the Rawlsian "difference" principle[46] that requires economic gains (in this case from a looser workplace standard and lowered employee disease costs) to be used to mitigate the disadvantage of the least well-off (the genetically most sensitive).

The more difficult choice is between allowing employers to test (and compensate) or allowing only employees to test if they wish to determine their own sensitivity and decide how to use that information. In the employer testing with compensation option, bargainers behind the veil could agree that employers would have the right to test individuals to determine their sensitivity

---

J. EMP. & LAB. L. 286, 313–14 (1999):
> If risk is conceptualized in terms of the personal habits or biology of individuals, it naturally appears beneficial to develop screening programs to identify people who take drugs or have genetic characteristics that may present a health hazard on the job. But if employers want to provide a safe workplace, they should tighten engineering controls, monitor exposurehazards, replace hazardous products, and collect scientific information on risks to populations; only such efforts can reveal whether working conditions are indeed safe. Priorities in health policy should be redirected toward reducing risk without falsely making it appear that genetically high-risk workers and drug users are the problem and without needlessly penalizing individuals perceived to be susceptible. Investing in improved management policies and working conditions could prevent disease more effectively than broad genetic screening. The search for high-risk individuals should not limit the use of effective strategies for reducing environmental hazards and disease that are already widely recognized but under funded.

45. This is a basic postulate of Rawls, accepting differences in the distribution of goods as long as they work to the advantage of all, including the least advantaged. *See* RAWLS, *supra* note 210, at 13 " … inequalities in wealth and authority are just only if they result in compensating benefits for everyone, and in particular for the least advantaged members of society." The application of this principle in the context of this essay is that as long as the savings from the reduced requirements of workplace toxics control systems are adequate to compensate the genetically susceptible, such a system is fair.

46. *See* text accompanying note 220.

and exclude those who are determined to be at greatest risk, as long as they either offer alternative, safer, equally paid employment or contribute a percentage of the "savings" from the more lenient exposure limits to a centrally-administered compensation fund. Persons excluded from a workplace by virtue of their genetic susceptibility would be compensated by the centrally-administered fund for their loss of opportunity. In this way employers would make an economically efficient decision to allocate funds either to clean up the plant environment or to test and compensate a subgroup of employees or employment applicants.

Under the employee freedom of choice option, a rule could require that workplaces be made reasonably safe for persons of ordinary sensitivity while prohibiting employers' use of tests that would identify those who are more than ordinarily sensitive. Under this alternative, only employers would be prohibited from gaining and using such knowledge; workers or applicants for employment would be free to self-test and make their own decisions as to whether or not to accept the risks associated with employment. The potential employees would then divide up into at least three groups: the risk-"ostriches" who simply do not want to know of their susceptibility; the risk-averse who test and self-exclude if at risk; and the risk-takers who test and ignore any risk.

The question of whether or not risk-"ostriches" and risk-prone individuals who take jobs despite knowing or avoiding knowledge of their genetic susceptibility should be allowed to recover in tort or worker's compensation for their resulting injuries is beyond the scope of this comment. However, the question of whether or not persons who are behind the veil would prefer a rule giving the employer the choice to screen and compensate rather than a rule giving employees freedom of choice for risk-"ostriches" and risk-taking individuals[47] is a very interesting question that comes close to the heart of Rawls's system.[48] When considering whether or not to allow persons to be risk-

---

47. Risk-takers and risk-ostriches can best be described by their position on the normal distribution of willingness to take risk. On the one extreme of risk-taking are those who would attempt to climb Mount Annapurna (with its 41 percent fatality rate, *see* http://www.mounteverest.net/story/stories/EvrstvsBigMa-killrstatistisMar22004.shtml); on the other extreme a risk-ostrich clearly includes someone who packs a parachute for a commercial flight in the United States. Somewhere in the middle of the distribution are those who exceed the posted speed limits by a few miles per hour in normal driving.

48. Richard Schmalbeck notes that Rawls posits extremely risk-averse individuals (because of their concern for the possibility that they will be at the very, very bottom of the actual distribution of social goods). Richard Schmalbeck, *The Justice of Economics: An Analysis of Wealth Maximization As a Normative Goal*, 83 COLUM. L. REV. 488, 513 (1983)

takers and voluntarily expose themselves to significant risk, the original position decision makers must take into account the impact of such freedoms on others, particularly those whose lot will be made significantly worse by such a freedom, such as minor children. After all, behind the veil it is necessary to consider not only the possibility that one's real world position would be that of a genetically susceptible job applicant, but also the possibility that one would be a loved one or dependent of a genetically susceptible job applicant.

The issue for the bargainers is to consider not only whether they would wish to be free to take on jobs that posed a special risk for them when there was no economic advantage to doing so, but also the impact such freedom would have on any surviving dependents or loved-ones of a risk-choosing individual. Bargaining for freedom under such circumstances is bargaining for the acceptability of such avoidable losses and the costs that they impose on others. Since for every non-risk-averse individual who suffers an avoidable loss there is likely to be more than one "aggrieved" loved one, rational bargainers may well determine that their self-interest would actually limit their future freedom to inflict avoidable losses on others to avoid being hurt by the avoidable loss of a loved one. Under Rawls's difference principle, the infliction of avoidable losses on those disadvantaged by the freedom to choose would seem to point toward limiting the freedom to choose risk, as long as reasonable compensation for the loss of opportunity were provided. [49]

Obviously, such compensation to the disadvantaged and excluded employees would lower the amount of profits to be distributed to investors or paid in wages to the genetically "normal." Despite the fact that most persons are more likely to be among the genetically "normal" compensators rather than the genetically sensitive compensated, the difference principle clearly requires compensation when the few sacrifice opportunity, albeit in part because of "brute luck,"[50] for the benefit of the many. The details of the compensation system are beyond the scope of this chapter, but employer testing and com-

---

(reviewing RICHARD A. POSNER, THE ECONOMICS OF JUSTICE (1981)): "**Rawls** is at one extreme, implicitly assuming complete risk aversion."

49. But cf., Draper, *supra* note 243, arguing that workplaces should be made safe for the genetically sensitive rather than removing them from exposure to the potential harm. Draper's argument appears to be premised on the assumption that such reductions in workplace risk could be made at reasonable cost. This article assumes that by attaching a compensation cost to discrimination against the genetically sensitive, rational employers would prefer to make changes in the workplace to reduce exposure where doing so would be cost-effective.

50. For a discussion of this term in its philosophical context, see Jules Coleman & Arthur Ripstein, *Mischief and Misfortune*, 41 McGILL L.J. 91, 120–22 (1995).

pensation seems to be the economically efficient and, from a Rawlsian perspective, ethically appropriate choice.

In conclusion, the EEOC's regulation, as upheld in Chevron v. Echazabal, ought to be interpreted as allowing genetic testing when the best form of evidence as to a direct threat to an employee's health would be provided by genetic testing. However, such a rule is ethically correct in the context of genetic testing only if there is compensation for the genetically disadvantaged.

§9.4 Embryonic Stem Cell Research

The controversy over the ethics of embryonic stem cell research has continued to plague the field, with advances on the scientific front bringing renewed opposition from those who see the research as being an inappropriate exploitation of a potential human life, i.e. that of the embryo.[51] First, to clarify and limit the ethical discussion, it is important to distinguish research on extra embryos that are created for purposes of assisted reproduction and then offered for research by the egg donor ("spare embryos"), from research on embryos deliberately created for research purposes. From a utilitarian perspective, research on the spare embryos is not a difficult ethical issue. Since the embryo must be one that the potential biological mother has decided not to implant (and, for purposes of this discussion is in excess of the number that could be used by persons who are seeking to be impregnated with the embryos of others), it loses nothing in future satisfaction while the living beneficiaries of the research or derived tissue stand to gain much. Similarly, from a Kantian perspective, it can only be a dilemma if the cluster of cells that form a blastocyst (on which embryonic stem cell research could be performed) can be posited to be a human being worthy of full respect. A necessary, if not sufficient, prerequisite for such Kantian autonomy is the ability to possess a will, that is, to have the ability to choose among ends.[52] It is likely that a Kantian would not accord equal human dignity to the cells' rights, in conflict with the

---

51. For an interesting discussion of this position as expressed by Leon Kass, the advisor on bioethics to President Bush, see Gary Rosen, *Who's afraid of Leon Kass?*, 115 COMMENTARY, 28–33, (2003).

52. While there are many situations in which persons are rendered unable to make their will known due to brain injury or other conditions, there are a number of doctrines that have been developed to essentially supply an answer to the question "What is this person's will?" *See* Richard S. Kay, *Section V: Criminal Law and Procedure: Causing Death for Compassionate Reasons in American Law*, 54 AM. J. COMPARATIVE L. 693 (2006). None of those doctrines are really applicable in the context of an embryo that could have never expressed

rights of the mother, the researcher, and the suffering patients.[53] When both utilitarianism and Kantianism agree, most assuredly Rawls would be in full accord. It is because of religious notions of personhood, not philosophical ones, that the current controversy has arisen. While the costs to the feelings of those offended on religious grounds are important, they are unlikely to tip the utilitarian balance nor do they have a Kantian right to impose their will on the parties directly involved. Leon Kass, the President's bioethics advisor, grounds his position not in any of the three philosophical traditions developed here, but rather on what he develops from a system that relies on moral intuition.[54] Unfortunately, his moral intuitions cannot be reconciled with any of the great philosophical traditions.

---

wishes, where the mother wishes to donate it to research and the choice is between being discarded or donated.

53. *See* Heather Johnson Kukla, *Note, Embryonic Stem Cell Research: An Ethical Justification*, 90 GEO. L.J. 503 (2002). Ms. Kukla would restrict the creation of embryos for research purposes but defends the use of spare (left over from assisted reproduction) embryos for research.

54. *See* Rosen, *supra* note 250.

CHAPTER 10

Special Regulatory Issues: Human Gene Therapy

§10.0 Introduction to Gene Therapy: At the Edge of a New Era in Medicine

Not all that many years ago a chapter on human gene therapy would have been more likely to appear in a work of science fiction than one concerned with current legal and business issues facing a growing industry. In recent years, gene therapy has progressed from being the concern of a relative handful of university researchers to being the core technology for a surprisingly large number[1] of biotech companies, such as Ceregene, Cardium Therapeutics, and Oxford BioMedica, and to being a clinical reality for hundreds of patients in trials of a variety of gene therapy techniques at research centers around the country.[2] Gene therapy is sufficiently different from other biotechnology applications to have its own regulatory path, reflecting both the technological and philosophical issues raised by the potential power of gene therapy. This chapter provides an overview of the technology as has developed and a brief examination of the regulatory framework and major features of the ethical debate about the need for greater control of future developments in human gene therapy.

---

1. A recent report listing such companies is summarized at http://www.mediligence.com/rpt-a605.htm.
2. Elizabeth Culotta, *New Startups Move in As Gene Therapy Goes Commercial*, 260 Science 914 (May 14, 1993).

§10.1 What is Gene Therapy?

Gene therapy, narrowly conceived,[3] is the attempt to treat diseases which have a genetic component at the genetic level, that is, by inserting a corrective gene into disease-related cells. The process is called transduction and the cells that then express the new gene are referred to as transduced. For instance, some of the first attempts at gene therapy on humans involved efforts to treat a severe immune system disease known as adenosine deaminase (ADA) deficiency or X-SCID (x-linked Severe Combined Immunodeficiency Disease). In this disease, the patients have two defective copies of the gene for ADA, an enzyme crucial to the normal function of T and B cells, the major immune system components. Lacking the enzyme, the patients are completely vulnerable to any bacterial or viral infection. Most such children die within months of birth. Gene therapy for this disease has focused on using a genetically engineered virus (viral vector) to infect the patients' B-lymphocytes with a correct copy of the ADA gene. In early attempts, lymphocytes were filtered from blood drawn from the patient, mixed with the virus in a culture inoculated with growth factors to stimulate multiplication of the lymphocytes, and then the lymphocytes transduced by the virus were reinfused into the patient. Because ordinary lymphocytes have a life span of only a few months, the procedure was, at best, only a temporary measure. The disease is now treatable by providing the enzyme itself to affected patients; nonetheless, in 1995 the NIH reported the possible success of a gene therapy trial for the disease, although patients continued to receive the enzyme replacement therapy as well.[4]

The first clearly "successful" gene therapy trials were based on a slightly more advanced approach to this problem, in a similar disease known as X-SCID. In X-SCID patients are also missing a working copy of a gene that results in the absence of functional T and B cells, however the problem is not the Adenosine Deaminase Enzyme, but the gene that encodes an important

---

3. I use the phrase "narrowly conceived," because the word therapy inherently signifies the attempt to ameliorate disease. It is clear, however, that much of the debate over gene therapy is driven by the difficulty in drawing the line between the treatment of disease and attempts at genetic "enhancement." An obvious example of such difficult line drawing is the distinction between treating dwarfism and treating patients who may simply be unhappy with their genetically determined height. As proponents of gene therapy are quick to point out, such ethical dilemmas are not unique to gene therapy. The dilemma already exists with respect to many "traditional" medical interventions, such as the use of human growth hormone injections.

4. http://www.genome.gov/10000521.

part of the cytokine proteins that provide key immune system activation signals. The gene is carried on the X chromosome, which means that it will most commonly affect males, whereas females who have a defective copy on one X-chromosome will be unaffected as long as they inherit a working copy from their other parent.[5]

Providing a "permanent" correction of the genetic defect required isolating the progenitor cells of B-lymphocytes and other blood cell types, so-called totipotent hematopoietic stem cells. These stem cells are long-lived and, if effectively transduced with the corrective gene, will pass it on indefinitely to their progeny of differentiated blood cells, including mature B-lymphocytes and T-cells. The extraordinary news that a number of children were successfully treated by this approach was followed two years later by the disheartening news that three of the children had developed leukemia resulting from the viral vector disrupting a normal gene pathway in the transduced stem cells.[6]

The attempts at treating ADA deficiency and X-SCID are representative of one of the two major branches of gene therapy, somatic cell gene therapy, and illustrate a number of the technical issues and approaches as well. In somatic cell gene therapy the therapeutic genetic material (RNA or DNA) is inserted into human body, or somatic, cells. B-lymphocytes, liver cells (for hypercholesteremia), and lung cells (to treat cystic fibrosis) were among the first somatic cells targeted. Because the genetic material is inserted into body cells, rather than reproductive cells, it is not heritable by the patient's offspring. It is distinguished from germ-line gene therapy, in which heritable genetic material is inserted into cells that include reproductive cells, so that the resulting genetic information would become a part of the patient's sperm or ova and could pass on to offspring. Germ-line gene therapy has been barred by NIH guidelines from federal research funding and is discussed below in §10.4(B).

Somatic cell gene therapy, aimed at body cells, can involve either an *in vivo* or *ex vivo* approach. The *in vivo* approach targets cells within the body, while the *ex vivo* approach (such as those described above for ADA and X-SCID deficiency) involves removing cells, manipulating them outside the body ("*ex vivo*") and reintroducing the manipulated cells into the patient. Both approaches require solutions to a number of technological challenges facing all somatic cell gene therapy: 1) the identification of a discrete gene which plays a crucial role in a disease process; 2) the identification of an appropriate tar-

---

5. *See X-linked SCID mutation database (IL2RGbase)* http://research.nhgri.nih.gov/scid/.
6. *See* Thomas H. Maugh, *Gene Therapy Experiments Put on Hold*, L.A. TIMES, March 4, 2005 at 16.

get cell set, capable of expressing the corrective genetic material at therapeutic levels; and, perhaps most important, 3) a method for delivering the genetic material in effective form to the target cells.

§10.1(A) Candidate Diseases for Gene Therapy

Although there may be as many as 4,000 human diseases that are genetically associated, a relatively small number of those in which a single gene defect has been identified, are now targeted for gene therapy trials.[7] Cystic fibrosis is the most common severe genetic disease for which a discrete genetic defect has been identified, and there have been numerous attempts at cystic fibrosis gene therapy. In cystic fibrosis, lung cells lack a functioning form of the protein known as the cystic fibrosis transmembrane protein, which mediates the intake of calcium ions into the cell. Since calcium ion concentration is one of the major determinants of the cell's regulation of cytoplasmic salt concentrations, the inability to maintain normal ion flow may be the cause of the extremely viscous secretions found in CF patients that plug up the airways of the lungs and result in the cystic fibrosis disease. Because animal models demonstrated that the insertion of a correct copy of the CF gene into lung epithelial cells would correct the disease, human CF gene therapy trials appeared promising; but obtaining clinically beneficial expression of the gene for a meaningful length of time has been very difficult.

While there are certainly other single-gene "genetic diseases" (such as hypercholesteremia and hemophilia) which are attractive targets for gene therapy, the number of patients affected by those diseases is relatively small. Thus it is not surprising that commercial gene therapy efforts have targeted bigger markets not traditionally conceived of as single gene "genetic diseases." These gene therapy targets include cancer and cardiovascular disease. The genetic target of these gene therapy candidates provides some insight into the variety of approaches gene therapy may take. For cardiovascular disease, Cardium Therapeutics' Generx is in advanced clinical trials for angina, the pain accompanying coronary artery disease caused by an insufficient supply of blood to the heart. Rather than attempting to clear the arteries, or target a gene that is a master switch for atherosclerosis, the Cardium approach is to use a viral vector carrying a gene intended to provide the genetic stimulus for the growth of new blood vessels (angiogenesis) and thus alleviate the insufficient blood flow.

---

7. *See* Richard C. Mulligan, *The Basic Science of Gene Therapy*, 260 SCIENCE 926 (May 14, 1993).

Gene therapy approaches for cancer use a much wider range of genetic approaches. Although particular genetic defects are now associated with some cancers (mutations in the p53 and RB oncogenes or tumor suppressor genes being the most well-understood), many of the cancer gene therapy trials to date have not focused on the oncogenes themselves. Instead, the gene therapy approaches have attempted to deliver other therapeutic genes to the tumors. One recent review article listed, among other approaches: oncogene inactivation (by delivering, for example, antisense or other "shutoff" signals aimed at the gene pathways driving cell division and proliferation; tumor suppressor gene replacement (providing additional copies of the genes that function to regulate cell division and proliferation); immunopotentiation (delivering genes to produce proteins that stimulate an immune response to the tumor cells) and; in the reverse of the Cardium approach to cardiovascular disease, anti-angiogenic gene therapy (using gene sequences to block the growth of new blood vessels, thus limiting tumor growth and survival).[8]

§10.1(B) Selecting Appropriate Target Cells

For some diseases, the nature of the disease dictates the target cells for gene therapy. For example, cystic fibrosis can only be treated in the affected lung cells, while ADA deficiency can only be treated in B-lymphocytes or their progenitor stem cells. For other diseases, where the pathology involves the loss of a circulating protein in the blood, a wider range of target cells is possible. For example, although the most common form of hemophilia involves the failure of bone marrow cells to produce Factor VIIIC, circulating factor VIIIC is therapeutic no matter what its source. Thus, if muscle cells, or liver cells, or even endothelial cells could be induced to secrete Factor VIIIC into the bloodstream in appropriate amounts, the desired therapeutic effect might be achieved. Because of rapid progress in tissue engineering, or the ability to grow large quantities of transplantable cells of a variety of types (including keratinocytes and fibroblasts), *ex vivo* transduction of those cell types is another approach to delivery.

§10.1(C) Methods for Delivering Therapeutic Genetic Material

For those diseases that require a highly specific, less accessible target cell type, there are significantly greater technological challenges for targeting the

---

8. R.M. Hughes, *Strategies for Cancer Gene Therapy*, 85 J. SURG. ONCOL. 28–35 (2004). (abstract available through PUBMED, PMID: 14696084).

cells with an appropriate vector. The most commonly used vectors are genetically engineered viruses used to carry the therapeutic gene into the target cell, whether *in vivo* or *ex vivo*. Construction and production of appropriate viral vectors is one of the major technological challenges for gene therapy. It is necessary to understand at least some of the technology of creating viral vectors to understand the regulatory, safety, and ethical issues. The first major types of virus used for gene therapy were retroviruses, particularly a strain of murine retroviruses that naturally infect mouse cells. These murine retroviruses, like all retroviruses, use RNA as their genetic material. Unlike ordinary RNA viruses, which simply use the host cells' cytoplasmic t-RNA and enzymes for replication, retroviruses employ a retroviral enzyme known as reverse transcriptase, which causes the "backwards" or "reverse" transcription of the viral RNA into DNA, which is then inserted into the host cells' nuclear DNA during cell division. This chromosomally integrated viral DNA is known as "integrated proviral" genetic material and will be permanently expressed by the host cell and inherited and expressed in progeny of the original host cell. However, retroviral infection of a cell is limited to cell types expressing an appropriate receptor for that retrovirus and integration of retroviral DNA can only occur during cell replication. Thus cells that lack the receptor or are primarily quiescent are not suitable targets for retroviral vectors.

More recently, RNA and DNA viruses, such as members of the adenovirus family, have been used in gene therapy. Such DNA viruses insert their DNA into the cytoplasm of host cells, where it is transcribed and translated by the host cells' m-RNA and t-RNA for viral replication (although occasional integration of adenoviral DNA has been reported). The expression of DNA viral vector gene sequences may be less persistent than those of retroviruses. On the other hand, the disadvantage of retroviruses is that the site of integration into the host DNA is random and there is some probability that integration of viral DNA will disrupt a vital host gene sequence, causing further disease, as may have been the case in the tragic French trial for X-SCID.[9] A more detailed analysis of the risks of gene therapy is discussed below in §10.2.

Retroviruses and adenovirus have been more thoroughly developed as gene therapy vectors than other viruses, although other viruses, including herpes viruses and adeno-associated viruses, are being explored as well. For purposes of this chapter, this basic discussion of retroviruses and adenoviruses sufficiently illustrates the general technological and regulatory issues confronting viral vectors.

---

9. *See* note 6 and accompanying text.

One of the major steps in viral vector gene therapy is the creation of replication disabled or "incompetent" viral vectors, whether the virus used in gene therapy is a retrovirus (such as the murine retroviruses), or a DNA virus (for example an adenovirus). This is done by using a so-called packaging cell-line, generally a mammalian cell culture, to produce the viral vector. In this approach, the viral genes necessary to produce complete new virus particles, for example genes that are responsible for the viral envelope proteins or their assembly, are inserted into the host cells as plasmid, non-nuclear, non-chromosomal DNA, while viral particles which are missing those genes but contain the therapeutic gene sequence are induced to infect the host cells. The host-cells (referred to as a packaging cell-line) then produce the envelope and assembly proteins, while the introduced viral particles replicate the rest of its genetic material. The packaging cell then assembles the parts to form new virus particles that can infect the target cells but still are replication incompetent (still lack the envelope and assembly genes inserted in the packaging cell). As long as the patients' cells lack the missing packaging genes, these viral vectors can infect those cells but cannot replicate.

Although most of the work in gene therapy has centered on viral vectors to carry genetic material to target cells, there are other, non-viral delivery systems being explored. For diseases that afford a wide range of target cell types, a correspondingly greater number of delivery systems are possible. Perhaps the most surprising gene therapy delivery system is the injection of DNA, in buffered solution, into muscle tissue. Scientists at Vical Corp. injected DNA sequences into muscle tissue as a control to test against a liposome delivery system Vical had been developing. Liposomes are microscopic fatty or lipid droplets that can be used to encapsulate nucleic acids or a wide variety of other pharmacological or therapeutic agents. Much to their surprise, the control animals showed greater uptake and expression of the control "naked" DNA than did the experimental animals that had received injections of the DNA in a liposomal vector. Subsequently it has been confirmed that muscle tissue will take up and express a small percentage of DNA injected into the muscle. The expression persists for some time, at least at low levels.

Direct injection of naked DNA is being explored primarily for gene therapy "vaccines." In these vaccines, the DNA for antigenic features of pathogenic organisms, or possibly even the DNA for cancer-specific antigens, is injected into muscle cells. The muscle cells then produce low levels of the antigenic proteins, generating an immunological response and, potentially, lasting immunity to the pathogen or even tumor cell, from which the antigen is derived. Vical currently has projects using this technology to develop vaccines for CMV, hepatitis B, malaria, HIV, anthrax and, most recently, avian flu.[10] The persistent, if low-level expression of DNA injected directly into muscle cells is also a potential delivery system for diseases for which a low-level of circulating protein would be therapeutic, such as hemophilia and pituitary dwarfism. Naked DNA is also being used to deliver immune-stimulating proteins, such as IL-2, in an effort to treat cancer.

Despite Vical's serendipitous failed experiment using liposomes against a control of "naked" DNA, scientists are still exploring liposomes as gene therapy vectors in vaccine protocols similar to those described above for naked DNA and for some cancer therapies as well. In one trial of liposomes, the DNA that encodes an immuno-stimulating protein known as HLA-B7 is encased in liposomes and then injected directly into melanoma tumors. The hope is that the tumor cells will take up the liposomes, express the HLA-B7 antigen, and stimulate a strong host-immune response against the tumor.[11]

Finally, scientists are also attempting to deliver DNA by linking it to small protein fragments (peptide ligands) that are recognized by specific receptors on target cells. This method could be adapted to a wide range of target cells for *in vivo* application; however, technical obstacles remain with respect to the stability of the DNA and the persistency of its expression. If those obstacles can be overcome, ligand-linked DNA could be a method of gene therapy with very low risk and very broad applicability.[12]

§10.2 Risks of Gene Therapy

No method of gene therapy is risk-free. There is a danger of contamination or inflammation even when "naked" DNA sequences are injected directly into muscle cells, and there is also a slight possibility that the genetic material inserted in gene therapy will drive the expression of protein at levels that may cause injury. Nevertheless, at this time most of the concerns about the scientific or medical risks of gene therapy (as opposed to the ethical and social risks discussed below) have focused on the risks created by the use of viral vectors in gene ther-

---

10. *See* Penni Crabtree, *For Bird Flu, Time Is of the Essence: S.D.'s Vical Is Working on Novel Approach: a DNA Vaccine*, San Diego Union-Trib., Nov. 11, 2005 at C-1.

11. A. T. Stopeck, et al., *Phase II study of direct intralesional gene transfer of allovectin-7, an HLA-B7/beta2-microglobulin DNA-liposome complex, in patients with metastatic melanoma*, 7 Clin Cancer Res. 2285–91 (Aug. 7, 2001) (available at PMID: 11489803).

12. R. Brokx and J. Gariepy, *Peptide- and polymer-based gene delivery vehicles*, 90 Methods Mol. Med. 139–60 (2004) (available at PMID: 14657563).

apy. Despite the much lower risks of liposomal delivery or direct injection of DNA, neither of those methods has yet demonstrated sufficient target-cell range or level of expression to obviate the need for the more difficult and risky viral vector approach. At some future date, the art of tissue engineering and stem cell-based technologies may have advanced sufficiently to provide a major pathway to gene therapy; but in the field of gene therapy as it is currently developing, the risk potential of viral vectors remains the chief concern.

§10.2(A) The Risk of Producing Replication-Competent Viruses

Analyzing the risks of a particular viral-vector-based gene therapy experiment requires determining two different variables. This subsection will examine the first variable, the risk that recombination of the viral vector will produce a replication competent virus that can be transmitted to other sites within the patient and, potentially, even to third parties. The second variable, the pathogenicity of the vector, (i.e. the consequences of infection by an inadvertent replication-competent virus or the adverse consequences of viral integration in the host) is examined in the next subsection.

While the process of creating replication-incompetent viral vectors is conceptually elegant and sound, replication-competent viruses could be produced through recombination of the viral vector's genes with the deleted packaging and envelope genes to produce a replication-competent virus. Such inadvertent recombination is a foreseeable risk in the packaging cell line with the helper packaging and envelope genes introduced into the packaging cells. Such recombination is known to occur, and one of the major parts of a gene therapy protocol deals with screening for such replication-competent virus.

For retroviral vectors, such recombination might also occur with so-called integrated proviral DNA, which are stretches of host cell genes that represent the genetic material introduced into an ancestor of the packaging cell line by a related virus. Such proviral genes are found in all mammalian cells. Similarly, recombination might occur with integrated proviral genes in the target cells of the patient or even with viral genes from wild-type viruses by which the patient is randomly infected during the course of therapy. For retroviral vectors, such risks cannot be completely eliminated, but the selection of appropriate retroviral strains for vectors (those least likely to be related to proviral sequences in the packaging cell line or patient) can reduce the risk of recombination to such low levels of probability that a risk-benefit analysis may justify proceeding with those therapies.

The risk of recombination to produce a replication-competent virus is probably greatest when the viral vector is closely related to common viruses that are found ubiquitously in the normal environment and in the cells of most normal adults, greatly increasing the opportunity for recombination. Furthermore, the ubiquity of viruses like adenovirus is a result of the relative ease with which such viruses are transmitted. For gene therapy experiments employing vectors like adenovirus, which are thus most likely to recombine to become both replication-competent and easily transmissible, the risks of the experiment to third parties can only be reduced by careful selection of the genetic sequences to be inserted into the vector.

§10.2(B) The Pathogenicity of Viral Vectors

As the previous subsection notes, the first concern in a risk-benefit analysis of a gene therapy procedure is the possibility of the inadvertent creation of replication-competent viruses. The second concern is the pathogenicity of any such virus, which is a function of its ease of transmission and the genes that it carries. For example, many of the early trials of gene therapy for cancer employ vectors containing genes for highly toxic anti-tumor agents such as TNF or IL-2.[13] These substances are potent immunomodulators and are highly toxic at relatively low doses (thereby making viral vector delivery to targeted tumor cells a potentially effective approach to cancer therapy). At the same time, the greater the toxicity of the new genetic material the greater the risk posed by a replication-competent virus.

This first concern, which is that the viral vectors would regain replication competence, is a function of the probability of a recombination (acquisition of the packaging genetic material) from the packaging cell or the host's target cells. That kind of recombination would most likely occur if the gene therapy viral vector infects a cell that also contains related, wild-type, replication-competent virus. The probability of such an event, that is the co-infection of the cell by a viral vector and a wild-type virus, is clearly a function of the how commonly the wild-type virus infects such cells. This in turn is determined, at least in part, by how easily transmitted the wild-type virus is. Retroviruses, though they may cause diseases as serious as HIV, are relatively difficult to transmit and are spread by sexual contact and blood. On the other hand, adenoviruses do not generally cause life-threatening diseases but can be much more easily transmitted by coughing, sneezing, and hand-to-mouth-to-hand contact.

---

13. *See* S.A. Rosenberg, *Gene Therapy for Cancer*, 268 J. AM. MED. ASS'N 2416–2419 (1992).

Obviously, the probability of a human liver cell being infected by a wild-type murine retrovirus or even closely related human retrovirus is very low, but the probability of encountering a wild-type adenovirus (many cold viruses are adenoviruses) in a human lung cell is much higher. Furthermore, a replication-competent retrovirus would not easily be transmitted to health care workers or others. However, a recombination that created a wild-type adenovirus carrying additional genetic material very toxic to normal cells, such as TNF, would be a serious problem. Thus, there is additional safety built into the choice of a virus which is not easily transmissible and which is extremely unlikely to recombine to become replication-competent, such as the retroviruses of murine origin used in the early TNF and IL-2 experiments.

To greatly reduce the risk of doing a gene therapy experiment with a common virus (albeit a replication incompetent virus), vectors have been developed that have multiple deletions affecting the genes necessary for packaging and assembly. These multiple deletions make it much less likely that recombination with a wild-type virus could restore competence to the vector, as each recombination is an independently low-probability event and numerous such events would have to occur in a single vector. Despite the risks of using a common virus to create a vector to deliver a toxic payload such as TNF, such a product is in clinical trials for cancer and is being developed by GenVec.[14] These experiments are ongoing and to date, all evidence from numerous researchers is that the vectors themselves are safe and do not recombine to regain the ability to replicate.

Adenovirus and adeno-associated virus vectors, which are closely related to common viruses frequently found in human tissue, have been used thus far in trials for diseases such as cystic fibrosis and familial hypercholesterolemia. Here the question of pathogenicity must focus on the function of the gene carried by the vector. For hypercholesteremia, the gene used in the early experiments encodes a lipoprotein receptor designed to help liver cells take more lipoprotein out of the blood stream. For a normal person, the harm produced by inadvertent infection with such a cholesterol-lowering agent would depend on his or her baseline level of lipid metabolism, but is unlikely to produce serious consequences at any likely level of infection. On the other hand, much less can be said with confidence about the function of the cystic fibrosis transmembrane protein. While we know that the absence of the functioning protein causes serious dysfunction in two affected cell types (lung and pancreas),

---

14. See *TNFerade(tm) for Pancreatic Cancer,* http://www.genvec.com/go.cfm?do=Page.View&pid=28.

it is difficult to know what would happen to normal cells of a variety of types with a substantially increased amount of the protein. It is here that experiments may well need to be done. Although treating animal models of the disease with the viral vector generates useful information, treating healthy animals would also generate useful information that is not currently required.

§10.2(C) Summary: Risks of Gene Therapy

In general, the two basic risk factors, the risk of recombination and the pathogenicity of a resulting virus can be restated in the form of a general model of risk assessment for gene therapy. The following formula demonstrates the relationship of the various factors that determine the risk of a viral-vector-based gene therapy procedure. R equals overall risk, Prcv represents the probability of recombination to a replication competent virus, and πrcv represents the pathogenicity of a replication competent virus (which is in turn a function of T, the ease of transmission of the wild type virus and V, the virulence or toxicity of the added genetic material together with the virulence or pathogenicity of the wild-type virus, thus:

$$\text{Risk} \approx \text{Prcv} \times \pi_{rcv} \text{ where } \pi_{rcv} = T \times V.$$

Unlike the risks of most experimental therapies, which are limited to the patient receiving the treatment, who may weigh the risks of such treatment against the risks of her serious underlying disease, viral vector gene therapy poses a very slight risk to third parties, through the inadvertent transmission of replication competent virus. This risk cannot be completely eliminated. Rather, it too can only be reduced by a protocol that not only minimizes the possibility of producing replication competent virus, but also minimizes the likelihood of inadvertent transmission to third parties, through rigorous procedures for viral vector handling and patient-care procedures as well as by appropriate monitoring of the patient herself.[15] It should be emphasized that while gene therapy does produce a risk to third parties, it is not the only medical procedure that does so. The currently used Sabin vaccine for polio (OPV), as well as other live virus vaccines, also present a small risk to third parties, and there are several cases per year of transmission of polio by the OPV vaccine to parents and other close contacts of vaccinated infants. As with all med-

---

15. The longer the viral vector may persist in the patient, the longer the follow-up necessary. For retroviruses, which persist indefinitely after integration, herpes vectors, which may also persist indefinitely, and some others, very long-term follow up may be imposed by the NIH or FDA.

ical risks, the risk of gene therapy to third parties must be weighed against the potential benefit to the patients.

§10.3 The Regulatory Framework for Gene Therapy

The risk-benefit analyses for proposed gene therapy experiments are done within a unique regulatory framework that is still evolving. For all researchers, approval is required from the institutional review board (IRB) of the hospital or other institution where the proposed gene therapy trial is to be performed. As with all experiments involving human subjects, IRB review and approval of the procedures for ensuring patient informed consent and other bioethical considerations of the proposed experiment are essential. In addition to the IRB review required for all experiments involving human subjects, because gene therapy involves the use of genetically engineered gene sequences, such protocols also require review from the Institutional Biosafety Committee (IBC) of the sponsoring institution. The IBC is principally concerned with the procedures for limiting the risks of inadvertent release of pathogenic viruses or toxic materials. This two-pronged institutional review is a pre-condition for external review by the NIH-RAC or the FDA or both.

§10.3(A) Notification of the NIH-RAC

For gene therapy trials conducted by researchers at universities and other not-for-profit institutions receiving federal funds, as well as commercially funded trials performed at facilities that also receive NIH funds, investigators must notify the Recombinant DNA Advisory Committee of the NIH (NIH-RAC). Up until 1992, the NIH-RAC delegated initial review of proposals for gene therapy to a Subcommittee on Human Gene Therapy. In February of 1992, the Subcommittee voted to discontinue itself and review of gene therapy protocols again became part of the NIH-RAC itself. To assist researchers in submitting protocols to the NIH-RAC to secure permission to do a gene therapy trial with human subjects, the NIH-RAC has published a document entitled "Points to Consider in the Design and Submission of Protocols for the Transfer of Recombinant DNA into the Genome of Human Subjects."[16] The NIH Gene Therapy Points to Consider, which are now Appendix M-1 of the

---

16. 55 Fed. Reg. 7438-01 (February 28, 1990).

NIH-RAC Guidelines,[17] sets forth in very synoptic form the major topics required for consideration and inclusion in the submission of a protocol.

In 1997, the NIH rules were changed yet again to require that universities and other NIH-fund recipients provide the NIH with notification of gene therapy experiments, retaining the Appendix M format for notification but eliminating the NIH's role in approval.[18] Approval is now solely in the hands of the Institutional Biosafety Committee (IBC) for gene therapy clinical trials, as well as the IRB and FDA, as with any other clinical trial. Although abandoning its formal role in approval, the NIH-RAC announced its intention to continue to review proposed experiments and to meet and consider any such proposed experiments that it considered raised novel safety or ethical concerns.

§10.3(B) Appendix M and the NIH's Principal Concerns

While the NIH no longer retains the right to block a gene therapy experiment, it does meet to review protocols that raise novel concerns and it reserves the right to express concerns to the FDA or Secretary of HHS. What sort of concerns might the NIH have? The answer to that question would appear to be found in the questions the NIH asks in Appendix M, the required form for notifying the NIH of a proposed gene therapy experiment. First, under the heading of objectives and rationale for the proposed trial, the NIH-RAC's principal concerns are clearly centered on risk-benefit. Part of the assessment of benefits is derived from the requirement that researchers respond to the issues of the existence of alternative, more conventional therapies, the intended results (i.e. cure, remission, palliation, etc.), and the degree to which the disease is one that will allow meaningful assessment of the results by clear, objective clinical markers of disease progression. At this point, gene therapy is not intended to be used as a substitute treatment when there are alternative therapies or little prospect of meaningful measurement of results.

The NIH-RAC's concern about the risks of viral vectors is clear:

> Two possible undesirable consequences of the transfer of recombinant DNA would be unintentional: (i) vertical transmission of genetic changes from an individual to his/her offspring, or (ii) horizontal transmission of viral infection to other persons with whom the individual comes in contact.

---

17. http://www4.od.nih.gov/oba/rac/guidelines_02/NIH_Guidelines_Apr_02.htm.

18. Recombinant DNA Research: Actions Under the Guidelines, 62 Fed. Reg. 59032 (October 31, 1997).

The data that must be included in a protocol submitted for NIH consideration returns to this concern. One key section of the NIH Gene Therapy Points to Consider (PTC) is the section entitled "Public health considerations."[19] Here the questions raised concern not only the general public health benefits that would accrue if a successful therapy were demonstrated, but also the risks of the "possibility that the added DNA will spread from the patient to other persons or to the environment." In particular, the NIH PTC asks, "What precautions will be taken against such spread (e.g. to patients sharing a room, health care workers, or family members)?" The section also raises the questions of what measures will be taken to mitigate the risks to public health and the risks to potential offspring of the patient and health care personnel.

Very clearly, the NIH PTC are intended to result in protocols which allow the NIH-RAC to weigh the potential for patient benefit, including the potential of benefit to other patients similarly afflicted, against the risks of creating a transmissible virus carrying the added genetic material (discussed above in §10.2). At the same time, the NIH PTC deals with the bioethical concerns common to all medical research: informed consent and the avoidance of unwarranted experimentation on human beings.

§10.3(C) FDA Review of Gene Therapy Protocols

After approval of a gene therapy protocol by the internal institutional IRB and IBC (and notification of the NIH when required), a proposed protocol must in all cases receive approval from the FDA (just as in the case of any other human trial of a new therapeutic or pharmaceutical). The FDA has a special Gene Therapy Advisory Committee to consider and recommend approval of gene therapy protocols prior to formal approval by the Commissioner of the FDA.

Like the NIH, the FDA has also published "Points to Consider in Human Somatic Cell Therapy and Gene Therapy" ("FDA PTC"). The availability of the FDA PTC was announced in the Federal Register on November 29, 1991 (56 FR 61022). In 1998, the FDA PTC was supplemented by a "Guidance for Human Somatic Cell Therapy and Gene Therapy" (http://www.fda.gov/cber/gdlns/somgene.pdf). Responsibility for the FDA's regulation of gene therapy was assigned to the FDA's Center for Biologics Evaluation and Research (CBER) and remains with CBER after the recent reorganization of biologics regulation (see §8.2(A)). CBER has published an additional regulatory guid-

---

19. *Id.* at Appendix M-II-B-4.

ance addressing the need to continue tracking and monitoring patients who took part in gene therapy clinical trials.[20] It is fair to say that CBER's approach reflects a very heavy emphasis on assuring that products produced by the use of viruses, serums, or cell culture are free from contamination. However, the FDA's approach to the regulation of gene therapy continues to evolve.

The most important trend in the FDA's regulation, from the perspective of companies developing gene therapy, is the FDA's increasing interest in the problem of long-term follow-up of gene therapy patients to monitor possible long-term adverse effects of some viral vectors. This concern arises because the three French X-SCID subjects developed leukemia some two years after treatment (see text accompanying footnote 6) and because it is known that replication competent virus can emerge after many years (if co-infection with a related wild-type virus occurs at any time after treatment so long as the original vector persists). For this reason, the FDA has considered but not adopted a proposal that would mandate long-term follow-up of all gene therapy patients, but currently requires long-term follow-up only for retroviral vectors,[21] while making the decision to require such follow-up for other vectors on a case-by-case basis. For retroviral vectors, the current requirement is pre-treatment, three months, six months, twelve months after treatment, and yearly thereafter.[22]

§10.4 THE ETHICAL DEBATE OVER GENE THERAPY

At the core of the ethical debate over gene therapy is a disagreement over whether or not human gene therapy, and particularly somatic cell gene therapy, represents something radically new and different. Those most concerned about the possible ramifications of gene therapy tend to view it as a technological revolution that represents a significant discontinuity with previous social and medical convention, on a par with such earlier technological revolutions as the Industrial Revolution. From that perspective, gene therapy is an

---

20. Gene Therapy Patient Tracking System, (http://www.fda.gov/cber/genetherapy/gttrack.htm#r5ret).

21. Supplemental Guidance on Testing for Replication Competent Retrovirus in Retroviral Vector Based Gene Therapy Products and During Follow-up of Patients in Clinical Trials Using Retroviral Vectors, http://www.fda.gov/cber/gdlns/retrogt1000.htm (October 18, 2000).

22. Id.

unprecedented step, a technology of power not only over Nature but over our own nature as well. This concern over the ethical issues raised by gene therapy is heightened by the perception that it has great potential significance for the entire future direction of humankind.

Proponents of gene therapy research who are less concerned about the portentous nature of the beginnings of gene therapy see gene therapy as presenting only new contexts for ethical dilemmas that we are already being forced to face by other medical technologies. For example, in response to concerns over the power that gene therapy offers to affect the development of human beings, the proponents of gene therapy ask whether it really is ethically different to correct a lipid metabolism deficiency with gene therapy than to treat it with a cholesterol-lowering drug. Similarly, in response to the concerns that gene therapy leads inexorably to gene enhancement, gene therapy proponents query whether the choice to intervene genetically to alter height presents a different ethical dilemma from that now posed by recombinant human growth hormone.[23] Finally, with respect to the potential power of gene therapy to allow parents to "treat" their embryos to choose the characteristics of their offspring, the proponents of gene therapy see little difference from the current technologies that allow pre-natal diagnosis and abortion, or even pre-implantation embryo screening by genetic testing.[24] A resolution to the debate over the uniqueness of the ethical issues posed by gene therapy is beyond the scope of this chapter, however, these issues can be grouped into questions about the general limitations of permissible gene therapy interventions, questions about the currently forbidden area of germ-line gene therapy, and ethical questions about the equality or inequality of future access to gene therapy.

§10.4(A) General Limitations on Somatic Cell Gene Therapy

Every new form of gene therapy, whether it involves a change in vector or a new therapeutic target, will undergo the review applicable to all medical therapies and the additional review discussed in §10.3. In addition, the NIH-RAC Guidelines adopt the position that gene therapy should only be aimed at therapy and not at enhancement. This position raises one of the central eth-

---

23. Compare M.M. Lee, *Idiopathic Short Stature*, 354 NEW ENG. J. MED. 2576 (Jun. 15, 2006) with S.E. Oberfield, *Growth Hormone Use in Normal, Short Children—A Plea for Reason*, 340 NEW ENG. J. MED. 557 (Feb. 18, 1999).

24. See Dontrich J. Jordaan, *Preimplantation Genetic Screening and Selection: An Ethical Analysis*, 22 BIOTECH. L. REP. 586 (2003); John Rennie, *Grading the Gene Tests*, SCI. AM., June 1994, at 88.

ical dilemmas for gene therapy. The dilemma is rooted in the perception that it would be wrong to use technology to substantially change "human nature" by a Brave-New-World-like effort to create a superior species.

It is easy to see that the line between enhancement and therapy will be a very difficult one to draw in a great many important cases. For instance, if a single gene could be identified as the cause of a form of profound retardation, there would be little quarrel over attempts to use gene therapy to treat an afflicted infant, or even fetus. If however, a single gene could be identified that has a significant beneficial impact on even one mental process associated with intelligence, such as the ability to do mental arithmetic calculations or even to memorize lists, there would be enormous resistance to using genetic techniques to enhance the abilities of "normal" individuals. Such resistance cannot be based on any "Kantian" notion of ethics (see Chapter 9) but is based on one of two arguments. The first argument is that a just society would not allow the rich to become not only financially but genetically ever-better-endowed than the poor, and that no society would choose to provide such benefits for all. The second argument is that it is wrong to tamper with nature, although why enhancement is tampering with nature while the eradication of disease is not is unclear. Further, once such a technology is possible, the problem of defining normalcy complicates the ethical issue. For instance, is an individual "normal" who is a gifted linguist but who cannot balance a checkbook and is in the very bottom ten-percent of the population in numerical ability? Is such an individual precluded from receiving treatment for his arithmetic deficiency (or dyslexia)? It would seem that it is only the limitations of current technology and the allocation of scarce resources for research that allow us to avoid such issues now. It is likely that we will not be able to avoid them indefinitely. Of course, as was noted in the beginning of this chapter, such problems may not be unique to gene therapy, but rather may simply increase in number as gene therapy develops.

§10.4(B) Ethical Issues Surrounding Germ-Line Gene Therapy

Germ-line gene therapy entails the insertion of new genetic material (DNA) into cells involved in reproduction, with the potential that the new DNA will be heritable. There are two major inter-related ethical problems raised by the prospect of germ-line gene therapy. The first is the concept of informed consent. Informed consent is a fundamental requirement for ethical relationships between physician and patient, and, to an extent, is derived from the basic Kantian ethical imperative that we treat others as full persons entitled to choose their own ends, rather than using others as means to our ends. Thus patients are entitled to choose freely whether to partake or not to partake of a given course of treatment, and they are not to be subjected to treatment merely because it meets the physician's concept of the good. To assure the autonomy of the patient and her meaningful participation in the decision making, bioethicists (and courts) have developed the doctrine of informed consent, which requires not only that the patient agree, but also that she be given all of the relevant information upon which to base a reasoned decision.

In germ-line gene therapy, regardless of the informed consent of the individual being treated, the creation of heritable traits effectively imposes the results of that decision on succeeding generations ad infinitum, without any possibility of the consent of those affected. Of course, the doctrine of informed consent as it is currently understood allows significant exceptions for the treatment of those unable to consent for themselves. Parents routinely consent for their children, and guardians or judges regularly consent for those who are judged incompetent and thus, like infants and children, are unable to choose meaningfully for themselves.

Is the difference between allowing a parent to choose for a child (even one in utero) and allowing an adult to choose for all of their descendants a qualitative difference or merely quantitative? Is there some ethical basis for imposing a limit on consent to one generation? Those who would argue for a permanent ban on gene therapy believe that there is such an ethical basis and that it is impermissible to choose for future generations. However, the distinction between choosing for one's child and choosing for one's grandchildren cannot be easily derived from the basic Kantian notion of autonomy and treating others with full respect for their own ends.

To fully appreciate the ethical complexity of the issue, imagine that scientists have developed a potential gene therapy for a devastating disease for which there currently is no alternative cure and which is invariably lethal (something along the order of Lesch-Nyhan, which is always devastating and usually fatal).[25] Assume that gene therapy for the disease entails a high probability (or even certainty) of affecting reproductive cells and thus creating a heritable trait (which would be the non-Lesch-Nyhan phenotype). Should therapy to the afflicted individual be denied because of the probable germ-line effect? To deny treatment to an individual because of concern about the potential impact on others is in fact a negation of the Kantian principle which gives rise to the original concern. It is requiring one person to suffer as a

---

25. http://www.ninds.nih.gov/health_and_medical/disorders/lesch_doc.htm.

means to protecting the ends of future individuals (even though those future individuals would likely wish the same treatment).

To explore the ethical issues further in the context of this hypothetical, we may need to abandon the Kantian-derived argument about informed consent and instead turn to a utilitarian argument. The essential utilitarian argument against germ-line gene therapy has two parts. The first part is simply the argument that there is insufficient knowledge about the full consequences of that may result from any intergenerational genetic changes. This argument is sometimes bolstered by references to disease-producing mutations (such as sickle-cell trait) that are subsequently discovered to confer an adaptive advantage under particular conditions (the sickle cell mutation in its heterozygous form does not cause disease but appears to confer a greater degree of malaria resistance). The second part of the utilitarian argument against gene therapy builds upon the first by focusing on the very great number of people potentially affected by any one germ-line change. Thus, if a genetic change does turn out to have unforeseen negative consequences, it serves the greater good if such a change is limited in scope, particularly to those individuals who are willing to bear the risk themselves. This utilitarian argument will be increasingly difficult to make as progress in understanding gene variation and protein function accumulates. At some point it will certainly be possible, for at least some genetic variations, to say that the benefit of a variation (e.g. to confer some resistance to a pathogen that has been largely eradicated, such as smallpox, or is routinely preventable, such as polio) requires preservation. It is hard to argue in such cases that one could be doing damage to the greater social welfare by removing from the genome a mutation that can produce devastating disease.[26]

It would seem difficult to make the case that in absolutely all cases the requirement of informed consent, more general Kantian notions or even utilitarian arguments can justify a prohibition against germ-line gene therapy. Although the technology is not yet developed to the point where germ-line therapy is a viable approach to the treatment of genetic diseases such as Lesch-

---

26. Patricia Carpenter Carson at the University of California at San Diego proposes that one such argument might be that if a less radical therapy could later be developed for such lethal mutations, such as a new pharmaceutical compound, it could then turn out that the otherwise lethal mutation might then in some instance confer a selective advantage. Nonetheless, if the argument is to be decided on utilitarian grounds, the probability of such instances must simply be calculated against the certain loss of utility that would occur if affected individuals are to be denied treatment until a less radical therapy is developed. Intuitively in such cases the utilitarian ethic demands treatment.

Nyhan, in the future if germ-line technology affords the potential for a cure while somatic cell therapies do not, it would be difficult to maintain a prohibition on germ-line therapy. Furthermore, if the technological capacity were to develop while a ban were imposed in the U.S. and EU, it is likely that some countries would not join in the prohibition and become destinations for those who need and could afford treatment for themselves or their offspring.

§10.4(C) The Ethical Problem of Access to Gene Therapy

In recent years, access to health care has been a major national political issue in the United States. Rapidly escalating health care costs have been the driving factor in the debate over health care access. A substantial component of both the rising costs and the issue of access has been the development of a number of very expensive medical technologies. Ranging from kidney dialysis to dual heart-lung transplants, modern medicine has created new but often extraordinarily expensive treatments for conditions that would otherwise be fatal. Such developments raise two related questions. First, to what extent should access to such expensive technologies be determined by the ability to pay, or at least to buy the best health insurance? Second, if not allocated by ability to pay, is there any other ethical basis for rationing expensive medical technology?

It may turn out that gene therapy, unlike other complex medical technologies, can be done at a reasonable and unarguably cost-effective price. If a rather standard set of safe and effective vectors could be developed, it might be that gene therapy for hemophilia or for cystic fibrosis could cost a few thousand dollars or less for a permanent therapy for diseases which now can cost tens of thousands per year. In that event, access to gene therapy is unlikely to present the sort of ethical problem that is presented by organ transplantation. On the other hand, gene therapy may well develop into the sort of expensive technology that has led to the current national debate.

The complicated issue of parental decision making for children's health care (also discussed above in connection with germ-line gene therapy) might also be considered as a question of access by parental control rather than by wealth. Currently the parent of a child who requires treatment ranging from a simple blood transfusion to an organ transplant may be opposed to the treatment, for any number of moral or religious reasons. In such cases, the courts have long been wrestling with the standard for determining when such parental opposition constitutes such neglect as to require state intervention to compel medical treatment.

The issue for gene therapy is not truly different from the problem raised by parental refusal of existing medical interventions.

For traditional medical therapies, state intervention is easily justified when the proposed therapy is life-saving, as in the case of a blood transfusion. It is somewhat more problematic when the condition of the child to be treated is one that "merely" impairs the quality of life, rather than one that threatens survival itself. Even in such cases, some courts have been willing to intervene to compel treatment when the difference in the prospective quality of life is marked, such as between normal mobility and significant disability. It is likely that the question of compulsory gene therapy for children will follow the same path, which may include the problem of line-drawing between therapy and enhancement discussed above in §10.4(A).

§10.5 Conclusion

Gene therapy is clearly a major target for biotechnology. With additional disease-producing mutations being identified at an extraordinarily rapid pace, the need to turn that knowledge into effective therapies will continue to push gene therapy research and development. In addition, the potential power of gene therapy for other uses, such as anti-viral therapies and vaccines, adds to the incentive for gene therapy development. Nevertheless, there are significant limitations in the current state of the art of gene delivery and significant risks to some of the current techniques, such as the use of retroviral vectors.

The NIH and FDA regulation of gene therapy has allowed research to go forward into the clinic while the regulatory pathway is still evolving. The ordinary issues of safety and effectiveness confronting the development of any biological are added to the unique risks surrounding the development of viral vectors carrying new genetic material and require very careful analysis and planning. For companies that wish to develop gene therapies, the desirability of choosing non-persistent vectors with relatively benign "payloads" is clear, in light of the financial, regulatory, and potential liability issues. Furthermore, choosing targets that the public clearly perceives as warranting treatment will limit future public controversy. Finally, the power of the technology, particularly the potential power of germ-line gene therapy, demands special caution. Nevertheless, the history of technological development teaches that the impetus to employ new tools is irresistible. Gene therapy is expected to be a major component of our medical arsenal. We can only hope that continuing scrutiny and debate will minimize the problems that its development will bring.

CHAPTER 11

Liability for Biotechnology Products

§11.0 Introduction

One of the issues of great concern to biotechnology investors and managers is the potential liability for personal injuries allegedly caused by a biotechnology product. In recent years there has been considerable public controversy about the impact of the tort liability system on American business. The "crises" in medical malpractice insurance and, to a lesser extent, in litigation over childhood vaccine injuries, have placed doctors and pharmaceutical manufacturers in the center of the public debate about tort reform.

It might be useful to consider the problem of liability for biotechnology products in the context of a hypothetical case. Assume that Biobiz Inc. has received FDA approval to market "Integrex" for the treatment of rheumatoid arthritis. The product is a genetically engineered protein produced in a mammalian cell fermentation process in Biobiz's Northern California Plant and administered to patients by injection. Shortly after its approval, there were several reports of adverse reactions in patients. Four patients suffered a significant loss of liver function within one week after receiving Integrex.

The first lawsuit for damages has been filed by an attorney representing one of the patients, Hedda Hurt. How will Biobiz's potential liability to Hedda be determined? Were Hedda's injuries due to an unforeseen side effect of the product itself or due to some contamination of the product with some other harmful substance? Did the label instructions accompanying the product contain any warning of the possibility of such a reaction? Was usage of the product contraindicated in patients with some identifiable characteristic such as high blood pressure or concurrent administration of some other drug, which should have excluded Hedda from receiving the product? It may also be important for the court to decide whether Biobiz's liability will be based on negligence or on strict product liability. Finally, it will of course be necessary to determine whether Biobiz's Integrex caused Hedda's injury.

This chapter examines the basic legal structure governing liability for biotechnology health care products; negligence, product liability, legal cause, and the Childhood Vaccine Injury Act of 1986. In addition, the chapter briefly examines the debate over the appropriate standard for liability for injuries caused by pharmaceutical and biotechnology products. There also is a brief discussion of liability for non-health care applications, such as pesticides, food products and industrial biotechnology products.

§11.1 A Basic Overview of Negligence

For at least the last century or more in most of the United States, liability for unintentionally inflicted personal injuries has been generally determined by the law of negligence. Negligence is often conceived of as relating liability to "fault" or blameworthiness, principally through the central negligence concept of "reasonable care." In other words, under negligence, the obligation of A to pay for injuries suffered by B, as a result of some activity on the part of A, is determined by an inquiry into whether or not A was at fault for not exercising reasonable care in the conduct of his activities. This of course requires a determination of what constitutes reasonable care in each case.

Perhaps the most accurate way to convey the meaning of reasonable care in negligence, a concept which has plagued generations of first-year law students, is to provide the full text of the jury instruction commonly used in California to guide juries of twelve lay-persons deliberating negligence cases.

> 401 Basic Standard of Care
>
> ...
>
> **A person can be negligent by acting or by failing to act. A person is negligent if he or she does something that a reasonably careful person would not do in the same situation or fails to do something that a reasonably careful person would do in the same situation.**
>
> ...

Is it all clear now? What does it mean to say that Biobiz did or did not use reasonable care, the care that the pharmaceutical manufacturer of ordinary prudence would use in order to avoid injury to themselves or others "in the same situation"? To answer that question, assume that Hedda Hurt's case has proceeded simply on the theory that the injury was due to contamination of the product by a virus, 'HVX', which entered into the production process undetected. In that case, there are three possible approaches, each of which might be used by Hurt's lawyer to convince the jury of Biobiz's negligence: violation of a statutory or regulatory requirement; deviation from industry custom; and, finally, the general concept of reasonable care, particularly as it may be derived from a cost-benefit analysis. Each of these three approaches to determining the standard of care for a negligence case is briefly examined in the following subsections.

§11.1(A) Violation of a Statute or Regulation as Negligence *Per Se*

One of the clearest and simplest ways for a plaintiff's lawyer, such as Ms. Hurt's, to prove negligence is to show that the defendant's actions violated a statutory or regulatory requirement. For example, the FDA has numerous regulations that define GMP (Good Manufacturing Practice). These regulations represent a minimal standard of reasonable care for the particular issues that the regulations cover. In our hypothetical case, if Hedda's lawyer can show that the FDA's good manufacturing practice rules required manufacturers like Biobiz to perform a particular test for the presence of HVX, then a failure of Biobiz to do that test would be negligence *per se*, as long as the requirement can be shown to have been intended to prevent injuries such as that which befell Hedda Hurt.

Thus the violation of a statutory or regulatory requirement resulting in an injury that the requirement is intended to prevent is clearly negligent and liability is almost certain to result. On the other hand, the issue of whether conformity to statutory or regulatory requirements is conclusive evidence of non-negligence, or reasonable care, is a much more complex issue. For example, if the FDA has a particular approved GMP procedure for guarding against HVX contamination, then following the FDA's GMP procedure is relatively certain to be considered reasonable care **if** the FDA's regulations should be considered to be both comprehensive and exclusive on the subject of HVX testing. On the other hand, if there are no GMP requirements relating to testing for HVX, Biobiz's compliance with general GMP requirements is not necessarily reasonable care. Finally the GMP procedures, as many other regulatory or statutory frameworks, leave room for additional procedures when appropriate for the individual circumstances of a particular type of manufacture. So even if Biobiz complied with GMP by performing the required test, Hurt's lawyer might point to the existence of an additional procedure that was not done but might have detected the problem and then argue that the additional procedure was required by reasonable care even if not specified in the GMP procedures. Thus a violation of GMP requirements is fairly conclusively negligent, while compliance with GMP may well simply require Hurt's lawyer to look to custom or general negligence for a way to build her case.

§11.1(B) The Relevance of Custom to Negligence

Since the central concept in negligence is the care that would be taken by the reasonable person (or company), it is logical that custom, if defined as that conduct or precautions used by an industry in general, would represent reasonable care. In other words, that which has become the general practice or custom is likely to represent reasonable care, the care that the reasonable person would take. So for example, if companies using mammalian cell cultures to produce human therapeutics follow an almost universal practice of using a very sensitive assay for HVX, then Hedda's lawyer will have a strong argument that Biobiz's failure to use such an assay was negligent, or fell short of reasonable care. On the other hand, the converse argument by Biobiz would be less powerful. That is, if Biobiz were to argue that the majority of similar companies, or even all other similar companies, **do not** use any precautions to screen for HVX contamination it would be helpful but not conclusive on the issue of whether or not the failure to take particular precautions against HVX contamination is negligence. The California standard jury instruction on the relation of custom to reasonable care underscores this point.

> 413 Custom or Practice
>
> **You may consider customs or practices in the community in deciding whether [name of plaintiff/defendant] acted reasonably. Customs and practices do not necessarily determine what a reasonable person would have done in.... They are only factors for you to consider.**
>
> **You should consider whether the custom or practice itself is reasonable.**

Why is a departure from industry custom very likely to be negligence while compliance with industry custom is only relevant to a determination of reasonable care but much less than conclusive? Simply because it is far easier to argue that the adoption of a safety precaution by industry is evidence that the precaution is necessary for reasonable care, while the failure of industry to adopt a precaution, particularly a newly developed procedure, may be easily ascribed to a lag between development and adoption. This point is generally thought to have been best expressed by Justice Learned Hand in *The T.J. Hooper*: [1]

> Indeed in most cases reasonable prudence is in fact common prudence; but strictly it is never its measure: a whole calling may have unduly lagged in the adoption of new and available devices.... [C]ourts must in the end say what is required; there are precautions so imperative that even their universal disregard will not excuse their omission.

Thus, if Biobiz can demonstrate compliance with both the regulatory requirements and industry custom, Ms. Hurt's lawyer has one last avenue by which to attempt to establish that Biobiz failed to exercise reasonable care.

§11.1(C) General Evidence of Reasonable Care

For activities within the common experience of most jurors, such as driving a car, it may well be possible to elicit from the jurors a consensus as to what constitutes reasonable care. For example, a jury needs no special guidance to determine that it is not reasonable care to divert attention from the road to adjust a car radio, even in the absence of a particular statutory violation, despite the fact that it is a behavior engaged in by the great majority of drivers at one time or another. However, the jury is almost certain to be lacking in any common experience which can guide them on the issue of the procedures that should be utilized by a reasonably prudent manufacturer in circumstances similar to those faced by Biobiz. Thus the stage is set for the introduction of that nearly universal feature of such product liability suits, the battle of experts.

The role of the expert, under the Federal Rules of Civil Procedure, is to testify when "scientific, technical, or other specialized knowledge will assist the trier of fact to understand the evidence or to determine a fact in issue" (Rule 702, Federal Rules of Civil Procedure). Clearly, a jury needs the help of experts to determine what constitutes reasonable care for a manufacturer of biologically produced therapeutics. What is less readily apparent is how the conflicting testimony of opposing experts can be focused in such a way as to be of practical assistance to a jury untrained in molecular biology or pharmacology. It is here that the trial lawyers earn their pay (or of course, fail to do so). What the lawyer must elicit from the expert, in a form that is understandable to the jury, is why a particular procedure is or is not "worthwhile," which is the last refuge in which reasonable care can be found.

---

1. 60 F.2d 737 (2d. Cir. 1932).

If reasonable care is the care that the ordinary prudent person or company would take to avoid a risk which is foreseeable and which can be reasonably avoided, then reasonable care will vary with the circumstances, particularly with the magnitude of any foreseeable risk. In a simpler context, it is likely to be reasonable care to wear a bicycle helmet while riding a bicycle but it is almost certainly more than is required when crossing the street on foot. In both cases a precaution is available, but the difference in the size of the risk varies considerably, rendering the precaution worthwhile (i.e. reasonable) in the one case but not in the other. This is precisely what is at issue in most far more complicated negligence cases. In the Biobiz hypothetical, the experts' disagreement may likely center on how big a risk is posed by HVX contamination, that is, how significant a problem it is in processes like Biobiz's and how feasible the steps are to avoid the risk.

Law professors and economists are fond of precisely quantifying the risk-benefit analyses in such cases. Such reductionism can yield deceptively simple equations that "prove" that the failure to take a precaution costing $50,000 is not negligent when the underlying risk is a one-in-100,000 chance of serious injury (assuming for the moment that no one would value a single serious injury or death at $5 billion, which is the product of $50,000 times 100,000). Juries are undoubtedly less given to such formal quantification, even when urged on by skillful defense counsel. Nevertheless, juries almost certainly struggle to resolve, based on the varying expert testimony on the size of the risk and the feasibility of precautions, how much the reasonable manufacturer should pay to minimize such risks. Thus Biobiz's defense of Hurt's negligence action will likely succeed only if the risk of HVX is sufficiently small to be insignificant, or the costs of avoiding HVX contamination sufficiently large as to be unreasonable.

§11.2 An Introduction to Strict Product Liability

The most significant exception to this dominant negligence framework has been the rise of "product liability" during the last thirty or so years. While negligence is tied to the concept of "fault," product liability is often referred to as "strict." This relative stringency derives from the fact that product liability does not rest on fault, at least insofar as fault is a failure to exercise reasonable care. Nevertheless, as the following discussion explains, strict product liability is far from automatic liability and, in many cases, the result in negligence and the result in product liability will be the same.

In 1944 a waitress, surnamed Escola, was injured when a Coca-Cola bottle exploded in her hand. In one of the most influential personal injury law decisions of the twentieth century, Justice Traynor of the California Supreme Court overturned 50 years of negligence jurisprudence and ruled that a manufacturer "incurs an absolute liability when an article that he has placed on the market, knowing that it is to be used without inspection, proves to have a defect that causes injury to human beings" Escola v. Coca-Cola Bottling Co. of Fresno, 150 P.2d 436 (Cal. 1944). Over the course of the next twenty years, courts and legislatures in virtually every state adopted Justice Traynor's reasoning. In 1966 these post-Escola developments formed the basis of §402A of the RESTATEMENT (SECOND) OF TORTS, which quickly became the primary focus of product liability law. §402A reads in part:

> (1) One who sells any product in a *defective condition unreasonably dangerous* [*emphasis added*] to the user or consumer or to his property is subject to liability for physical harm thereby caused to the ultimate user or consumer.... [(2)] although ... [(a)] the seller has exercised all possible care in the preparation and sale of his product....

Since its adoption by the influential American Law Institute, Restatement §402A has been at the center of the growing debate over manufacturers' liability, the United States' competitiveness, and the product liability system. Before attempting to look at the way courts have most recently applied §402A to the problem of human therapeutics and vaccines, it is necessary to examine the general scheme of §402A, particularly the ways in which liability for products under that section differs from negligence liability.

§11.2(A) Defective Condition Unreasonably Dangerous

The heart of product liability law is the concept of defect. In negligence, liability is based on the failure to exercise reasonable care. In product liability, liability is based on the selling of a defective product that is unreasonably dangerous. The distinction is a subtle one. There are two kinds of defects, defects in design and defects in manufacture. In addition, a product may be unreasonably dangerous because of a failure to provide an adequate warning. Each of these three bases for liability; manufacturing defect, design defect, and warning are discussed in this subsection.

§11.2(B) Manufacturing Defect

The simplest kind of product liability case for a plaintiff to prove and win is one in which a manufacturing defect can be established. Quite simply, a "**product contains a manufacturing defect if the product differs from the man-**

ufacturer's design or specifications or from other typical units of the same product line." Cal. Civil Jury Inst. 1202. Thus, the hypothetical Ms. Hurt has a slam-dunk product liability case (in jurisdictions which apply §402A to pharmaceuticals) if her lawyer can establish that HVX viral contamination of the lot from which Ms. Hurt's dose was taken made the product different from the manufacturer's specifications or from other lots of the same drug. This is true no matter how careful Biobiz's quality control process was and is probably the single most significant consequence of applying product liability developed under Restatement §402A rather than negligence. The implicit assumption in the treatment of manufacturing defects is that the manufacturer's design or specifications or identical products set a standard, an injurious departure from which is always defective and unreasonably dangerous.

§11.2(C) Design Defect

The more difficult cases, and the cases which generate the greatest controversy for products liability law, are the design defect cases. In design defect, the plaintiff claims that the product was unreasonably dangerous even though made precisely to the manufacturer's specifications. In large part, the product liability concept of defective and unreasonably dangerous design is very close to the negligence concept of reasonable care. In California, as in a number of other states, there are two alternative approaches to the issue of design defect, "consumer expectations" and "risk/utility."[2] In the first approach, a product is defective if it fails to meet the reasonable expectations of the ordinary consumer: that is, the defendant's product was more dangerous than would ordinarily be expected for products of that kind.[3] This type of consumer expectations test approaches negligence because in most cases reasonable consumer expectations are derived from the level of safety attained by other manufacturers' products, which in negligence would be strong evidence of the standard of reasonable care. Ms. Hurt also has a strong case against Biobiz based on the consumer expectations standard of defect, since it is easy to make a persuasive argument that liver damage is far beyond the level of adverse effect that the ordinary consumer would expect from an arthritis treatment (unless of course, the product was accompanied by a warning of just such a risk, as discussed in §11.2(D)).

---

2. Barker v. Lull Engineering Co., 573 P.2d 443 (Cal. 1978).
3. Barker v, Lull Engineering Co, *Id. See also* Conde v. Velsicol Chem. Corp., 816 F.Supp. 453, (S.D. Ohio1992).

On the other hand, a particular product that meets the expectations of the ordinary consumer is not necessarily protected against liability for a product design that causes injury. In a number of states, including California, the jury is permitted to find liability **either** on the basis of the consumer expectations test **or** the risk/utility test (and some states use only the risk/utility test). For example, even where consumer expectations of a product are that it may cause injuries, as may be the case with some vaccines, if the plaintiff can demonstrate that those side effects are now reasonably avoidable through some technological advance that has been made since the injurious product was developed, then products not made with the newer technology may be defective.

In this alternative risk/utility approach, a product risk that makes the product defective and unreasonably dangerous (and therefore renders the manufacturer liable for that risk) is very much like a foreseeable risk that a reasonable manufacturer would have avoided through reasonable care.[4] Indeed, the negligence concept of reasonable care and the product liability concept of unreasonably dangerous is much the same in many jurisdictions, particularly those that do not use the consumer expectations test. In other jurisdictions product liability may differ from negligence primarily when the product turns out to have an unforeseen danger, which would exceed the expectations of the ordinary consumer. In negligence, reasonable care need only be taken to avoid those risks that were foreseeable by the reasonable manufacturer at the time the product was sold. In strict liability, a manufacturer can be held liable for having failed to take precautions against a risk that only became apparent in hindsight with the passage of time, particularly under the consumer expectations test. In most jurisdictions using a risk-utility analysis for design defect, the risks to be considered do not include those unforeseeable risks that could only be discovered after the product has been on the market.[5]

To amplify this point, consider an alternative hypothetical, in which Ms. Hurt's injury was allegedly produced not by a manufacturing defect such as

---

4. Voss v. Black & Decker Mfg. Co., 450 N.E.2d 204, 208 (1983):
The test is whether the product as designed was "not reasonably safe"—that is, whether it is a product which, if the design defect were known at the time of manufacture, a reasonable person would conclude that the utility of the product did not outweigh the risk inherent in marketing a product designed in that manner.

5. This is also the general rule for design defect liability taken by the Restatement (Third) of Torts: Product Liability §2(c). The authors of the Restatement (Third) suggest much greater limits on design defect liability for prescription drugs in Restatement (Third) of Torts: Product Liability §6 but it is not at all clear that those additional limitations will be adopted by many states.

viral contamination, but a "design" defect such as an interaction with another drug, for example, an antidepressant. Further, assume that the interaction had not been observed during the clinical trials and would not have been predicted on the basis of the structure or mechanism of action of the two compounds.

To prove that a design defect was created by such a previously undetected risk, a plaintiff would have to show that this additional risk makes the product's risks as a whole outweigh the product's benefits, particularly if there is an alternative product, in this case another arthritis drug, that provides similar benefit without that risk. Potential liability for later-discovered defects, along with the consumer expectations analysis, are really the biggest differences between "strict" product liability and negligence, but these additional grounds for liability now increasingly are being restricted by state courts and state legislatures. In California, since the decision in *Brown v. Superior Court*, 751 P.2d 470 (Cal. 1988), it is clear that negligence rules apply to all prescription drugs (see §11.5). In other states, the courts are divided on whether or not to retain this strict liability standard for drugs with after-discovered defects or to limit design defect liability for prescription drugs to foreseeable risks and therefore bring them within a negligence standard.

§11.2(D) Failure to Warn

This third possible basis for a product liability case is established when a plaintiff asserts that her injury was caused by the manufacturer's failure to provide an adequate warning of the product's risks (to enable the plaintiff to make an informed choice as to whether or not to purchase the product) or proper instructions as to its use (to enable the plaintiff to avoid injury). For example, Ms. Hurt would allege that the manufacturer was liable for not warning her (or, more precisely, her physician) that Integrex could cause liver failure if taken in conjunction with the antidepressant used by Ms. Hurt. Here the question of whether the risk was known or reasonably knowable at the time of the sale becomes crucial. While in some jurisdictions, later-discovered risks, even those which had been unforeseeable prior to their occurrence, may be weighed on the risk/benefit scale of product defect, in the great majority of jurisdictions liability for failure to warn is limited to defects which were known or reasonably knowable. However, the risks for which a manufacturer is liable for failure to warn are not just those that would make the product unreasonably dangerous as a whole, but simply those that a reasonable physician/patient would need to know to make an informed decision as to whether and how to use the product. The California civil jury instruction is, once again, a reasonable statement of the law in most jurisdictions:

1205 Strict Liability—Failure to Warn—Essential Factual Elements

...

1. That [name of defendant] [manufactured] the [product];
2. That the [product] had potential [risks/side effects/allergic reactions] that were [known] [or] [knowable by the use of scientific knowledge available] at the time of [manufacture];
3. That the potential [risks/side effects/allergic reactions] presented a substantial danger to users of the [product];
4. That ordinary consumers would not have recognized the potential [risks/side effects/allergic reactions];
5. That [name of defendant] failed to adequately warn of the potential [risks/side effects/allergic reactions];

...

The warning must be given to the prescribing physician and must include the potential risks, side effects, or allergic reactions that may follow the foreseeable use of the product....[6]

Note both the risk and manner of use must be foreseeable and the warning must, in general, be given to the physician and not the patient/consumer. In the Hedda Hurt hypothetical if administration of Integrex to a patient who is also taking a particular type of antidepressant is foreseeable and reasonable pre-market testing would have discovered the adverse effects of taking both drugs, then a serious or near fatal interaction with the other drug would almost certainly be considered a defect that is unreasonably dangerous, at least in the absence of a warning.

§11.2(E) Preemption of Liability by FDA Approval: A Complicated Tale

Preemption is the legal term applied when a federal law precludes a state law on the same subject. Congress, under its power to regulate commerce, has the power to pass laws (subject to the President's signing) that exempt a reg-

---

6. Judicial Council of California Civil Jury Instructions CACI 1205.

ulated industry from specific state requirements or even from liability for damages under state personal injury law.[7] Manufacturers have also attempted to argue that because the FDA approved their products or the labeling and warnings accompanying their products, liability for a defective product or a failure to warn would conflict with those federal regulatory decisions and should be preempted. In *Medtronic Inc., v. Lohr*, 518 U.S. 470 (1996), the Supreme Court rejected the preemption argument for medical devices that were approved under the 510(k) substantial equivalence standard (see §8.12), despite an express provision of the Medical Device Amendments of 1976, 21 U.S.C. §360(k). The reasoning in *Medtronic* would appear to apply to drugs as well, particularly because the statutes pertaining to drugs contain no express preemption provision comparable to that in the Device Amendments. However, the Supreme Court left open the question of whether devices subject to the more rigorous PMA review process and the answer to that question is still unclear.[8] Most courts have ruled that approval under the full PMA process protects medical devices against tort liability. Because the specific statutory preemption requirement of 21 U.S.C. 360(k) applies only to devices, the general consensus of state courts has been that state tort liability for drugs is not preempted at all.

Although the majority of state courts have not given preemptive effect to FDA approvals of drugs, to complicate matters the FDA took action in January 2006 to assert a preemptive effect for its new labeling rules for drugs and biologics.[9] According to the FDA, any state tort action based on the inadequacy of the warning in the labeling of an FDA-approved drug or biologic would interfere with its statutory authority and is preempted. The power of the FDA to make such a determination in the absence of any statutory change

---

7. Such protection was most recently extended to avian flu vaccine manufacturers to encourage the accelerated development of vaccines that could protect against an avian flu pandemic. *See* Emily Heil, *Senators Seek Assurances About Pandemic Flu Protection*, NAT'L J. CONG. DAILY, January 31, 2006 (available on Lexis).

8. Gregory J. Scandaglia & Therese L. Tully, *Express Preemption and Premarket Approval Under the Medical Device Amendments*, 59 FOOD & DRUG L.J. 245 (2004).

9. *See* Requirements on Content and Format of Labeling for Human Prescription Drug and Biological Products, 71 Fed. Reg. 3922 at 3933–3936 (comments 12 and 13), January 24, 2006; *also* Guidance for Industry: Clinical Studies Section of Labeling for Human Prescription Drug and Biological Products—Content and Format; *and* http://www.fda.gov/cder/guidance/5534fnl.pdf; *and* Guidance for Industry: Labeling for Human Prescription Drug and Biological Products Implementing the New Content Format Requirements, DRAFT GUIDANCE, http://www.fda.gov/cder/guidance/6005dft.pdf.

and the adequacy of the regulatory procedure by which it did so will undoubtedly be challenged by plaintiffs in product liability actions. Its ultimate resolution will almost certainly be a matter for the Supreme Court sometime in the next two to four years. In the meantime, any prudent manufacturer of a drug will ensure compliance with its post-marketing duty to report any unexpected and serious adverse events (see §8.4(d)) and discuss proposed labeling changes with the FDA as soon as any pattern of such adverse events would indicate that such a change might be desirable.

§11.3 Legal Cause: A Problem in Both Negligence and Strict Liability

Regardless of whether a court will apply negligence or strict product liability to Ms. Hurt's case, it will be necessary for her to prove that Biobiz's product, Integrex, was the legal cause of her injuries. For a plaintiff prove that the defendant's action was the legal cause of her harm, she must convince the jury that it was more likely than not that the defendant's conduct or product was a substantial factor in bringing about plaintiff's injury: "**A substantial factor in causing harm is a factor that a reasonable person would consider to have contributed to the harm. It must be more than a remote or trivial factor. It does not have to be the only cause of the harm.**" [10]

The doctrine of legal cause, also known as proximate cause, is one that has long been a source of difficulty for lawyers. In the context of a medical product liability case the source of the difficulty is typically factual, that is whether or not the defendant's conduct was a cause-in-fact, rather than the legal complexity of determining whether or not one of several causes was a "substantial" factor. The problem is that it is terribly difficult in many cases to establish a probable cause-and-effect relationship between a particular agent and a particular physical sequelae. People suffer from cancer, heart attacks, strokes and many other serious diseases with unfortunately high frequency in the general population. Thus it can be quite difficult to establish that a particular case of cancer, heart attack, stroke, or other serious disease was produced by an adverse reaction to a drug rather than to some other, more general cause.

Scientists, lawyers, and executives in the biotechnology industry may understand the problem of establishing legal cause more easily if it is viewed as closely related to the problem all biotechnology companies must face, which

---

10. Judicial Council of California Civil Jury Instructions CACI 430.

is to prove that their products are safe and effective. Like proof of efficacy, proof that an injury is legally caused is primarily approached through a statistical analysis. In the typical torts case this requires comparing the epidemiology of the particular disease or injury suffered by the plaintiff (e.g. Ms. Hurt's liver failure) in the general population with the epidemiology of the disease in the group from which the plaintiff is drawn (patients taking Integrex or patients taking Integrex in conjunction with the antidepressant).

Just as in the drug approval process, where companies often re-analyze data to show efficacy in particular subgroups (e.g. males between the ages of 18 and 35, rather than all those who received the test drug), plaintiffs' lawyers will strive to re-analyze data to show an unusual incidence of adverse reactions in a particular subgroup to which a particular plaintiff belongs, such as patients with high blood pressure or smokers. Once the plaintiff demonstrates that a particular subgroup is at risk, it is relatively easy to argue that a warning was required for that subgroup, and, in the absence of a warning, that the product was defective and unreasonably dangerous for that group.

Epidemiology is not the only source of evidence about causation in a product liability torts case, although the defense bar and some courts would push in that direction *See "Agent Orange" Product Liability Litigation*, 611 F. Supp. 1223 (E.D.N.Y. 1985). Just as the case for FDA approval of a new chemical entity is strengthened by a strong scientific rationale, which may be based on *in vitro* data, animal models, structural analysis, and other sources of support for the efficacy of a drug, lawyers in a product liability case will seek to introduce such other sources of support for their positions on the issue of causation. For example, in *Richardson v. Richardson-Merrell*, 649 F. Supp. 799 (D.D.C. 1986), one of the many cases in which the teratogenicity of Bendectin was at issue, much of the disputed evidence came from *in vitro* and animal studies. The problem of determining the relevancy and reliability of such evidence has been a continuing one for courts and is frequently confronted in the context of the decision as to whether or not a particular expert is qualified to testify as to his opinion on the issue of causation.

§11.3(A) Expert Testimony in Product Liability Cases

The complexity of the issue of causation in most drug product liability cases has led to an additional battle. The issue of how cause-in-fact can be proved is inevitably linked to the question of what sorts of witnesses are qualified to testify as experts and what restrictions should be placed on the kinds of testimony expert witnesses may offer.

The issue of scientific evidence on causation was addressed by the United States Supreme Court in a pharmaceutical liability case, also involving Bendectin, *Daubert v. Merrell Dow Pharmaceuticals, Inc.*, 509 U.S. 579 (1993). In *Daubert*, the district court granted summary judgment for the defense, based on the testimony of defendant's expert epidemiologist that of more than 30 published studies involving over 130,000 patients, no study had found Bendectin to be a human teratogen (*Daubert* at 2791). In granting summary judgment, the district court held that the testimony of Plaintiffs' experts was largely inadmissible and therefore Plaintiff had not provided sufficient evidence to call into question the defendant's experts' testimony that Bendectin had not caused the plaintiffs' injuries.

The testimony by plaintiffs' experts, which the district court had either excluded or given very little weight, concerned various *in vitro* analyses, animal studies, and the reanalysis of previously published epidemiological studies (*Daubert* at 2791–92). The district court based its refusal to consider plaintiffs' experts' testimony on the traditional "Frye" rule (derived from *Frye v. United States*, 293 F. 1013, 1014, (1923)) that requires expert scientific opinion to be inadmissible unless the principle on which it is based is "sufficiently established to have general acceptance in the field to which it belongs." The district court reasoned, and the court of appeals affirmed, that in light of the overwhelming weight of the published epidemiological studies that found no link between Bendectin and birth defects, the plaintiffs' experts' unpublished studies and re-analyses of others' data could not meet the Frye standard of general acceptance within the scientific community.

All nine justices of the United States Supreme Court voted to vacate the judgment of the district court and court of appeals, ruling that the Frye standard had been replaced by the Federal Rules of Evidence Rule 702, which provides:

> If scientific, technical, or other specialized knowledge will assist the trier of fact to understand the evidence or to determine a fact in issue, a witness qualified as an expert by knowledge, skill, experience, training, or education, may testify thereto in the form of an opinion or otherwise....

All the justices agreed that Rule 702 now controls the admissibility of expert scientific witnesses in the federal courts and that the lower courts had erred in excluding plaintiffs' experts on the basis of Frye and the lack of publication or peer review of plaintiffs' experts' studies and analyses.

The Court was not unanimous in joining those sections of Justice Blackmun's in which he attempted to provide guidance on the issue of the way that Rule 702 should be applied to future cases. Justice Blackmun, joined by six other justices, reasoned that Rule 702, together with Rules 401 and 402 which require that evidence be relevant, obligates a trial judge to take an active role in evaluating whether a proffered scientific expert is proposing to testify to "scientific knowledge that will assist the trier of fact to understand" or determine a fact in issue. In elaborating on the trial judge's role in such cases, Justice Blackmun stressed that Rule 702's "scientific knowledge" standard required a trial judge to determine whether or not the proffered expert's testimony was grounded in the scientific method of hypothesis testing, empiricism, and falsifiability. In addition, Blackmun stated that whether or not the testimony had been embodied in a peer-reviewed publication, its "known or potential rate of error" and general acceptance by the scientific community were relevant to, but not determinative of, the question of admissibility.

Justices Rehnquist and Stevens dissented from this portion of Justice Blackmun's opinion, arguing that, having decided that Rule 702 replaced the Frye standard, there were no further issues properly before the Court in the *Daubert* case and that therefore Justice Blackmun's elaboration on the meaning of the "scientific knowledge" requirement was inappropriate speculation about future cases.

In light of *Daubert*, the use of a variety of experts in drug product liability cases seems certain to provoke continued controversy. Plaintiffs' lawyers will undoubtedly read *Daubert* as an invitation to bring in an even greater range of expert testimony on the issue of causation. Defense lawyers will certainly respond by vigorously contesting the issue of whether or not particular plaintiffs' expert testimony is appropriately based in scientific knowledge and the scientific method. Judges' responses to *Daubert* will undoubtedly vary greatly. For example, under Justice Blackmun's opinion, how should proffered testimony about the structural similarity between Bendectin and other teratogens be handled? Is such testimony falsifiable and grounded in the scientific method? Certainly the narrow question of its chemical structure in relation to the chemical structure of other compounds is "falsifiable," but how is testimony as to its significance to be judged?

Ultimately the "falsifiability" and grounding in the "scientific method" of such testimony depends upon what epidemiology teaches, at the very least, about the teratogenicity of all of the other structurally-related compounds for which epidemiological data is known. But what is a judge to do if 30 percent of such compounds are known to be teratogens, but teratogenicity cannot be related to any particular structural variations within that group of compounds? Some judges, but not all, might decide that the jury should be permitted to hear the evidence, along with all other evidence, with the weakness of the structure/activity association being a factor for the jury to weigh in assessing its weight. Similarly, if animal toxicology data is used, the weight of any inference to be drawn derives from the epidemiological studies that demonstrate the reliability of that animal model for predicting human carcinogenicity. A very few scientists who enjoy combat with trial lawyers in the courtroom may indeed relish these "Daubert" hearings, but it seems very likely that most trial judges will find them torturous.

At some point in the future, the relationship between a particular pharmaceutical (or environmental agent) and a particular illness may be better understood, as molecular biology and biotechnology develop ever-more-powerful tools for understanding the precise pathways by which various substances act on a molecular level. Thus someday the difficult debate about the epidemiology of Bendectin, for example, may be largely resolved by knowledge of the intracellular targets (if any) of Bendectin in a developing fetus, its specificity and affinity for those targets, and the strength of its effect. Until then, the problem of causation and expert testimony in products liability is likely to loom large in suits against the manufacturers of biotechnology-derived human health care products.

§11.4 The National Vaccine Injury Compensation Program: 42 U.S.C. §300aa-1 *et seq.* (2006): One Answer to the Causation Conundrum

The difficulty of proving causation in product liability cases involving vaccines, with the resulting expense and uncertainty for both manufacturers and plaintiffs, is a problem of such proportions that Congress has attempted to provide a federal legislative solution. Vaccines were singled out for federal intervention both because of the number of cases involving seriously injured children (primarily alleging reactions to the Pertussis component of the widely used Diphtheria, Pertussis, Tetanus "DPT" vaccine) and because the cost of litigation and product liability insurance was cited by a number of manufacturers as the reason for their withdrawal from the business of developing and producing vaccines, particularly childhood vaccines.

The resulting statutory program, known as the Vaccine Injury Compensation Program, 42 U.S.C. §300aa-1 *et seq.* (2006), "the Act," is a relatively straightforward effort to replace the uncertainty and expense of the traditional product liability system with a more certain, but smaller, administrative remedy for injured persons. This is accomplished; first, by providing a table of compensable injuries which are presumptively vaccine-related if they occur within particular time periods after vaccination; and, second, by limiting non-compensatory damages (damages for non-monetary losses such as pain and suffering) to a maximum of $250,000. At this time the Act's scope is limited, as the vaccine injury compensation table lists compensable adverse reactions only for the principal childhood vaccines: DPT: vaccines for measles, mumps, or rubella; the two major polio vaccines; plus the influenza type b vaccine, the trivalent influenza vaccine developed each flu season, Hepatitis-B, Varicella (chicken pox), and Rotavirus vaccines. The CDC can recommend additional vaccines as they are developed and more are likely to be added to the program in the future.

If a recipient of one of the listed vaccines manifests one of the listed consequences within the time limit (for example any child who manifests encephalitis within three days of DPT vaccination), may file for compensation with the United States Claims Court, instead of filing a state tort action. Proof, by a preponderance of evidence, of manifestation of a listed reaction within the specified time period is sufficient for the claimant to receive an award unless the respondent (the United States, represented by the Secretary of Health and Human Services) proves by a preponderance of evidence that there was some other cause for the reaction. Thus, where the actual cause of the encephalopathy is unknown, the claimant would recover actual monetary damages, including past and future medical expenses; pain and suffering up the limit of $250,000; and, reasonable attorney's fees.[11]

The Act does not preclude an injured plaintiff from pursuing a state tort action, but there are significant limitations placed on state tort claims. Procedurally, a state tort action cannot be filed until after a claim for compensation under the federal program is filed and a disposition of the claim is made. After a disposition of the federal claim, a claimant must file a notice of election if she wishes to forego any award made under the program and pursue a civil action in state court. Furthermore, although plaintiffs who received an award under the program might be tempted to waive the award and proceed in state court to circumvent the program's ceiling on non-economic damages, there are also strong incentives for accepting the program award and foregoing the state tort action.

First, as noted above, state actions may well raise substantial hurdles for the proof of causation that are dispensed with in the Vaccine Injury Compensation proceeding. Second, although the Act provides that states cannot

---

11. *See* Davis v. Sec'y. of HHS, 20 Cl. Ct. 168 (1990).

deny injured persons the right to bring civil actions for vaccine injuries, state product liability actions for vaccine injuries covered by the Act are further restricted. Most notably, the Act provides:

> §300aa-22. Standards of responsibility
> .... (1)No vaccine manufacturer shall be liable in a civil action for damages arising from a vaccine-related injury or death associated with the administration of a vaccine after the effective date of this part *if the injury or death resulted from side effects that were unavoidable even though the vaccine was properly prepared and was accompanied by proper directions and warnings.*
> [*and, for vaccines manufactured in conformity with federal regulations, unless the plaintiff shows*]
> (B) by clear and convincing evidence that the manufacturer failed to exercise due care *notwithstanding its compliance with such Act and section (and regulations issued under such provisions).*
> [emphasis and bracketed, italicized language added.]

Thus, state courts are generally precluded from imposing strict liability for vaccine injuries. The Act further limits recovery in state actions based on the manufacturer's failure to warn the vaccine recipient of possible adverse reactions, in the absence of manufacturer fraud or misconduct in the reporting of data to the FDA.

To analyze the impact of the Act on state court liability, it is helpful to examine separately the various potential causes of a vaccine injury. The three logically most significant causes are:

1) the inherent risks of a particular vaccine, which the Act terms "unavoidable,"
2) injuries due to flaws in the manufacturing process, such as contamination of a particular lot; and, finally,
3) injuries arising from a poorly designed vaccine.

State tort claims for the first category, unavoidable side effects, are clearly barred by the Act. State claims for improper manufacture or batch defects are allowed, but the plaintiff is required to prove the manufacturer's negligence (lack of "due care") by "clear and convincing evidence" (§300aa-22, *supra*) rather than the ordinary civil liability standard of a preponderance of the evidence.

State tort claims of the third type, which assert that a particular manufacturer's vaccine, although properly prepared, carried risks not present in other manufacturers' versions of the same vaccine are not as clearly dealt with by the Act. One such commonly argued instance is the relative superior safety of the

Salk Inactivated Polio Vaccine over the more commonly used Sabin Oral (attenuated live) Polio Vaccine. Such arguments have generally been unsuccessful.[12] It would seem that the Act permits a plaintiff to assert that such adverse reactions are not "unavoidable" and therefore not barred from state tort relief. Nevertheless, such claims might be subject to the "clear and convincing evidence" requirement of "lack of due care." In such a case, the plaintiff would face the difficult challenge of establishing that the manufacture of a particular type of vaccine, again for example a live attenuated polio virus vaccine, is in itself a lack of due care, given the availability of safer types of polio vaccine.

Although an argument that a particular type of vaccine is defectively designed may be allowed under the Act, it is likely to be substantially restricted under the developing consensus in state product liability law. The problem is very similar to the issue of whether or not a vaccine is "unavoidably unsafe" under Restatement 402A comment k (see §11.5 *infra*). Since virtually every recent case considering the liability of a vaccine manufacturer for design defects has ruled the questioned vaccine unavoidably unsafe, it is very unlikely that "design defect claims" may be successfully pursued regardless of the legislative intent of the Vaccine Injury Compensation Act.

In summary, the hostility of state courts to vaccine design-defect claims, the Act's restrictions on state failure-to-warn claims, and state liability for unavoidable adverse reactions, together with the "no-fault, no-causation" structure of the Act, will likely persuade most eligible plaintiffs to opt for recovery under the Act.

§11.5 Drugs, Vaccines, and Other Human Therapeutics as Unavoidably Unsafe: Restatement (Second) of Torts §402A Comment k

The basic formulation of product liability law set forth in §402A contains numerous comments explaining the views of the Restatement's drafters on a number of important points. No comment in the RESTATEMENT OF TORTS (SECOND) has been the subject of more attention than comment *k*, "unavoidably unsafe products." The heart of Comment k is the argument that products such as vaccines and drugs, which have significant potential value to society, should not be burdened with strict liability:

---

12. *See e.g.* Reyes v. Wyeth Laboratories, Inc. 498 F.2d 1264 (5th Cir. 1974).

RESTATEMENT (SECOND) OF TORTS §402A comment k.

> There are some products which, in the present state of human knowledge, are quite incapable of being made safe for their intended and ordinary use. These are especially common in the field of drugs.... It is also true of many new or experimental drugs as to which, because of lack of time and opportunity for sufficient medical experience, there can be no assurance of safety, or perhaps even purity of ingredients, but such experience as there is justifies the marketing and use of the drug notwithstanding a medically recognizable risk. The seller of such products ... is not to be held to strict liability for unfortunate consequences attending their use....

The question comment k poses for biotechnology-derived pharmaceuticals is whether the comment intended to exempt all pharmaceuticals from strict liability, or if some identifiable subset should be automatically exempted, or if such decisions should be made case-by-case by considering facts concerning the dangers and possibility of avoidance on a manner for each drug. The majority of state supreme courts that have considered the issue have held that each product must be dealt with individually, permitting the trial court to determine that a product was "unavoidably unsafe" only if the manufacturer could establish that there was no alternative to the product in question that was safer and equally effective. Recently, as represented by the decision of the California Supreme Court in *Brown*, some state courts are coming to the conclusion that all prescription drugs, vaccines, and FDA-approved medical devices should be considered unavoidably unsafe and therefore subject to negligence liability only.[13]

The debate over the scope of comment k and its application to human therapeutics is a significant part of the larger debate over the impact of product liability law on the United States' economy and competitiveness. To understand the arguments used by both sides, it is necessary to understand the reasoning behind Justice Traynor's original formulation of product liability law in *Escola*. Justice Traynor provided three major rationales for his holding: first, that stricter liability for manufacturers would minimize the absolute number and severity of product-related injuries (loss minimization); second, that strict product liability would be more fair because of manufacturers' superior ability to obtain insurance and pass on the cost of injuries as a cost of business (loss spreading); and, third, that strict liability would obviate the difficulty of proving negligence in many cases.

---

13. *See also* Grundberg v. Upjohn, 813 P.2d 89 (Utah 1991).

The first two rationales provided by Traynor are essentially economic hypotheses, and the burgeoning field of law and economics literature has produced an enormous body of literature analyzing the economic impact of strict liability versus negligence. For example, compare the elegant argument, grounded in economic theory, for strict liability made by Calabresi and Hirschoff, *Toward a Test of Strict Liability in Torts*, 81 YALE L.J. 1055 (1972), with the defense of the negligence system by another giant of law and economics, Richard Posner in his article *Calabresi's The Costs of Accidents: A Generation of Impact on Law and Scholarship: Guido Calabresi's The Costs of Accidents: A Reassessment*, 64 MD. L. REV. 12 (2005). The vast body of literature about the economic issues of tort law may be summarized quite simply: reasonable experts disagree, and to a large extent, the relative advantages of each system present empirical questions for which we have no good experimental evidence or even design.

Although economists may not agree on the abstract question of whether strict liability or negligence produces a socially optimum allocation of resources, it is incontrovertible that increased liability does make doing business more expensive and, at the margin, discourages investment in businesses where the potential for liability is both uncertain and large. As a result, the reasoning of *Brown* is one that is strongly endorsed by industry in general and the biotechnology industry in particular. With investment capital the lifeblood of biotechnology, the added burden of potential liability for unforeseeable injuries is a potentially crippling if it reduces the willingness of investors to fund an already risky enterprise.

§11.6 RESTATEMENT OF TORTS (THIRD) AND PRODUCT LIABILITY REFORM

The RESTATEMENT (THIRD) OF TORTS: PRODUCTS LIABILITY was released by the American Law Institute in 1998 and is an effort by a group of lawyers, judges, and academic experts to both summarize developments in the law of product liability since the Second Restatement's §402A and, to an extent, influence the development of product liability by favoring some developments over others. It is too early to tell whether or not the views of the Third Restatement will ultimately supplant the substantial body of state court law that has developed under 402A, but there is a significant possibility that it will have a considerable influence on the future course of the law. For purposes of this brief introduction to product liability law, the most important aspects of the Third Restatement are that the drafters added a special section just to deal with the liability of manufacturers and sellers of prescription drugs and medical devices and that the Third Restatement unequivocally takes the position that prescription drugs should be judged only on a negligence standard, for their foreseeable risks, and be considered defective only if their risks are such that a reasonable physician would not prescribe it for any subgroup of patients:

§6 Liability of Commercial Seller or Distributor for Harm Caused by Defective Prescription Drugs and Medical Devices

(a) A manufacturer of a prescription drug or medical device who sells or otherwise distributes a defective drug or medical device is subject to liability for harm to persons caused by the defect. A prescription drug or medical device is one that may be legally sold or otherwise distributed only pursuant to a health-care provider's prescription.

(b) For purposes of liability under Subsection (a), a prescription drug or medical device is defective if at the time of sale or other distribution the drug or medical device:

(1) contains a manufacturing defect as defined in §2(a); or

(2) is not reasonably safe due to defective design as defined in Subsection

(c); or

(3) is not reasonably safe due to inadequate instructions or warnings as defined in Subsection (d).

(c) A prescription drug or medical device is not reasonably safe due to defective design if the foreseeable risks of harm posed by the drug or medical device are sufficiently great in relation to its foreseeable therapeutic benefits *that reasonable health-care providers, knowing of such foreseeable risks and therapeutic benefits, would not prescribe the drug or medical device for any class of patients* [emphasis added].

(d) A prescription drug or medical device is not reasonably safe due to inadequate instructions or warnings if reasonable instructions or warnings regarding foreseeable risks of harm are not provided to:

(1) prescribing and other health-care providers who are in a position to reduce the risks of harm in accordance with the instructions or warnings; or

(2) the patient when the manufacturer knows or has reason to know that health-care providers will not be in a position to reduce the risks of harm in accordance with the instructions or warnings.

(e) A retail seller or other distributor of a prescription drug or medical device is subject to liability for harm caused by the drug or device if:

(1) at the time of sale or other distribution the drug or medical device contains a manufacturing defect as defined in §2(a); or
(2) at or before the time of sale or other distribution of the drug or medical device the retail seller or other distributor fails to exercise reasonable care and such failure causes harm to persons.

§11.7 Liability for Biotechnology in Other Contexts: Beyond Product Liability for Drugs, Vaccines, and Medical Devices

Although much of the debate about product liability and its impact on the biotechnology industry has focused on the liability of manufacturers of drugs, vaccines, and medical devices, liability for injuries in other contexts should also be a concern for the industry. Biotechnology companies of all types may have liability to their workers if they are injured as a result of their employment, while suppliers of biotechnology products to other companies must consider their liability to the employees of those companies that use the products that they manufacture and sell.[14] In addition, manufacturers of other biotechnology products such as biopesticides, genetically engineered plants, and industrial chemicals such as enzymes and toxic waste control agents must also consider the potential for liability if their products cause injury to persons or property. This section briefly considers the issues raised in these other, non-health care contexts.

§11.7(A) Liability in the Workplace

It would be misleading to discuss the topic of liability in the biotechnology workplace without prefacing the discussion by acknowledging the extraordinary safety record of the industry to date. Although the many companies and research institutions involved in biotechnology research have long engaged in

---

14. *See, e.g.* Andrulonis v. U.S., 724 F. Supp. 1421 (N.D. N.Y. 1989), *aff'd in part* and *rev'd in part*, 952 F. 2d 652, *cert. denied* 505 U.S. 1204 (1992). The procedural history of the case on appeal is quite complicated, but concerns only the relative share of damages to be born by the U.S. government as opposed to other defendants. The finding on liability was essentially unquestioned. *Andrulonis* is discussed in detail in the next subsection, §11.71.

research programs involving a number of highly pathogenic organisms, such as HIV, hepatitis virus, anthrax, and others, **to date there is no record of even a single incident of serious worker infection or disease in biotechnology research in the U.S.** That is one reason why the case with the most interesting implications for biotechnology workplace liability, *Andrulonis*, actually predates the advent of the modern biotechnology industry. Nevertheless, a close reading of the case helps to illuminate how the fundamental liability standard of "reasonable care" (see §11.1, *supra*) will likely be applied to biotechnology research and development, when and if a laboratory accident should unfortunately occur.

In 1977 Jerome Andrulonis was a 34-year-old senior bacteriologist with the New York State Department of Health (NYSDOH), working in a state laboratory on a project to develop an orally-active rabies vaccine that could be administered to wildlife via bait in the wild. The project soon centered on the use of a tablet-coating machine, recommended by the Wisconsin Alumni Research Foundation (at one time the technology-transfer enterprise of the University of Wisconsin). The machine was used by NYSDOH to coat sugar pills, first with a rabies virus strain and then with an "enteric" coating that would permit the coated tablet to pass through the target animals' digestive systems before dissolving and being absorbed through the intestine. The machine was essentially a chamber into which sufficient air was blown to suspend capsules, while through an opening various solutions could be aerosolized into the chamber and onto the air-suspended tablets.

It was apparent to Andrulonis and others that the chamber of the coating machine was not airtight, as the colored enteric coating was observed being deposited as a film on objects in the vicinity of the machine's chamber. Nevertheless, no effort was made to use a physical containment system (such as placing the machine under a laminar flow hood or inside a glovebox chamber) to avoid worker contact with the airborne virus used in the process. The principal, in fact the only, precaution taken was the vaccination of the laboratory workers, including Andrulonis, and the monitoring of blood-antibody levels to ensure that the vaccination was still producing an effect.

In his opinion, Judge Munson, of the federal district court for the Northern District of New York, detailed the state of knowledge about rabies at the time of the experiments, as well as the process by which rabies vaccines are produced and the way in which the particular strain of rabies virus used in the NYSDOH research was developed.

First, although all of medical history disclosed only a very small number of reported cases of airborne transmission, it was believed by experts that rabies could be transmitted internasally, by airborne or aerosol transmission.[15] Following two incidents in the late 1950s, in which persons died after apparently being exposed internasally in a poorly ventilated cave filled with millions of bats, a variety of controlled experiments confirmed the possibility that aerosol transmission of the virus was possible.

Second, although after the early 1960s experts believed that aerosol transmission was possible, it was unknown whether or not the common rabies vaccine, manufactured by Eli Lilly and used prophylactically on Andrulonis and others, was protective against an airborne transmission. The label accompanying the vaccine, however, did not make any mention of the possibility that the vaccine was ineffective against aerosol or airborne virus.

Third, the virulence for humans of the particular strain used in the experiments was unknown. The strain (ERA-BHK/21) had been created by the traditional vaccine research technique of cultivating succeeding generations of the virus in a number of cell-types or tissue types, from species other than the target species for which the vaccine was intended. The strain of virus known as ERA, licensed as a veterinary vaccine, had been developed by successive cultivation, or passaging numerous times, in hamster kidney cells, chicken egg embryos, and porcine kidney cells. To increase the concentration of the virus for the NYSDOH experiments, a scientist from the federal Center for Disease Control (CDC) further modified the virus by incubating the ERA strain in a cell-system derived from hamsters. This manipulation did increase the concentration of the virus, now denominated ERA-BHK/21, while affecting its other characteristics in unknown ways.[16]

Trying the case without a jury, Judge Munson's application of the negligence standard to these facts and to the actions of the NYSDOH and the CDC is quite informative, particularly in regard to the issue of what constitutes the foreseeable risks that a reasonable person should be required to take care to

---

15. *See* W. Winkler, *Airborne Rabies*, in G. Baer (ed.), THE NATURAL HISTORY OF RABIES, at vol. 2, pp. 115–17, (1975), cited in *Andrulonis* at n. 97.

16. As the *Andrulonis* court stressed, the traditional means of attenuating of a virus for one or more species (the ERA vaccine was licensed as safe and effective for six species of domestic animals) does not necessarily attenuate the virus for other species. For example, by forcing the virus to grow in chicken eggs, the virus's ability to infect horses may be lost (although perhaps not its immunogenicity), but its ability to infect humans may be unaffected or even increased. Although the original ERA strain was deemed nonpathogenic to humans, the further manipulation performed at the Center for Disease Control created a new viral strain with unknown properties. The resulting strain was pathogenic for the unfortunate Andrulonis.

avoid. First, in choosing to work with a particular pathogen (the rabies virus), the defendants were held to the knowledge of the foreseeable risks that would be possessed by one of "the community of rabies experts" of about 25 to 50 scientists in the United States.[17] Thus the risk of aerosol transmission was a foreseeable risk, although apparently it was not actually foreseen by the defendants. Similarly, the defendants' standard of care assumed knowledge of the fact that aerosol transmission, via direct exposure to the olfactory nerve ending in the nasal cavity, might not provide an effective opportunity for the serum antibodies produced by the Lilly vaccine to attack the virus, although at the time this was only a theoretical risk and had never been experimentally demonstrated. Additionally, the defendants were required to know that the strain of rabies virus used in their experiments was of unknown pathogenicity to humans.[18] Finally, the defendants were held to a standard of good laboratory safety practices, particularly the standards of safe handling for potential pathogens of unknown virulence. In specific, the court faulted the defendants for their failure to utilize any physical containment apparatus to control airflow of the virus.[19]

---

17. The defendants were presumed to know all that reasonably should have been known, as of March 29, 1977, by scientists venturing beyond routine laboratory procedures involving the rabies virus and proceeding to plan and supervise experimental research projects involving that virus. Specifically, Drs. Debbie, Baer, and Winkler, whose conduct or representations are relevant to the claims made in this lawsuit, are charged with knowledge of the principles and theories discussed in the pages that follow, as well as familiarity with the research upon which those principles and theories were based. In 1976 and 1977, these men were either members, or engaged in research projects that should only have been undertaken by members, of a select group referred to at trial as the "community of rabies experts," which in the United States numbered between twenty-five and fifty scientists. Andrulonis, 724 F. Supp. at 1434.

18. The *Andrulonis* opinion noted that: "In a 1976 publication the CDC recommended that such virus strains 'should be regarded as potentially virulent for purposes of managing the treatment of exposed humans.'" *Andrulonis*, 724 F. Supp. at n.279, *citing* DEP'T. OF HEALTH, ED. AND WELFARE, *Recommendations of the Public Health Service Advisory Committee on Immunization Practices: Rabies*, 25 MORBIDITY AND MORTALITY WEEKLY REP. 403, 406 (1976).

19. The *Andrulonis* opinion provided a detailed discussion of precisely how the defendant's conduct fell short of reasonable care:

> In 1977, proper risk assessment required consideration first of what was known and not known about the organism used. It was important to know whether the organism was a pathogen capable of causing disease and, if so, what concentration of the organism was ordinarily necessary to cause disease, whether the organism could produce illness in humans, the severity of the illness the organism could cause, and the routes of infection through which the organism could in-

It should also be noted that, in general, worker suits against employers for workplace injuries are covered by a state worker compensation scheme, in which the principal issue is causation "within the scope of employment." Most state worker compensation schemes provide injured workers with a limited recovery without the need for proving negligence, in exchange for eliminating the worker's right to sue the employer in negligence. However, as in *Andrulonis*, the apparent preemption of negligence lawsuits against employers is frequently circumvented. If a claim of negligence can be made against a third party, such as the supplier of a reagent or device used in the laboratory, than the employer is likely to be dragged into the lawsuit by the third party, and the employer's negligence will in fact be in issue. In *Andrulonis* the injured worker alleged negligence on the part of a number of third parties, particularly the supplier of the virus with which he was infected, with the result that his own employer was indeed entangled in the negligence suit.

Again, although the laboratory safety record of the biotechnology industry is exemplary, Judge Munson's opinion in *Andrulonis* sends a clear message. The negligence standard is far from toothless when applied to laboratory research. When embarking on a new project involving a pathogen, reasonable care requires both good laboratory practice and a sufficient search of the literature to provide the level of knowledge of the absolute experts in the field of research on that particular pathogen. Despite the trend of recent years towards more relaxed practices in the handling of recombinant organisms and pathogens (see Chapter 7 §7.2), any departure from prescribed laboratory practices can be considered a basis for liability and, in the case of potential

---

vade a host animal. Next, the researcher was required to determine whether the procedures and practices within the laboratory were adequate given the nature and characteristics of that organism, with emphasis being placed on the familiarity of laboratory workers with clearly defined procedures designed to minimize exposures. Finally, proper risk assessment required inquiry into whether the equipment and facilities used for a particular laboratory activity were adequate, an inquiry that usually turned on whether the level of containment provided for an etiologic agent was commensurate with the level of hazard that reasonably should have been perceived for a given work activity. *Andrulonis*, 724 F. Supp. at 1477 (citations omitted).

Yet no efforts to contain the aerosols created by the machine were undertaken. Reasonably available and widely used technology existed in March 1977 through which such containment could have been achieved without unduly frustrating the purposes of the Uni-Glatt experiments. Andrulonis at 1478.

pathogens of undetermined virulence, some degree of physical containment is absolutely required.

§11.7(B) Liability for Biotechnology Pesticides

Liability for the traditional chemical poisons used as pesticides has, not surprisingly, been the subject of considerable litigation. One legal development of import in this area was the controversy among the federal courts of appeal as to whether or not the federal statute regulating pesticide registration and labeling, the Federal Insecticide, Fungicide and Rodenticide Act, 7 U.S.C. §§136–136y, preempts liability for inadequate warnings accompanying pesticides.[20] FIFRA contains this provision expressly preempting state laws that might impose different or conflicting labeling requirements on pesticides: "(b) Uniformity. Such State shall not impose or continue in effect any requirements for labeling or packaging in addition to or different from those required under this Act." 7 U.S.C. §136v(b).

In 2005, the U.S. Supreme Court issued an opinion in *Bates v. Dow Agrosciences*, 544 U.S. 431 (2005), which provided significant guidance to the interpretation of the FIFRA preemption clause, although leaving a central issue to future resolution on a state basis. The court decided that state tort claims for pesticide-related injuries were not broadly preempted by FIFRA, including state claims based on a failure to warn that calls into question the pesticides registered label. The court reasoned that the imposition of liability in tort for breach of such common law standards as "reasonable care," which might "induce" a manufacturer to alter its conduct, is not in general the imposition of a "requirement" preempted by federal regulation unless Congress has clearly expressed its intent to preempt state tort liability. However state lawsuits that are brought for false advertising, or misrepresentation, are permitted only so long as the state's law is based on a duty that is equivalent to the "misbranding" duty imposed under FIFRA. In other words, since FIFRA

---

20. Papas v. Upjohn, 985 F.2d 516, (11th Cir. 1993) is one of the most recent cases holding that FIFRA preempts state tort actions based on inadequate pesticide warnings. *Accord*, Worm V. American Cyanamid, 970 F.2d 1301, (4th Cir. 1992). The leading case on the other side is Ferebee v. Chevron Chem. Company, 736 F.2d 1529, (D.C. Cir. 1984). Burke v. Dow Chem. Co., 797 F. Supp. 1128 (D.Ct. E.D.N.Y. 1992) held that a wide range of state warning claims were not preempted, while Chem. Specialties Mfr.'s Ass'n. v. Allenby, 958 F. 2d 941 (9th Cir., 1991), FIFRA did not preempt state suits based on "point-of-sale" warning signs.

imposes on manufacturers a continuing duty to seek EPA approval to update their labels in light of any new safety data that would render the current label inadequate and "misbranding," a state tort law that merely provides civil damages for breach of an equivalent duty is not preempted. Thus a court faced with a suit against a pesticide manufacturer for failure to warn must examine the actual content of the state's common law claim to see whether or not it is different from the FIFRA duty to provide adequate labeling instructions to avoid misbranding.

Beyond the issue of the warning or misrepresentation of product safety, the question of the standard of liability for manufacturers of pesticides is clearly one of state law. Most states have placed pesticide manufacturers within the general category of strict product liability under Restatement §402A (see §11.2 of this chapter). Although a manufacturer of pesticides can raise the issue of whether or not its product is unavoidably unsafe under the RESTATEMENT (SECOND) OF TORTS §402A comment k, it is a far more difficult argument and far less the subject of widespread popular sentiment than it is in the context of human therapeutics (see §11.5 of this chapter). Thus, the principal defense of a manufacturer, whose pesticide has caused damage to crops or personal injuries to exposed persons, relies on the adequacy of the label and whether or not it can be argued that the damage was the result of clear misuse of a product in light of labeled directions. In cases where the problem is simply the manufacturer's inability to adequately advise of the propensity of a product to cause injury or damage, liability is almost certain to result.

Given the extremely low vertebrate toxicity of the pesticidal proteins, such as Bt, engineered into plants, the most significant litigation may well be over "genetic pollution." In genetic pollution suits the producers of non-modified or organic crops may assert claims against manufacturers and farmers for any inadvertent changes in their crops caused by cross-pollination with neighboring genetically engineered crops. Such cases have yet to produce a significant body of state law, but are indeed an area in which to watch for future developments.

§11.7(C) Liability for Food Products Produced by Biotechnology

Food products will be affected by biotechnology in a number of ways. In addition to food plants that are treated with or which produce biopesticides (see §11.7(B)), food-producing animals will be treated with biotechnology veterinary products and both food animals and food plants will be genetically engineered to alter their characteristics. It is almost inevitable that these developments will lead to allegations that such products have caused illness or injury and that product liability litigation will result.[21]

Litigation over food products is a very old field of law yet, surprisingly, there is still some controversy over the appropriate rules for determining liability for injuries from food products. Most states follow a rule of strict liability for food that is contaminated or adulterated by some foreign, hazardous substance. Thus if a particular batch or lot of food contains pathogenic bacteria or some foreign substance such as shards of glass or metal, the issue is very much like that of a manufacturing defect (see §11.2(B)) and liability is certain, regardless of a manufacturer's precautions to avoid such contamination.

Somewhat more controversial is the question of the standard of liability of a seller for substances that are a "natural" part of a particular food. There are many cases in which suits have been brought over the injuries caused by chicken bones, oyster shells, cherry pits, and even a peppercorn. Most states follow the rule that liability for such substances turns on whether the object complained of would have been expected by the reasonable consumer.[22] In those states, an unusually large bone, piece of shell, or cherry pit that would not be expected is a basis for liability. California and a small number of other states follow a different rule for such "natural" components, holding sellers of such products only to a negligence standard rather than a strict liability standard.[23]

Under the prevailing framework for product liability for foods, there will be relatively stringent liability if injuries are attributable to "added" substances such as biopesticides or biotechnology veterinary-product residues. A more difficult question will arise if injuries are attributable to a genetically engineered food product or to a genetically engineered ingredient in a food product (such as the Bt (bacillus thuringiensis) toxin, the endogenously-produced pesticide in genetically engineered corn and other crops). In this context, the question may be whether the genetically engineered constituent is a "natural"

---

21. In the closest brush to date with food-based product liability to date, Aventis Cropscience unintentionally released to food makers genetically modified corn containing "Starlink" ("Cry9c"), a variety of the insecticidal Bt protein that had not yet been approved for human consumption. See §7.85. In the wake of publicity over Starlink corn products, a number of persons reported to the FDA that they had experienced illness after eating Starlink corn products. However, investigators concluded that none of the persons reporting illness were likely to have been reacting to the Starlink Cry9c. See Melinda Fulmer, *Testing Finds No Link Between Gene-Modified Corn, Illnesses*, L.A. TIMES, June 14, 2001, at Business p.3.

22. *See, e.g.*, Yong Cha Hong v. Marriott Corp., 656 F.Supp. 445 (D. Md.1987).

23. Mexicali Rose v. Superior Ct., 822 P.2d 1292 (Cal. 1992).

ingredient or a foreign or contaminating ingredient. If the substance is considered to be a natural ingredient then liability for harm will be determined by whether its danger exceeded the reasonable expectations of the ordinary consumer (or alternatively, whether its manufacture constituted negligence). If the substance is considered to be a foreign or contaminating ingredient liability for its harm is more absolute.

It may be helpful to examine this problem in the context of a hypothetical, in which a particular protein, for instance "Bt" (see §7.4), is engineered into corn and causes an allergic reaction in a relatively small percentage of the population. One such allergic person, the hapless Ms. Hedda Hurt, undergoes the extreme allergic reaction of anaphylactic shock and brings suit against Ag Biobiz to recover damages for her injuries.

It is unlikely that Bt would be considered a "natural" constituent of corn, even a genetically engineered corn. Even if Bt is subjected to a formal approval by the FDA for food additive status or is used pursuant to a manufacturer's self-determination of GRAS (generally recognized as safe, see §7.7(C)), it still could not be considered a "natural" constituent, because corn or corn preparations (such as corn meal) do not necessarily contain it. Nor does GRAS or food additive status provide any statutory preemption of state tort liability. Although GRAS approval or food additive status would be strong evidence that the manufacturer was not negligent in selling products containing Bt, it is unlikely that negligence would be the standard or liability. In such a case, a product liability determination becomes a more difficult battle.

Under the "ordinary consumer expectations" test, even if allergenicity is limited to a very small percentage of the population, the product may well be defective (as the ordinary consumer may well not expect severe allergic reactions to arise in people accustomed to eating corn safely). The drafters of Restatement 402A provided their views on food allergy liability in comment (j), reasoning that liability ought to exist when a substantial number of people are allergic to the ingredient, the ingredient would not be expected in the particular product, and the allergenicity was known or knowable. This would seem to limit the liability for failure to warn when later-discovered allergenicity was not reasonably "knowable" in advance. On the other hand, if the plaintiff succeeds in arguing for the application of the risk-utility prong of design defect liability, a jury may well find that the risk as it ultimately becomes known outweighs the utility of the genetic modification that introduced Bt into the food supply. Given the limits of pre-market testing to demonstrate the absolute safety of any new food substance, biotechnology companies that introduce new compounds into the food supply face a significant likelihood of being required to defend their products against very difficult liability suits. The clear need is for as exhaustive a set of premarket allergenicity tests as are scientifically feasible.[24]

§11.7(D) Liability for Other Biotechnology Products in Industry and the Environment

Biotechnology will likely be used to produce organisms and reagents for a wide variety of industrial and environmental applications, such as the detoxification of chemical waste sites. Again, it is very likely to be an exceedingly safe enterprise, because of the organisms selected for such uses and because of the kinds of substances that would be produced in this way. Nevertheless, any manufacturing enterprise faces some risk of liability suits, and it is unlikely that the biotechnology producers of industrial or environmental reagents will be an exception. The legal system will probably respond to such lawsuits by handling them in much the same way as product liability lawsuits, generally using the two-prong *Barker* test (see footnote 281).

One possible additional difficulty may arise if plaintiffs' lawyers attempt to have the environmental release of genetically engineered plants or microorganisms judicially determined to be an abnormally dangerous activity and therefore subject to extremely stringent standards of liability (RESTATEMENT (SECOND) TORTS §§519–520). Under the Restatement §520, among the principal factors that are to be considered in determining whether an activity is abnormally dangerous are:

(a) existence of a high degree of risk of some harm ...
(b) likelihood that the harm that results from it will be great;
(c) inability to eliminate the risk by the exercise of reasonable care;
(d) extent to which the activity is not a matter of common usage;
(e) inappropriateness of the activity to the place where it is carried on; and
(f) extent to which its value to the community is outweighed by its dangerous attributes.

In most jurisdictions, the decision as to whether or not a defendant's activities should be considered abnormally dangerous is made on a case-by-case basis, because of the very situation-specific nature of the inquiries into whether or not an activity is a matter of common usage and inappropriate to the place where it was done. In the case of some inadvertent harm from the

---

24. The importance of allergenicity testing was emphasized by an inter-agency conference. *See* Assessment of the Allergenic Potential of Genetically Modified Foods, 66 Fed. Reg. 56839-03, (Nov. 13, 2001).

release into the environment of a genetically engineered microorganism, there will likely be a ferocious battle of experts over the underlying degree of risk and the ability to eliminate it by reasonable care. Molecular biologists and microbial ecologists have been at odds over the issue since the beginning of the age of genetic engineering, and the judicial outcome of this battle is difficult to predict.[25]

§11.8 Conclusion

The biotechnology industry, like any other major source of goods and services, will undoubtedly be subjected to liability suits alleging a wide range of injuries in a variety of contexts involving the entire range of biotechnology applications. The United States is already in the midst of a significant national debate over the appropriate basis for compensating injured persons and that debate will undoubtedly have a substantial impact on the development of biotechnology. In the health care products field, recent judicial decisions such as *Brown v. Superior Court* and the RESTATEMENT THIRD OF TORTS, PRODUCT LIABILITY (see §11.5 *supra*, are undoubtedly of great comfort to the industry. In other contexts, such as the workplace, the extraordinary safety record of the biotechnology industry should provide comparable reassurance. In the pesticide, food product, and industrial applications of biotechnology, the legal framework is unlikely to be as accommodating. Companies developing products for these end uses will undoubtedly face the most difficult legal battles in the event of the unfortunate occurrence of personal injuries or property damage.

---

25. *See generally* BIOTECHNOLOGY RISK ASSESSMENT: ISSUES AND METHODS FOR ENVIRONMENTAL INTRODUCTIONS, J.R. Fiksel and V.T. Covello (eds.), 1986).

INDEX

AAAS, 45
abnormally dangerous activity, 331
accelerated approval (drugs), 198, 227, 229
accredited investors, 132
adeno-associated viruses, 282
adenovirus, 282–283, 286–287
ADEs (adverse drug experiences), 226–227
adulterated, 183–185, 189, 249, 251–252, 329
adverse drug experiences (ADEs), 226–227
adverse reactions, 40, 89, 208, 299, 312, 316–318
advertising, 241–243, 245, 247, 327
Advisory Panel, 208
agricultural biotechnology, 12–14, 172, 178, 194–195, 197
allergenics, 203
allergic reaction, 330
Alzheimer's disease, 10, 82, 213
American Type Culture Collection (ATCC), 91
amino acids, 8, 20–21, 28, 43, 85, 87, 199
ANDA (Abbreviated New Drug Application for generic drugs), 240
angel, angel investors, angel investor clubs, 129, 132–134, 137–138
angiogenesis, 34, 81, 280
animal biologics, 171, 179
animal models, 115–118, 125, 199, 207, 209, 212–213, 280, 288, 312
annual reports, 120, 156
antibodies, 7–14, 29–34, 40–43, 54–55, 65, 75, 86, 88, 92, 124, 179, 203, 212, 233, 325

antibody diagnostic, 11–12, 40, 206
anti-dilution, 140, 142, 145
antigenic determinant, 30–33
antigens, 12, 30, 33, 35, 41–42, 180, 283
antisense, antisense oligonucleotides 8, 36, 39, 110, 188, 281
antisense therapeutics, 36
antitoxins, 171, 179, 203
antivenins, 203
approvable letter, 223–225
Archaea, 23
Ariad Pharmaceuticals, 101
articles of incorporation, 130–131, 140, 150
Asilomar, 25, 164
Association of University of Technology Managers (AUTM), 45
A THEORY OF JUSTICE (by John Rawls), 258, 260–261
authorized shares, 139–140
auto-antigen, 117
Bacillus thuringiensis toxin, (Bt),(Bt toxin), 172, 192, 329
bacteria, 8, 11–12, 21, 23–26, 30, 32, 34–35, 163–164, 329
Baltimore, David, 100
base pair, 20
batch to batch, 204
Bayh-Dole, 46–48, 50, 57–59
B-cells, 30, 32–34
Berg, Paul, 25, 163–165
best mode (§112), 75, 89, 92–93, 101
biocatalysts, 13, 42
biodiversity, 97–99
bioequivalent, 234
bioethical decision making, 257

biologics, 65, 162, 171, 179, 197–255, 291, 310
bioprospecting, 97–99
bioremediation, 176, 178
biosafety, 165–167, 289–290
Biosafety Level, 167
biosensors, 13
biosynthetic materials, 13
Biotechnology Industry Organization (BIO), 7, 165
biotechnology life-cycle, 4, 14, 16, 19
bioterrorism, 228, 257, 263–264
blastocysts, 10
Blue Sky Laws, 155
Boston University, 15
BRCA1, 23
BRCA2, 23
Bt (Bacillus Thuringiensis), Bt toxin, 172, 192, 329
burn rate, 120, 159
business plan, 5, 62–64, 67, 109–110, 112, 115–116, 119–120, 125, 135, 137–139, 141, 147, 159, 244
by-laws, 140
Calgene, 188
CANDA (Computer Assisted New Drug Application), 209
capital structure, 131, 134, 139
carbohydrates, 21, 110, 187, 190–191, 238
catalysts, 13
categorical imperative, 260–261, 264
cause-in-fact, 311–312
CBER (Center for Biologics Evaluation and Research), 198, 200–203, 205–206, 221, 239, 254, 291–292
CDC (Center for Disease Control), 200, 316, 324–325
CDER (Center for Drug Evaluation and Research), 198, 200–202, 205, 221, 228, 239–241, 246–247, 254, 310
CDRH (Center for Devices and Radiological Health), 200, 205, 221
confidentiality and non-disclosure agreement, CDA, 53
cDNA, 28, 79, 85–88, 259
Celebrex, 211

Center for Biologics Evaluation and Research (CBER), 198, 200–203, 205–206, 221, 239, 254, 291–292
Center for Devices and Radiological Health (CDRH), 200, 205, 221
Center for Devices and Radiological Health, 200, 205
Center for Drug Evaluation and Research (CDER), 198, 200–202, 205, 221, 228, 239–241, 246–247, 254, 310
Centers for Disease Control (CDC), 200, 316, 324–325
Centocor, 9, 41
central dogma, 20–21
chemokine receptor CCR-5, 81
childhood vaccines, childhood vaccine injuries, 315–316
Chinese hamster ovary (CHO) cells, 8, 25, 93
chromosome, 22, 25, 79, 81, 258, 279
Class I devices, 205
Class II devices, 205
Class III devices, 206
clinical hold, 169, 222
clinical research associates (CRA), 221
CNS (central nervous system), 118, 199, 210, 213
coding, 21–22, 28, 75, 79, 87
codons, 20–21, 87
combinatorial library, combinatorial library screening, 8
commercial speech, 242–243
common shares, 130–131, 139–140
compassionate use, 229
Computer-Assisted New Drug Application (CANDA) 209
constant region, 31, 33
consumer expectations test, 306–307, 330
Contract Research Organizations, CRO, 221
convertible preferred (stock), 131
Coordinated Framework for the Regulation of Biotechnology, 15, 161
core technology, 37, 67, 116, 136, 138, 143, 145–146, 199, 277
CRA (clinical research associate), 221
Crick, Francis, 5, 19

cytokines, 95, 203
date of conception (invention), 136
DDMAC, 241, 247
dedicated biotechnology companies, 6, 109
detailers, 243–244
Developing Sponsored Research Agreements: Considerations for Recipients of NIH Research Grants, 56, 58
device (medical), 170, 205–206
device 510(k), 206
Dietary Supplement Health Education Awareness Act, DSHEA, 247–252
dietary supplements, 241, 247–248, 250–252
difference principle, 262, 271, 273
direct to consumer advertising, 245
DNA polymerase, 28–29
DNA sequencing, 27
doctrine of equivalents, 87, 91, 102–104
dose-ranging, 208
dotcom bubble, 134
double helix, 19
Dow Agrosciences, 173, 327
drug label, 242
DSHEA see Dietary Supplement Health Education Awareness Act
due diligence, 134–135, 137, 152–153
E. coli, 8, 25–26, 28, 93, 237
early access, 227, 229–230
efficacy, 67, 114, 117–118, 126, 169–170, 174, 181, 204, 207–208, 213, 215, 222, 225, 228–230, 235–238, 240–241, 247, 254, 312
Eli Lilly, 41, 101, 324
embryonic stem cell, embryonic stem cell research, 10–11, 96, 258, 260, 274–275
EMEA (European Medicines Evaluation Agency), 237, 240, 254–255
enablement (§112), 91, 93
endpoint, 117, 206, 211–212, 214–217, 228
environmental impact assessment, 182
EPA (Environmental Protection Agency), 15–16, 161–162, 168, 172–178, 181, 183, 192, 328

epidemiology, 312, 314–315
epitope, 12, 30–32
Erythropoietin, 34, 65, 80, 93–94, 125, 147, 237–238, 240
eukaryotes, eukaryotic cells, 21, 23–26
European Medicines Evaluation Agency (EMEA), 237, 240, 254–255
EUP (Experimental Use Permit), 173
ex vivo gene therapy, 39–40
exclusive licenses, 64–65
exons, 21–22, 28, 79
Experimental Use Permit, 173
expert witnesses, 312
expression, 8–9, 22, 28, 33, 40, 53, 101, 133, 164, 237, 260, 280, 282–285
expression system, 8, 28, 33
fabs (fragments of antibody binding regions), 33
failure to warn, 308–310, 317, 327–328, 330
farmers exemption, 100
fast-track drug, 229
FDA (Food and Drug Administration), 5, 8, 10, 15–16, 34, 36, 40–42, 71, 112–114, 119, 125–127, 153, 158–163, 168–171, 173–174, 179–180, 183–191, 193, 197–255, 261, 288–292, 298–299, 301, 309–312, 317, 329–330
FDA Modernization Act of 1997, FDAMA, 227, 229, 235, 242
FDA's Statement of Policy for New Varieties of Food Plants, 186
FDCA (Federal Food Drug and Cosmetic Act), 162, 171, 187, 192, 205
Federal Fungicide, Insecticide and Rodenticide Act, FIFRA, 162, 172–175, 183, 192, 327–328
field of use, 64–67, 144
first option to negotiate, 69
first to invent, 106
first-in-human, 207
first-inventor-to-file, 106
food additive, 183–187, 190, 330
Food Safety Inspection Service (FSIS), 179
Food, Drug and Cosmetic Act, 162, 168, 187, 202

Foods Derived From New Plant Varieties Created by Genetic Engineering, 187
Form 10-K, 156
Form 10-Q, 156
formulary, formularies, 14–16, 202, 209–211, 216, 241, 243, 245
formulation, 201, 206–207, 211, 216, 235, 250, 318–319
forward looking statements, 157
FTE (full time employee equivalent), 145
gel electrophoresis, 26–27, 28
Gelsinger, Jessie, 40
gene, 5, 7–8, 10–11, 20–23, 27–28, 31, 35–36, 38–40, 43, 49, 75–76, 78–82, 85–88, 90–91, 94–95, 101, 103, 107, 110–111, 124, 166–167, 172, 179, 184–185, 187–188, 192–193, 201, 203, 258–259, 265–267, 269–270, 277–298
gene cloning, 28, 39
gene probes, 11, 35–36
gene therapy, 5, 7–8, 10, 38–40, 43, 110, 166, 201, 203, 259, 277–298
Gene Therapy Advisory Committee, 291
Genentech, 10, 87, 102, 104, 110, 213, 259
general partners, 135
generally recognized as safe, GRAS, 183–186, 188, 189, 191, 330
generic, 75, 126, 155, 198, 233–240, 250, 265
generic biopharmaceuticals, 198, 237, 239
genetic pollution, 328
Genome, 12, 22–24, 26–27, 80–81, 87–88, 106–107, 258–259, 266, 269, 278, 289, 296
genomics, 8, 80, 111, 233
genotype, 23
geographic territory, 64–67, 146
germ-line gene therapy, 38, 279, 293–298
globalization, 253, 257
glyphosphate resistant, 193
GMO (genetically modified organism), 193
GMP, Good Manufacturing Practice, 301–302
Guidance for Human Somatic Cell Therapy and Gene Therapy, 291

HAMA (human anti-mouse antibodies), 33
harmonization, 107, 253–254
Harvard oncomouse, 78
Hatch Waxman Act, 231
Hatch, Orin, 233
hazardous waster remediation, 13, 176, 178
health insurance, 211, 297
heavy-chains (antibody region), 30–31
HEPA (high-efficiency particulate air filter), 167
herpes viruses, 282
HHS (Health & Human Services), 54, 200–201, 235, 250, 252, 290, 316
hits, 9, 86, 125
Howard Hughes Medical Institute (HHMI), 62
human gene therapy, 5, 7, 10, 38, 166, 277–298
Human Genome Initiative, 106–107
Human Genome Project, 22, 27, 258–259, 269
Human Genome Sciences, 80–81, 107
human growth hormone, 8, 26, 34, 43, 114, 237, 240, 278, 293
human insulin, 8, 10, 26, 34, 43, 79, 103, 240
human tissue, 80, 87, 93, 95–96, 287
Huntington's disease, 38
hybridization, 12, 35, 90, 191
hybridoma, 32–33, 92
hyper-variable region (antibody region), 31
IDE (investigational device exemption), 206
IDEC (now Biogen Idec), 10, 41–42, 120, 213, 218–219
IL-1 (Interleukin-1), 41
IL-2 (Interleukin-2), 34, 284, 286–287
ImClone, 157
Immunomodulators, 203, 286
In vivo gene therapy, 39–40
IND (Investigational New Drug), 66, 201, 206–207, 220–222, 228–231, 234
indication, 16, 53, 112–114, 117–121, 126, 169, 198–199, 206–207,

INDEX 337

209–213, 216–217, 220, 232, 234, 244, 246
indigenous people, 98
infringement, 61, 87, 92, 100, 102–104
initial public offering, IPO, 15, 120–122, 126, 130, 140, 149, 151–152, 154, 156, 158
Institutional Biosafety Committee, IBC, 165–167, 289–291
integration (viral), 7, 24, 76, 253, 282, 285, 288
interference, 76, 83, 106, 265
International Conference on Harmonisation, ICH, 254–255
International Trade Commission, 94
introns, 21–22, 28, 79
invention assignment and disclosure agreement, 48
Investigational Device Exemption (IDE), 206
Investigational Review Board (IRB), 220–221, 230, 289–291
investigator brochures, 220
IRB (Investigational Review Board), 220–221, 230, 289–291
ISIS Pharmaceuticals, 36
Japanese Pharmaceutical and Medical Devices Agency, 254
Kantian, 257–258, 260–265, 270, 274–275, 294–296
Kass, Leon, 275
know-how, 61–62, 75, 135
Kohler and Milstein, 6, 32
La Jolla Cancer Research Foundation, 15
label indication, 199
LD50, 166
leads, 9, 63, 81, 92, 110, 112, 124–125, 138, 147, 199, 210, 293
legal cause, 300, 311
letter of intent, 151, 154
licensed technology, 61–64, 67, 144, 147
licensee, 51–52, 55–56, 58–67, 101, 137
licensor, 61–64, 66, 137
ligand, 8, 22, 110
ligase, 26
light chains (antibody region), 31
limited liability, 130

limited partners, limited partnerships, 130, 135
lipids, 21, 190–191
liposome, liposomes, 39, 283–284
liquidation preference, 131
Locke (John), 262
lock-up period, 156
lymphocytes, 30, 33, 54, 278
managed care, 16, 216, 245
marketing, 13, 15–16, 53, 55–56, 66–67, 114, 143–146, 148–149, 154, 174, 179, 181, 183, 189, 198, 210, 226, 230, 235, 241–243, 245, 253, 307, 319
Markush (patent claim), 124
medical device, 205–207, 211, 310, 321–322
medical malpractice insurance, 299
Medicare Part A, 211
Medicare Part D, 211, 216
messenger RNA (m-RNA), 9, 20, 22, 24, 28–29, 36, 79, 86, 87, 282
Microbial Products of Biotechnology: Final Regulations Under the Toxic Substances Control Act, 178
milestones, milestone payments, 55, 57, 62, 66–67, 114, 120–121, 124, 129, 133, 141, 144, 146, 150, 159
misbranding, 241, 244, 327–328
monoclonal antibodies, 7–9, 11, 29, 33–34, 40–43, 88, 92, 203
Monsanto, 6, 192
morning after pill, 201
m-RNA (messenger RNA), 9, 20, 22, 24, 28–29, 36, 79, 86, 87, 282
Mullis, Kerry, 29, 49
multi-center studies, 208
Mycogen, 173
naked DNA, 12, 35, 39, 283–284
National Academy of Sciences, 194
National Institutes of Health (NIH), 7, 10, 14, 45, 54–59, 68–69, 133, 161–162, 164–166, 200, 236, 258–259, 278–279, 288–291, 295, 298
National Institutes of Health Recombinant-DNA Advisory Committee Guidelines

(NIH-RAC Guidelines), 162–167, 290, 293
National Veterinary Services Laboratory, 181
negligence, 300–302, 304–308, 311, 317, 319–321, 324, 326, 329–330
negligence per se, 301
negotiation, 56, 59–61, 66, 69, 98, 121, 125, 262
net present value, 139, 141
neurodegenerative diseases, 11, 82, 112–113, 115–118
neutralizing antibodies, 32, 54
new chemical entity, 223, 312
new dietary ingredient, 249–250
new drug, 36, 114, 120, 127, 159, 197–198, 201–202, 204, 206, 208–210, 215, 219–220, 222–223, 225–227, 230, 248, 253–254, 259, 261
New Drug Application, NDA, 36, 67, 122, 126, 127, 145, 201, 206, 208–210, 220–226, 229, 232, 234, 240, 242, 254
new foods, 183
NIH see National Institutes of Health
NIH Gene Therapy Points to Consider, 289, 291
NIH-RAC, 25, 162–167, 289–291, 293
NIH-RAC Guidelines, 162–167, 290, 292
NIH Sponsored Research Guidelines, 58–59
non-confidential marketing, 53
nonobviousness (§103), 75, 80, 84–89, 91, 99, 105–106
not approvable letter, 223–225
Novartis, 6, 57, 60, 148
novelty (§ 102) 75, 76, 82–83
Nozick, Robert, 279
nuclear magnetic resonance spectroscopy, 37
nucleotide, nucleotides, 20, 26–29, 49, 79, 90, 177, 258
Office of Technology Transfer (OTA), 48, 50–51
off-label, off-label use, 16, 114, 210, 235, 241–244

oncogenes, 9, 281
optimal indication, 209–210, 212–213
options (stock), 131, 140–141
organelles, 23
Orphan Drug Act, 198, 231–233
Orphan Drug exclusivity, 232–233, 235
packaging cell-line, 283
Parallel Track, 230
Paris Convention, 76–77
Parkinson's, 11, 118, 123, 127, 199, 214, 216, 244
Patent and Trademark Office (PTO), 76
Patent Cooperation Treaty, 76
Patent Term Extension, 126, 231, 234, 239
Patent Term Restoration Act, 198, 233
patenting of whole animals, 97
PCR (polymerase chain reaction), 29, 49
pediatric term-extension, 198
pediatric use, 226, 231, 236
peptide, 21, 54, 199, 238, 284
peptides, 54–55, 199
person of ordinary skill in the art, 74, 76, 82, 84–85, 87–88, 91–92, 101–102
personal injuries, 299–300, 328, 332
pesticide residue, 183, 192–193
phage, 24, 33, 163
Pharmaceutical Research and Manufacturers Association, 254
pharmacoeconomic, 216
pharmacokinetics, 87, 207, 236
Phase I clinical trials, 207, 220, 234
Phase I/II clinical trials, 121, 125, 207–208, 229
Phase II (clinical trials), 111–112, 116–118, 120–122, 126, 145, 148, 208, 215, 217, 219, 229–230, 284
Phase III (clinical trials), 67, 118, 120, 122, 126, 145, 160, 208, 214–215, 217–221, 229–230, 241
Phase IV (post-marketing studies), 209, 225–226
phenotype, 23, 293
PhRMA (Pharmaceutical Research and Manufacturers of America), 254
PHS (Public Health Service), 200
piggyback registration rights, 150

PIP (Plant Incorporated Protectant), 173, 175, 192
PIPEs (Private Investment in Public Equities), 149
pivotal trial, 208
PLA (Product License Application), 206, 208, 222, 232
Plan B, 201
plant incorporated protectant (PIP), 173, 192
plant pest, 162, 174, 179, 181–183
Plant Protection Act (PPA), 162, 179, 181
plant regulators, 162, 173
plasmids, 25
platform technologies, 111
platform technology, 61, 146
pluripotent, 10, 96
PMA (Pre-Market Approval), 206, 308
PMDA (Japanese Pharmaceutical and Medical Devices Agency), 254–255
poliovirus, 35, 259
Pollack, Robert, 164
polyclonal antibodies, 32
Polymerase Chain Reaction (PCR), 29, 45
polypeptides, 21, 30, 188, 199
post-marketing reports, 226
post-marketing, Phase IV studies, 209, 225
post-money valuation, 137
potency, 181, 204
PPA (Plant Protection Act), 162, 179, 181–182
preclinical toxicology, 112, 126, 207, 209, 214, 254
predicate device, 206
preemption, 309–310, 326–327, 330
preemptive rights, 140, 150
preferred shares, 130–131, 139
preferred stock, 129, 131
Premanufacture Notice (PMN) for New Chemical Substances, 177
premarket approval, 206, 310
premarket notification (dietary supplements), 250
premarket notification (medical devices), 206
premarket notification (new, biotechnology-derived foods), 187, 189–190

pre-money valuation, 137
Prescription Drug User Fees Act, PDUFA, 224, 223–224
primary endpoint, 211–212, 216
priority review, 227, 229
private placement, 132, 145
pro forma operating statement, 141
product liability, 16, 189, 300, 303–308, 311–312, 314–315, 317–320, 322, 328–332
Product License, Product License Application 181, 206, 208, 222
product of nature objection, 73, 79–80
prokaryotes, prokaryotic cells, 21, 23–25
proof of concept, 207
prospectus, 135, 153–155
proteins, 8–9, 13, 21–22, 24–30, 32, 34–37, 41–43, 73, 79–80, 85–88, 95, 173, 179–180, 187–190, 201, 203–204, 233, 237–239, 279, 281, 283–284, 328
proteomics, 8, 111–113, 115, 124, 233
provisional applications, 77, 81, 106
PTO (Patent and Trademark Office), 76
PTO's 2001 Guidelines for Written Description, 86
Public Health Service (PHS), 200, 203, 248, 325
purity, purity, safety and potency, 168, 181, 204–205, 238, 319
quarterly reports, 156
quiet period, 155–156
rational drug design, 8–9, 36–37, 82
Rawls, John, 258, 260, 261–262, 264, 266, 269, 273, 275
Rawlsian, 257, 263–264, 266–267, 270–271, 274
reach through (license clause), 62
reach through (patent claim), 100, 107
reasonable care, 300–307, 322–323, 325–327, 331–332
red herring, 153–154
registration statement, 153–155
reimbursement, 15–17, 151, 209–211, 216, 245
Remicade, 9, 118–119
ReoPro, 41

restriction enzyme, 26
retroviruses, 24, 282–283, 286–288
reverse equivalents, 104–105
reverse transcriptase, 24, 28, 36, 79, 282
Rituxan, 42, 219
RNA, 6, 9, 19–22, 24, 28–29, 35–36, 86, 176, 258–259, 279, 282
RNA viruses, 24, 282
road shows, 154
Roundup (glyphosphate herbicide), 172
Rousseau, 262
route of administration, 116, 206
royalties, 56, 62–64, 66–67, 111, 137, 144
Sabin oral polio vaccine, 34
safe harbor, 156–157, 246
sales of biotechnology products, 3
Salk Institute (The), 5, 14, 15, 45, 53
Sandoz, 57–58, 60, 95
Sanger process, 27
Sarbanes-Oxley, 151, 158
Schwing & Albers, 194
scientific evidence, 222, 241, 245, 313
Scripps Research Institute (The), 4, 33, 45, 56–57, 60
secondary evidence of nonobviousness, 88–89
signal strength, 215
single patient access protocols, 229
single Patient IND, 229–230
Slovic, Fischhoff & Lichtenstein, 194
social contract, 262–263
somatic cell gene therapy, 38, 279, 292–293
sponsor, 51, 59, 67–69, 186, 222, 225–227, 229–230, 232, 234–236, 238, 244
sponsored research agreements, SRA, 56, 58, 67–68
STAT (signal transduction and activation of transcription), 22
stem cell therapies, 201
stem cells, 10–11, 39, 96, 260, 279, 281
stock options, 131, 140, 266
strategic alliance, 142–149, 158
structure/function claims, 251
Subchapter S corporations, 130

subgroup analysis, 208
sublicense, 63
substrate, 37
sub-unit vaccines, 35
surrogate clinical endpoints, 227
survival index, 159
Sutton, Willie, 46, 134
SV-40, SV-40 tumor virus, 25, 163–164
Syntro, 180
Syrrx, 37
Takeda Pharmaceuticals, 37
T-cells, 24, 30, 32, 34, 81, 213, 279
Tech Coast Angels, 133
terminator, 27
thrombolytics, 203
tissue plasminogen activator, tPA, 34, 87
TNF (tumor necrosis factor), 9, 32, 41, 212–213, 286–287
tolerance, 42, 172, 175, 184, 192
toolbox companies, 111
tort reform, 299
totipotent, 10, 279
Toxic Substances Control Act (TSCA), 162, 175–178, 181
toxicants, 188
toxicity, 166, 169, 175, 188, 207, 230, 286, 288, 328
transcriptase, 24, 28, 36, 79, 282
transcription factors, 22
transgenic animals, 13
transgenic plants, 12, 167, 192
transmembrane receptor, 22
Traynor, Justice, 305
Treatment INDs, 229
t-RNA, 20, 24, 28, 282
TSCA (Toxic Substances Control Act), 162, 175–178, 181
tumor suppressor, 281
unavoidably unsafe, 318–319, 328
underwriter, 151–154
United Nations Convention on Biodiversity, 98
University of Pennsylvania, 40
unmet medical need, 118, 212, 227
USDA (United States Department of Agriculture), 15, 161–163, 168, 171–174, 178–183

USPTO's 1999 Revised Utility Guidelines, 80
utilitarian, 257, 261, 263, 265, 270, 274–275, 296
utilitarianism, 258, 260–261, 263, 265, 275
utility (patent) see also USPTO's Revised Utility Guidelines, 75, 80–82
vaccines, 7, 12, 15–16, 34–35, 42–43, 54–55, 168, 171, 178–181, 201, 203, 263, 283, 288, 298, 305, 307, 310, 315–319, 322–323
valuation, 134, 137–139, 141, 145, 159
variable region, 31, 33
veil of ignorance, 262, 264, 269–270
Venter, Craig, 258
Veterinary Biological License (U.S.), 181

viral vector, viral vectors, 39–40, 278–280, 282–286, 288, 290, 292, 298
viruses, 6, 8, 19, 23–24, 33, 35, 39–40, 54, 163, 171, 179–180, 204, 258, 282, 285–287, 289, 292
Virus-Serum-Toxin Act, VSTA, 171, 179–180
Watson, James, 5–6
Waxman, Henry (Congressman), 233
Whitehead Institute, 14, 45, 100
written description (§112), 82, 86, 89–91, 93, 101, 107
Wyden, Ron (Congressman), 57
X-ray crystallography, 37
X-SCID, 40, 278–279, 282, 292
Zevalin, 10, 218–219